T0298339

Introduction to Renewable Energy Conversions

Introduction to Renewable Energy Conversions

Sergio C. Capareda

CRC Press
Taylor & Francis Group
Boca Raton London New York

CRC Press is an imprint of the
Taylor & Francis Group, an **informa** business

CRC Press
Taylor & Francis Group
6000 Broken Sound Parkway NW, Suite 300
Boca Raton, FL 33487-2742

© 2020 by Taylor & Francis Group, LLC
CRC Press is an imprint of Taylor & Francis Group, an Informa business

No claim to original U.S. Government works

Printed on acid-free paper

International Standard Book Number-13: 978-0-367-18850-4 (Hardback)

Library of Congress Cataloging-in-Publication Data

Names: Capareda, Sergio C., author.
Title: Introduction to renewable energy conversions / Sergio C. Capareda.
Description: First edition. | Boca Raton, FL : CRC Press/Taylor & Francis
Group, 2019. | Includes bibliographical references and index.
Identifiers: LCCN 2019018479 | ISBN 9780367188504 (hardback : acid-free paper) |
ISBN 9780429199103 (ebook)
Subjects: LCSH: Renewable energy sources—Mathematics. | Renewable energy
sources—Problems, exercises, etc. | Force and energy—Mathematical
models. | Engineering mathematics—Formulae.
Classification: LCC TJ808.3 .C37 2019 | DDC 621.042—dc23
LC record available at https://lccn.loc.gov/2019018479

Visit the Taylor & Francis Web site at
http://www.taylorandfrancis.com

and the CRC Press Web site at
http://www.crcpress.com

eResource material is available for this title at https://www.crcpress.com/9780367188504.

I dedicate this book to my beloved Mama, Adoracion Canzana Capareda, for her selflessness and daily prayers for our family's health, safety, and success, until her recent demise. She has been my inspiration to keep working hard and be a good provider for my family. She is forever in my heart and mind as she now looks after us from above.

Contents

List of Figures xix
List of Tables xxv
Foreword xxvii
Preface... xxix
Acknowledgments xxxi
Author.. xxxiii

1. Introduction to Renewable Energy 1
    1.1   Introduction 1
    1.2   Advantages and Disadvantages of the Use of Renewable
          Energy Resources...2
          1.2.1   Advantages..2
          1.2.2   Disadvantages...2
    1.3   Renewable Energy Resources..3
          1.3.1   Solar Energy..3
          1.3.2   Wind Energy..6
          1.3.3   Biomass Energy 7
          1.3.4   Hydro Power...9
          1.3.5   Geothermal Energy...11
          1.3.6   Salinity Gradient 15
          1.3.7   Fuel Cells 15
          1.3.8   Tidal Energy..16
          1.3.9   Wave Energy 18
          1.3.10 Ocean Thermal Energy Conversion Systems.......................19
          1.3.11 Human, Animal, and Piezoelectric Power............................20
          1.3.12 Cold Fusion and Gravitational Field Energy........................21
    1.4   Renewable Energy Conversion Efficiencies 22
    1.5   Renewable Energy Resources—Why? 23
    1.6   Summary and Conclusion..24
    1.7   Problems...24
          1.7.1   Carbon Dioxide Required to Make Carbohydrates...............24
          1.7.2   Kinetic Energy of a Mass of Wind 25
          1.7.3   Carbon Dioxide Production during Ethanol Fermentation 25
          1.7.4   Theoretical and Actual Power from Water Stream 25
          1.7.5   Theoretical Thermal Conversion Efficiency of Rankine Cycle............25
          1.7.6   Fuel Cell Efficiencies 25
          1.7.7   Tidal Power Calculations 25
          1.7.8   Solar Water Heater Conversion Efficiency............................25
          1.7.9   OTEC Energy Conversion..26
          1.7.10 Solar PV Conversion Efficiency..26
    References 26

2. Solar Energy..**29**
   2.1   Introduction.. 29
   2.2   The Solar Constant and Extraterrestrial Solar Radiation 30
   2.3   Actual Solar Energy Received on the Earth's Surface.......................... 31
   2.4   Solar Energy Measuring Instruments... 32
   2.5   Solar Time.. 33
   2.6   Geometric Nomenclatures for Solar Resource Calculations............... 35
   2.7   Extraterrestrial Solar Radiation on a Horizontal Surface................... 40
   2.8   Available Solar Radiation on a Particular Location 42
   2.9   Solar Energy Conversion Devices 45
          2.9.1   Solar Thermal Conversion Devices 45
                   2.9.1.1   Solar Refrigerators... 45
                   2.9.1.2   Solar Dryers... 47
                   2.9.1.3   Solar Water Heaters.. 49
          2.9.2   Solar Photovoltaic (PV) Systems 50
          2.9.3   Solar Thermal Electric Power Systems.................................... 52
          2.9.4   Solar Thermal Power Systems with Distributed Collectors.............. 53
          2.9.5   Solar Thermal Power Systems with Distributed Collectors
                 and Generators 53
          2.9.6   High-Temperature Solar Heat Engines 54
   2.10   Solar Collector System Sizing.. 55
   2.11   Economics of Solar Conversion Devices 57
   2.12   Summary and Conclusions.. 59
   2.13   Problems 60
          2.13.1   Extraterrestrial Solar Radiation... 60
          2.13.2   Solar Time.. 60
          2.13.3   Solar Declination Angle 60
          2.13.4   Angle of Incidence 60
          2.13.5   Hour Angle, Time of Sunrise, and Number of Daylight Hours........ 60
          2.13.6   Theoretical Daily Solar Radiation, H_o.................................... 60
          2.13.7   Theoretical Hourly Solar Radiation... 61
          2.13.8   Clearness Index to Estimate Beam and Diffuse Radiation 61
          2.13.9   Sizing Solar PV Panels... 61
          2.13.10 Economics of Solar Energy 61
   References 61

3. Wind Energy..**63**
   3.1   Introduction.. 63
   3.2   Basic Energy and Power Calculation from the Wind............................ 65
   3.3   The Worldwide Wind Energy Potential.. 69
   3.4   The Actual Energy and Power from the Wind 69
   3.5   Actual Power from the Wind.. 72
   3.6   Windmill Classification.. 73
          3.6.1   Classification according to Speed 73
                   3.6.1.1   High-Speed Windmills.. 73
                   3.6.1.2   Low-Speed Windmills.. 73
          3.6.2   Classification according to Position of Blades....................... 73
                   3.6.2.1   Upwind Windmills 73
                   3.6.2.2   Downwind Windmills... 74

	3.6.3	Classification according to Orientation of Blade Axis	74
		3.6.3.1 Vertical Axis Windmills	74
		3.6.3.2 Horizontal Axis Windmills	74
3.7	Wind Speed Measuring Instruments		75
3.8	Wind Power and Energy Calculations from Actual Wind Speed Data		77
	3.8.1	The Rayleigh Distribution	77
	3.8.2	The Weibull Distribution	80
3.9	Wind Design Parameters		84
	3.9.1	Cut-In, Cut-Out, and Rated Wind Speed	84
	3.9.2	General Components of Horizontal Axis Windmills for Power Generation	84
	3.9.3	Wind Speed Variations with Height	86
	3.9.4	Wind Capacity Factor and Availability	87
3.10	Comparative Cost of Power of Wind Machines		88
3.11	Conclusion		89
3.12	Problems		90
	3.12.1	Kinetic Energy from Wind	90
	3.12.2	Power from the Wind	90
	3.12.3	Power Differential as Wind Speed Is Doubled	90
	3.12.4	Actual Power from Windmill	90
	3.12.5	Rayleigh Distribution Estimate	90
	3.12.6	Estimating Average Wind Speed from Rayleigh Distribution	91
	3.12.7	Average Wind Velocity for a Given Site and Hours of Occurrence	91
	3.12.8	Estimate Weibull Parameters k and c from Linear Regression Data	91
	3.12.9	Wind Speed at Different Elevation	91
	3.12.10	Payback Period for Wind Machine	91
References			91
4. Biomass Energy			**93**
4.1	Introduction		93
4.2	Sources of Biomass for Heat, Fuel, and Electrical Power Production		95
	4.2.1	Municipal Solid Wastes	95
	4.2.2	Municipal Sewage Sludge	96
	4.2.3	Animal Manure	96
	4.2.4	Ligno-Cellulosic Crop Residues	97
4.3	Biomass Resources That May Have Competing Requirements		97
	4.3.1	Oil Crops	97
	4.3.2	Sugar and Starchy Crops	98
	4.3.3	Fuel Wood	98
	4.3.4	Aquatic Biomass	98
4.4	Various Biomass Conversion Processes		99
	4.4.1	Physico-Chemical Conversion Processes	99
		4.4.1.1 Biodiesel Production	99
	4.4.2	Biological Conversion Processes	102
		4.4.2.1 Bio-Ethanol Production	102
		4.4.2.2 Biogas Production	104

          4.4.3   Thermal Conversion Processes...107
                   4.4.3.1   Pyrolysis...107
                   4.4.3.2   Gasification 108
                   4.4.3.3   Eutectic Point of Biomass.....................................110
                   4.4.3.4   Combustion Processes 112
     4.5   Economics of Heat, Fuel, and Electrical Power Production from Biomass............113
          4.5.1   Biodiesel Economics 113
          4.5.2   Ethanol Economics 114
     4.6   Sustainability Issues with Biomass Energy Use....................................115
     4.7   Conclusion..116
     4.8   Problems...116
          4.8.1   Area Required to Build a Power Plant....................................116
          4.8.2   Electrical Power from MSW 117
          4.8.3   Feedstock Requirement for a 3 MGY Biodiesel Plant........................117
          4.8.4   Sugar Needed to Produce Ethanol 117
          4.8.5   Biogas Digester Sizing..117
          4.8.6   Residence Time for Biomass Conversion in
                   Fluidized Bed Reactors 117
          4.8.7   Chemical Formula for Biomass...117
          4.8.8   Air-to-Fuel Ratio (AFR) Calculations....................................118
          4.8.9   Eutectic Point of Biomass 118
          4.8.10   Area Needed for Wood Power...118
     References 118

5. **Hydro Power**..**121**
     5.1   Introduction 121
     5.2   Power from Water 123
     5.3   Inefficiencies in Hydro Power Plants 125
     5.4   Basic Components of a Hydro Power Plant..127
     5.5   Water Power–Generating Devices 129
          5.5.1   Water Wheels and Tub Wheels..130
          5.5.2   Turbines..130
          5.5.3   Specific Speeds for Turbines..131
          5.5.4   Turbine Selection...132
     5.6   Hydraulic Ram 134
          5.6.1   Construction and Principles of Operation 134
          5.6.2   Hydraulic Ram Calculations...136
          5.6.3   Design Procedures for Commercial Rife Rams.........................138
          5.6.4   Specifying Pipe Sizes and Discharge Pipe Lengths 139
          5.6.5   Starting Operation Procedure for Hydraulic Rams..................140
          5.6.6   Troubleshooting Hydraulic Rams 141
     5.7   Types of Hydro Power Plant...141
          5.7.1   On the Basis of Operation..141
          5.7.2   Based on Plant Capacity...142
          5.7.3   Based on Head..142
          5.7.4   Based on Hydraulic Features 142
                   5.7.4.1   Conventional...142
                   5.7.4.2   Pumped Storage Systems.......................................142
          5.7.5   Based on Construction Features...147

5.8		Environmental and Economic Issues	149
5.9		Conclusions	150
5.10		Problems	150
	5.10.1	Theoretical Power from Water	150
	5.10.2	Actual Efficiencies of Micro Hydro Units	151
	5.10.3	Hydro Power Plant Calculations	151
	5.10.4	Pump Specific Speed	151
	5.10.5	Volumetric Efficiency of Hydraulic Rams	151
	5.10.6	Energy Efficiency of Hydraulic Rams	151
	5.10.7	Specifying Drive Pipe Size and Lengths	151
	5.10.8	Specifying Drive Pipe Size Using Rife Ram	151
	5.10.9	Pumped Storage Power Production	152
	5.10.10	Pumped Storage Power Production Water Use	152
	References		152

6. Geothermal Energy **153**

6.1		Introduction	153
6.2		Temperature Profile in Earth's Core	154
6.3		Geothermal Resource Systems	158
	6.3.1	Liquid-Dominated Systems	158
	6.3.2	Vapor-Dominated Systems	159
	6.3.3	Hot Dry Rock Systems	159
	6.3.4	Geo-Pressure Systems	159
6.4		Geothermal Resource Potential in Texas	160
6.5		Geothermal Power Cycles	161
	6.5.1	Analysis of the Thermodynamic Cycle (Exell, 1983)	162
	6.5.2	Energy Flows or First Law Analysis	163
6.6		Geothermal Heat Pumps	168
	6.6.1	Geothermal Heat Pump (Opposite of Refrigeration)	171
6.7		Geothermal Power Cycles	174
	6.7.1	Non-Condensing Cycle	174
	6.7.2	Straight Condensing Cycle	175
	6.7.3	Indirect Condensing Cycle	177
	6.7.4	Single Flash System	177
	6.7.5	Double Flash System	177
	6.7.6	Binary Fluid Cycle	178
6.8		Geothermal Power Applications	180
6.9		Levelized Cost of Selected Renewable Technologies	182
6.10		Environmental Effects of Geothermal Power Systems	183
6.11		Conclusion	184
6.12		Problems	185
	6.12.1	Well Selection	185
	6.12.2	The Ideal Rankine Cycle	185
	6.12.3	Efficiency of Ideal Geothermal Cycle	185
	6.12.4	Changes in Efficiency and Power Output	186
	6.12.5	COP of Ideal Refrigeration Cycle	186
	6.12.6	Ideal Vapor Refrigeration System	186
	6.12.7	Power Consumed in Heat Pump	186
	6.12.8	Cost Comparison	187

6.12.9   Number of Households Served by Geothermal Facility 187
6.12.10   ROI of Geothermal Heating and Cooling 187
References 187

7. Salinity Gradient **189**
7.1   Introduction... 189
7.2   The Solar Pond.. 190
       7.2.1   Advantages.. 193
       7.2.2   Disadvantages... 194
7.3   Energy of Sea Water for Desalination 194
7.4   Pressure-Retarded Osmosis (PRO) 195
       7.4.1   PRO Standalone Power Plants
                (Statkraft, Netherlands, 2006) 197
       7.4.2   Statkraft Prototype (Norway, Co.) 197
7.5   Reverse Electro-Dialysis (RED) 198
7.6   Specific Applications or Locations... 201
7.7   Limitations and Factors Affecting Performance and Feasibility..................... 202
7.8   Performance and Costs... 203
7.9   Potential Energy and Barriers to Large-Scale Development............................ 204
7.10   Environmental and Ecological Barriers 205
7.11   Conclusions 206
7.12   Problems 206
       7.12.1   Sensible Heat from Solar Pond... 206
       7.12.2   Theoretical Carnot Cycle Efficiency 207
       7.12.3   Osmotic Pressure Calculations 207
       7.12.4   Work Done against Pressure 207
       7.12.5   Energy Required to Boil Seawater... 207
       7.12.6   Size of PRO Unit to Generate Given Power 207
       7.12.7   Amount of Membrane to Use to Generate Power
                for a Household 207
       7.12.8   RED Salinity Gradient System 207
       7.12.9   Cost of RED Power Plants 208
       7.12.10 Simple Payback Period for Salinity Gradient Power Plant.............. 208
References 208

8. Fuel Cells **211**
8.1   Introduction... 211
8.2   The Various Types of Fuel Cells 215
       8.2.1   Proton Exchange Membrane Fuel Cells.................................. 215
       8.2.2   High-Temperature Proton Exchange Membrane Fuel Cell.............. 216
       8.2.3   Direct Methanol Fuel Cell... 216
       8.2.4   Alkaline Electrolyte Fuel Cell 217
       8.2.5   Phosphoric Acid Fuel Cell... 218
       8.2.6   Solid Oxide Fuel Cell, High Temperature 219
       8.2.7   Solid Acid Fuel Cell... 220
       8.2.8   Molten Carbonate Fuel Cell, High Temperature 220
       8.2.9   Regenerative Fuel Cell 221
       8.2.10   Solid Polymer Fuel Cell 222
       8.2.11   Zinc-Air Fuel Cell.. 222

| | 8.2.12 | Microbial Fuel Cell | 223 |

8.2.12   Microbial Fuel Cell 223
8.2.13   Other Fuel Cells: Biological, Formic Acid, Redox Flow
         and Metal/Air Fuel Cells 224
8.3   Data for the Different Major Types of Fuel Cells 224
8.4   Various Fuels Used for Fuel Cells and Issues 225
8.5   Advantages and Disadvantages of Fuel Cells 226
8.6   Balance of Plant 227
8.7   Existing and Emerging Markets for Fuel Cells 228
      8.7.1   NASA Helios Unmanned Aviation Vehicle 229
      8.7.2   Naval Research Lab Spider Lion 229
      8.7.3   The PEMFC Commercial Fuel Cell Module by Ballard
              (NEXA TM 1.2kW) 231
      8.7.4   Heliocentris Fuel Cell System 232
8.8   The Future of the Fuel Cell 233
8.9   Conclusions 233
8.10 Problems 234
      8.10.1   Conversion Efficiency of a Direct Methane Fuel Cell 234
      8.10.2   Maximum Conversion Efficiency for a
               Direct Methane Fuel Cell 234
      8.10.3   Heat Energy Losses in a Direct Methane Fuel Cell 234
      8.10.4   Hydrogen Needed (in kg) to Produce a Liter of Water 235
      8.10.5   Potassium Carbonate Produced in an Alkaline Fuel Cell 235
      8.10.6   Ideal Water and Carbon Dioxide Produced for a
               Direct Methane Fuel Cell 235
      8.10.7   Zinc Needed for Every Tonne Zinc Oxide Produced in a
               Zinc-Air Fuel Cell 235
      8.10.8   Practical Conversion Efficiency for a Direct Methanol Fuel Cell 235
      8.10.9   Efficiency of a Spider Lion Fuel Cell 235
      8.10.10 Efficiency of a Commercial Fuel Cell 236
References 236

9. Tidal Energy 239
9.1   Introduction 239
9.2   Worldwide Potential of Tidal Energy 242
9.3   How Tidal Energy Works 245
9.4   Tidal Power Generation Schemes 247
      9.4.1   Single-Basin Ebb Cycle Power Generation 248
      9.4.2   Single-Basin Tide Cycle Power Generation 250
      9.4.3   Single-Basin Two-Way Power Generation 252
      9.4.4   Double-Basin Systems 253
9.5   Other Tidal Power Generating Methods 256
9.6   Cost of Tidal Energy Systems 258
9.7   Environmental Concerns 259
      9.7.1   Beneficial 259
      9.7.2   Non-Beneficial 259
9.8   Conclusions 260
9.9   Problems 260
      9.9.1   Variation of Tide Level with Time Using Sine Curve 260
      9.9.2   Reservoir Volume Calculation 261

9.9.3   Time to Release Water from Reservoir..261
9.9.4   Power from Tidal Reservoir..261
9.9.5   Energy from Tidal Reservoir..261
9.9.6   Matching Household Energy Requirements.....................................261
9.9.7   Water Level Decline with Time for a Given Basin..........................261
9.9.8   Power Generated from Small Basin..262
9.9.9   Power and Energy from Double-Basin System................................262
9.9.10   Cost to Recover Initial Investment..262
References...262

10. Wave Energy 265
10.1   Introduction..265
10.2   Power from Wave...266
10.3   World's Wave Power Resource...268
10.4   Various Generic Wave Energy Converter Concepts.......................269
       10.4.1   Point Absorber Buoy..269
       10.4.2   Surface Attenuator...274
              10.4.2.1   Wave Contouring Rafts (Cockerell Rafts).......................276
       10.4.3   Oscillating Wave Surge Converter....................................276
       10.4.4   Oscillating Water Column..276
       10.4.5   Overtopping Device..277
       10.4.6   Submerged Pressure Differential......................................278
10.5   Other Common Types of Currently Deployed
       Wave Energy Converters..279
       10.5.1   Hose Pump...279
       10.5.2   Salter's Duck..280
       10.5.3   Masuda Buoy...281
10.6   Typical Hydraulic Circuit for Wave Generators.............................281
10.7   Approximating Wave Height Using Significant Wave
       Height, H_s...284
10.8   Beneficial and Non-Beneficial Environmental Impacts of
       Wave Power..286
       10.8.1   Advantages..286
       10.8.2   Disadvantages...286
10.9   Year-Round Distribution of Wave Energy......................................286
10.10   Economic Aspects and Potential Locations......................................287
10.11   Countries with Wave Energy Studies (IRENA, 2014).....................288
       10.11.1 United Kingdom...289
       10.11.2 Australia..289
       10.11.3 Denmark..289
       10.11.4 United States...290
       10.11.5 Belgium..290
       10.11.6 Sweden..290
       10.11.7 Ireland...290
       10.11.8 Israel..291
10.12   Conclusion...291
10.13   Problems..292
       10.13.1 Determine the Constant for Wave Power Equation.......292
       10.13.2 Basic Power from Wave...292

10.13.3 Wave Power in Storms..292
10.13.4 Total Power from Wave 292
10.13.5 Hydraulic Power Developed from Buoys 292
10.13.6 Hydraulic Power..293
10.13.7 Hydraulic Jack Power (Metric).....................................293
10.13.8 Hydraulic Jack Power (English System) 293
10.13.9 Piston Power for Surface Attenuator 293
10.13.10 Significant Wave Height (H_s)...................................294
10.13.11 Capital Cost of Wave Converters 294
References 294

11. Ocean Thermal Energy Conversion (OTEC) Systems 297
11.1 Introduction..297
11.2 The Basic OTEC System 299
11.3 OTEC Components and Temperature Profiles 299
11.4 Other Applications of OTEC 302
11.5 Uses of OTEC Systems 305
11.6 Basic Thermodynamic Cycle: Rankine Cycle 306
11.7 OTEC Power Generation Systems...................................309
11.7.1 Closed Cycle..309
11.7.1.1 Efficiency Calculations.................................312
11.7.2 Open Cycle 315
11.7.3 Hybrid Systems..316
11.8 Projects Under Way for OTEC Systems (IRENA, 2014)......317
11.8.1 Natural Energy Laboratory of Hawaii Authority
(NELHA) 317
11.8.2 OTEC Projects in Japan 318
11.8.3 OTEC Facility in India..318
11.8.4 Other OTEC Projects Around the World........................318
11.9 Technical Limitations and Cost (IRENA, 2014)................319
11.10 Conclusion..321
11.11 Problems...322
11.11.1 Heat Capacity of the Ocean 322
11.11.2 Ideal Carnot Cycle Efficiency 322
11.11.3 Volume of Water Needed for a 100 kW of Power...........322
11.11.4 Calculating Water Pumping Power 322
11.11.5 Base Load Power Calculations 322
11.11.6 OTEC Closed Cycle Calculations.................................323
11.11.7 Actual OTEC Cycle Examples 323
11.11.8 Heat of Evaporation Calculations 323
11.11.9 Estimating the Number of Households Served by OTEC...........324
11.11.10 Estimating the Initial Capital Cost of OTEC 324
References 324

12. Human and Animal Power, and Piezoelectrics..........................327
12.1 Introduction..327
12.2 Animal Power..330
12.2.1 Draft Animal Performance Compared with
Mechanical Tractors...331

12.2.2   Draft Horsepower Capability of Various Animals 331
12.2.3   Unique Perspectives of Animal Power.. 333
12.3   Human Power.. 334
12.3.1   Advantages of Humans for Energy Use 335
12.3.2   Disadvantages of Humans for Energy Use 336
12.3.3   Human Factors in Energy and Power:
The Ergonomic Factors 336
12.4   Piezoelectrics 338
12.4.1   Applications of Piezoelectricity 340
12.4.2   High-Voltage Power Sources... 340
12.4.3   Use of Piezoelectric Devices as Sensors... 343
12.4.4   Piezoelectric Devices as Tiny Actuators 343
12.4.5   Piezoelectric Motors... 344
12.4.6   Potential Future Applications of Piezoelectricity.......................... 344
12.5   Conclusions... 345
12.6   Problems.. 345
12.6.1   Power from Animals.. 345
12.6.2   Power from Humans.. 345
12.6.3   Various Units of Power 346
12.6.4   Power from Groups of Animals 346
12.6.5   Energy Output of a Cow in the Form of Milk................................. 346
12.6.6   Power of Humans over Longer Periods of Time............................. 346
12.6.7   Power from Arms and Legs of Humans... 346
12.6.8   Basic Piezoelectric Power from Numerous
Repeated Cycles... 346
12.6.9   Piezoelectric Power from Single Tap 346
12.6.10   Charging a Cell Phone with Piezoelectric Power.......................... 347
References 347

13. Cold Fusion and Gravitational Energy **349**
13.1   Introduction... 349
13.2   The Cold Fusion Theory... 350
13.3   Calorimetry... 353
13.4   Cold Fusion by Other Names 355
13.5   Key Figures in Fusion Energy Research... 356
13.5.1   Randell L. Mills, Brilliant Light Power, New Jersey 356
13.5.2   Michael McKubre, Energy Research Center,
SRI International 357
13.5.3   David J. Nagel, George Washington University 357
13.5.4   Rossi's E-Cat.. 357
13.5.5   International Thermonuclear Experimental Reactor...................... 358
13.6   The Gravitational Power Potential.. 359
13.7   Tachyon Field Energy 361
13.8   Len's Law and Faraday's Law.. 363
13.9   Other Scientists Investigating Gravitational Field Energy and Other
Renewables... 364
13.9.1   Dr. T. Henry Moray, American Physicist 365
13.9.2   Professor Shinichi Seike, Director, Gravity Research
Laboratory, Japan.. 365

	13.9.3	Bruce De Palma's N-Machine	366
	13.9.4	Paramahamsa Tewari of India and His Space Power Generator	367
13.10		Non–Energy-Related Applications of Gravitational Field Energy	371
13.11		Conclusions	371
13.12		Problems	372
	13.12.1	Energy Balance in Electrolysis Setup	372
	13.12.2	Heat Capacity of Calorimeters	372
	13.12.3	Heat Released from Combustion of Chemicals	373
	13.12.4	Energy Balance in N-Machine or N-Generator	373
	13.12.5	Determining Magnetic Fluxes, Voltages, and Current in Conducting Coils	373
	13.12.6	Estimating Gravitational Forces at Various Elevations	373
	13.12.7	Calculating Acceleration due to Gravity at Various Elevations	374
	13.12.8	Calculating Voltages, Current, and Magnetic Fluxes in Coils	374
	13.12.9	Calculating Input and Output Power in an Electric Motor	374
	13.12.10	Improving the PF of Resistive Motors	374
	References		374

14. Environmental and Social Cost of Renewables **377**
14.1	Introduction	377
14.2	Technical Advancement of Renewable Energy Technologies	378
14.3	Balance of Systems	382
14.4	Overall Economics and Levelized Cost of Renewable Energy	384
14.5	Life Cycle Analyses of Renewables	387
14.6	Pollutant Emissions of Some Renewable Energy Technologies	390
14.7	Sustainability Issues of Renewables	394
14.8	The Social Costs of Renewables	396
14.9	Conclusion	398
14.10	Problems	399

	14.10.1	Area Required for Solar PV Systems	399
	14.10.2	Algal Oil Production and Yield Calculations	399
	14.10.3	Size and Cost of PV Systems for Large Commercial Applications	399
	14.10.4	Balance of System Cost as Percentage of PV Cost	399
	14.10.5	SO_2 Daily Emissions Rate for Coal Power Plants	399
	14.10.6	SO_2 Daily Emissions Rate for Biomass Power Plants	399
	14.10.7	Ozone and SO_2 Concentration Units from NAAQS Standards	400
	14.10.8	Net Energy Ratio (NER) for Biofuels	400
	14.10.9	Net Energy Balance (NEB) for Biofuels	400
	14.10.10	Return on Investment for the Production Cost of Solar PV Systems	400
	References		401

Appendix A: Table of Conversion Units **405**

Index **407**

List of Figures

Figure 1.1   The eight thermodynamic pathways for solar energy conversion 4

Figure 1.2   Low-cost micro–hydro power units for rural villages 11

Figure 1.3   Map showing the world's geothermal provinces 12

Figure 1.4   The ideal Rankine Cycle used in geothermal power systems 14

Figure 1.5   Attractive forces between the earth and the moon that cause the variations in water levels along the shores 17

Figure 2.1   A simplified sketch of the distribution of solar radiation absorbed by the earth on a typical day in the tropics 32

Figure 2.2   A photo of a simple research-grade solar pyranometer 33

Figure 2.3   Nomenclatures in the estimate of solar radiation received on a given surface, oriented at various directions on the earth; some important geometric angles relating a solar collector with sun's angular position 36

Figure 2.4   Photo of a research-grade solar pyranometer mounted on a solar panel.. 43

Figure 2.5   Average solar radiation received in College Station, Texas, for a year 45

Figure 2.6   Schematic diagram for a solar refrigeration system 46

Figure 2.7   Schematic of a simple solar refrigerator 47

Figure 2.8   Practical pressure-temperature profile in a solar refrigerator....................... 48

Figure 2.9   Typical performance of a simple solar refrigerator 48

Figure 2.10   Simple solar dryer for agricultural products 49

Figure 2.11   Various types of solar water heaters 50

Figure 2.12   Schematic of a simple solar PV home system 51

Figure 2.13   Schematic of a solar village power system.. 51

Figure 2.14   Schematic of a solar thermal electric power generation system 52

Figure 2.15   Solar thermal power systems with distributed collectors 53

Figure 2.16   Solar thermal power system with distributed collector and engine............. 54

Figure 2.17   Average prices of solar thermal collector systems 54

Figure 3.1   Typical generation of wind movement 64

Figure 3.2   Utility-scale land-based 80-meter wind map of the United States 70

Figure 3.3   Power coefficient curve following the Betz coefficient..................................70

Figure 3.4   Power coefficient for some windmill designs...71

Figure 3.5   Photo of various units of windmills for power generation in the
              northern part of the Philippines...75

Figure 3.6   Photo of cup anemometers as well as a sonic 3D anemometer.....................76

Figure 3.7   Wind speed histogram and Rayleigh distribution curve fit.........................78

Figure 3.8   Variations of Rayleigh distribution mode parameter c 78

Figure 3.9   Histogram of actual wind speed data in College Station, Texas, for
              the 2005 at 10-meter height...80

Figure 3.10   The wind speed histogram using raw data and the Weibull
              distribution curve fit..82

Figure 3.11   Diagram showing the relationship between cut-in, cut-out,
              and rated wind speed..85

Figure 4.1   Various biomass conversion pathways 99

Figure 4.2   The physical depiction of the transesterification process and the
              governing mass balance...100

Figure 4.3   The action of catalysts on the ester linkage to generate glycerin
              and the ester of the vegetable oil 101

Figure 4.4   Ethanol production process flow chart...103

Figure 4.5   Batch digestion setup for designing anaerobic digestion reactors..............105

Figure 4.6   Pilot anaerobic digester with floating gas holder...106

Figure 5.1   The hydrologic cycle and the transport of water in a
              continuous cycle...122

Figure 5.2   Top hydro power–producing states in the United States in 2011.................123

Figure 5.3   Basic components of a hydro power plant...127

Figure 5.4   The Bhumibol Dam and hydro power plant in Thailand..............................129

Figure 5.5   A schematic of Turgo wheel action...131

Figure 5.6   Blade configuration comparison for different specific speeds.....................133

Figure 5.7   Turbine application chart: a monograph for turbine type selection............133

Figure 5.8   Basic components of a hydraulic ram 135

Figure 5.9   Typical hydraulic ram installation...135

Figure 5.10   Nomenclatures for the design and calculation of hydraulic ram
              efficiencies..137

Figure 5.11   Nomenclatures for the design of commercial Rife rams 138

Figure 5.12   Schematic for the general arrangement of a pumped storage
              system showing the higher level pool and the lower level pool..................143

Figure 5.13 A pumped storage facility in the Philippines: the Kalayaan pumped storage hydro power plant located in Kalayaan, Laguna 144

Figure 5.14 Photos of the Kalayaan pumped storage hydro power plant in the Philippines 145

Figure 5.15 Schematic of the three basic types of pumped storage power plants.. 146

Figure 5.16 Schematic of the different types of hydro power plants based on their construction features 148

Figure 6.1 The world's geothermal provinces along the Ring of Fire in the Pacific Ocean 155

Figure 6.2 Temperature profile in the earth's core 156

Figure 6.3 Temperature profile of earth's surface...................................... 157

Figure 6.4 Geothermal resource areas in Texas... 160

Figure 6.5 Regions in Texas with proven geothermal resource bases 161

Figure 6.6 Basic geothermal power cycle 162

Figure 6.7 The TS diagram for geothermal systems 163

Figure 6.8 Values of thermodynamic properties at each major point in the thermodynamic cycle... 166

Figure 6.9 The TS diagram for the vapor refrigeration cycle...................... 169

Figure 6.10 Vapor refrigeration cycles... 169

Figure 6.11 Ideal refrigeration cycle 170

Figure 6.12a–c Basic geothermal power cycles... 175

Figure 6.12d–f More geothermal power cycles 176

Figure 6.13 The geothermal power plant in the Philippines at Tiwi, Albay 179

Figure 6.14 The delivery of steam through complicated pipelines 179

Figure 6.15 Levelized cost of renewable energy technologies 182

Figure 7.1 A solar pond facility in Pecos, Texas 190

Figure 7.2 Typical salinity gradient/solar pond conversion system................ 192

Figure 7.3 Basic PRO system schematic design... 196

Figure 7.4 A simplified PRO system designed and envisioned by MIT engineers in the United States.. 199

Figure 8.1 Typical design of a fuel cell 212

Figure 8.2 Basic schematic of a hydrogen-oxygen fuel cell and electrolytes used.. 213

Figure 8.3   Another illustration of fuel cell reactions...214

Figure 8.4   The NASA Helios Unmanned Aviation Vehicle (UAV)..................................229

Figure 8.5   Photo of the Naval Research Laboratory's Spider Lion.............................230

Figure 8.6   The PEMFC module by Ballard installed in a vehicle................................231

Figure 9.1   Primary areas around the world with high potential for
              tidal energy..240

Figure 9.2   Variations in height of time as a function of time......................................240

Figure 9.3   Possible sites for tidal power stations worldwide.....................................243

Figure 9.4   Best places to observe high tides that vary in height from 12 to 16
              meters [40 to 52.5 ft] in a day..244

Figure 9.5   Simple schematic of tidal power generation system...................................245

Figure 9.6   Diagram of the single-basin system...248

Figure 9.7   Diagram of the events in a single-basin ebb cycle system.........................249

Figure 9.8   Diagram of the single-basin tide cycle system and the corresponding
              events..251

Figure 9.9   Diagram of the single-basin double-cycle system and the
              corresponding events..252

Figure 9.10   Diagram of the double-basin system for continuous power
              generation..254

Figure 9.11   Operating regime for double-basin continuous power generation
              system..254

Figure 10.1   World's wave power potential (kWh/m)..268

Figure 10.2   Various wave energy converter concepts (1 = point absorber,
              2 = attenuator, 3 = oscillating wave surge converter, 4 = oscillating
              water column, 5 = overtopping devices, 6 = submerged pressure
              differential)..269

Figure 10.3   Hydraulic power from a wave point absorber..270

Figure 10.4   Piston hydraulic motion as the wave changes..275

Figure 10.5   The Cockerell wave contouring raft...276

Figure 10.6   The Pendulor is an example of an oscillating wave surge
              converter..277

Figure 10.7   The oscillating water column...277

Figure 10.8   The Isaac wave energy converter...278

Figure 10.9   The hose pump (a heaving buoy device)...279

Figure 10.10 The Salter's duck...280

Figure 10.11 The mechanism of the Masuda buoy...281

Figure 10.12a   The hydraulic circuit for a typical piston-based wave energy converter showing downward stroke of piston cylinder (wave receding) 282

Figure 10.12b   The hydraulic circuit for a typical piston-based wave energy converter showing upward stroke of piston cylinder (wave peaking) 282

Figure 10.13   Wave power distribution for a year in India (59°N, 19°W) 287

Figure 11.1   The world's ocean thermal energy conversion (OTEC) resources 298

Figure 11.2   The basic OTEC thermodynamic cycle....................................299

Figure 11.3   Component parts of an OTEC system 300

Figure 11.4   Applications of OTEC Systems 302

Figure 11.5   Basic data for an ideal OTEC system 308

Figure 11.6   Ideal T-s diagram for an OTEC system....................................308

Figure 11.7   Heat and work energy of the ideal thermodynamic cycle.....................309

Figure 11.8   Schematic of an operational closed OTEC system.....................310

Figure 11.9   Ideal thermodynamic cycle for OTEC systems 311

Figure 11.10   A closed-cycle OTEC system using liquid with low boiling point 313

Figure 11.11   The OTEC open-cycle system 315

Figure 11.12   Relationship between initial capital cost ($/kW) and power output of OTEC systems...320

Figure 12.1   Harnessing factor relationship derived from Cambell's (1990) reported table 333

Figure 12.2   Deformation of a piezoelectric material and the generation of electric voltage or power..339

Figure 13.1   Schematic diagram of the original experiment by Pons and Fleischmann 351

Figure 13.2   A typical power triangle for improving motor PF.....................370

Figure 14.1   Best conversion efficiencies of solar PV cells as reported by the USDOE...379

Figure 14.2   Unsubsidized levelized cost of energy comparison between conventional and renewable technologies 385

Figure 14.3   Life cycle analysis (LCA) flow chart for solar PV systems 387

Figure 14.4   Graphical estimation of National Source Performance Standards (NSPS) for SO_2 emissions from coal-powered plants. 392

List of Tables

Table 2.1   Table to Easily Estimate the Value of n for a Given (Non–Leap-Year) Month 30

Table 2.2   Solar Radiation Data in College Station, Texas 44

Table 3.1   Surface Roughness Coefficients for Various Terrain Descriptions 86

Table 4.1   Proposed and Final Renewable Fuel Volume Requirements for 2018-2020 in the United States 115

Table 5.1   Performance Chart for Rife Hydraulic Ram 139

Table 5.2   Recommended Values of Drive Pipe Diameter and Supply Head for a Given Range of Volumetric Flow Rates 139

Table 6.1   Summary of State Points in the Ideal Rankine Cycle Example 167

Table 6.2   Characteristics of the Largest Geothermal Power Plant in the Philippines 180

Table 7.1   Theoretical and Technical Potential of Salinity Gradient 205

Table 8.1   Data for Different Types of Fuel Cells 225

Table 9.1   Worldwide Power Potential from Tidal Energy 243

Table 9.2   Estimated Costs for Existing and Proposed Tidal Barrages 259

Table 11.1   Electric Power Consumption per Capita for Selected Countries 306

Table 12.1   Sustainable Power of Various Work Animals Using Maximum Reported Weights and Constant Work Speed 328

Table 12.2   Sustainable Power of Draft Animals and Humans at Similar Work Speed 329

Table 12.3   Drawbar Pull and Power Developed from Various Work Animals 332

Table 12.4   Reworked and Expanded Harnessing Factors for Work Animals 332

Table 12.5a   Reworked Man-Hour Requirements per Hectare [Acres] of Various Farm Activities 338

Table 12.5b   Man-Hour Requirements per Hectare [Acre] of Various Farm Activities 338

Table 13.1   Various Input/Output Power Formulas or Efficiency for Different Types of Electric Motors 368

Table 14.1   Breakdown of Installed Cost for a Residential Solar Photovoltaic System 383

Table 14.2 Levelized Cost of Electrical Power from Various Energy Sources.................384

Table 14.3 LCOE Calculator Results for Various Factors.......................................386

Table 14.4 Early LCA Studies on Various Solar PV Systems...388

Table 14.5 Revised LCA Studies on Various Solar PV Systems......................................388

Table 14.6 Recent LCA Studies on Various Solar PV Systems..389

Table 14.7 GHG Emissions for Various Renewables and Other Conventional
Power Sources 389

Table 14.8 The National Ambient Air Quality Standards (NAAQS).............................391

Foreword

Renewable energy technologies will become very important in the near future for two main reasons: first, the global energy–related carbon dioxide emissions are still increasing despite attempts to slow them down, affecting our environment; second, there will be an inevitable economic growth for many countries outside of China, the United States, and Europe. Hence, we can expect a much greater increase in fossil-fuel energy usage and in turn worsening its negative effect on climate change.

Many developing countries are now slowly investing on renewables than many developed countries relative to their percentage use of fossil fuels. Perhaps the primary reason is the pressure from environmentalist groups about the overall global effect of harnessing renewable energy. However, many developing countries are simply not interested in increasing the renewable energy mix in their energy usage due to numerous technical and economic reasons. Therefore, renewable energy textbooks such as this one are very important in the training of young engineers who would take the lead in revising this trend of continued increase in fossil fuel use to lessen carbon dioxide emissions and in turn minimize the current climate changes occurring worldwide.

This textbook is then a valuable contribution to the goal of educating many engineers in the proper implementation of renewable energy projects. This textbook presents a large number of renewable energy technologies available for the public, seemingly covering more than others of its kind. Chapters 2 to 5 discuss what I would consider the major renewables: solar (Chapter 2), wind (Chapter 3), biomass (Chapter 4), hydro power (Chapter 5), and geothermal energy (Chapter 6). These resources should be the primary focus of many countries planning to improve their renewable energy mix. The remaining chapters are what I call the emerging renewables that will be of importance to many other countries, especially those countries with access to or near water bodies. In fact, aside from fuel cells (Chapter 8), the other chapters deal with water bodies: salinity gradient (Chapter 7), tidal (Chapter 9), wave (Chapter 10), and ocean thermal energy conversion systems (Chapter 11).

I like the inclusion of human and animal power (Chapter 12). They are still in widespread use in many developing countries, and in a large majority of these countries, agricultural energy needs still rely on human and animal power sources. The concept of piezoelectrics, included in this chapter, is perhaps in the right place due to the smaller amount of power that could be generated over a long period of time. The chapter on cold fusion and gravitational field energy (Chapter 13) is an eye-opener for me. I have heard of cold fusion but never really thought of it as a major player in the future, but I may be wrong, and I wish my assumption is incorrect. How I wish we could harness gravitational field energy as some scientists have been advocating for so many years. But yes, as the author has warned the reader, please have an open mind getting into these fields.

The last chapter of the book is definitely noteworthy. It brings together the justification for investing on renewables. This goes back to my initial reaction on renewables; that is, countries have to invest in energy technologies that will not be harmful to our environment and at the same time will be of use by hundreds of generations to come. Hence, I am hoping that this textbook will provide engineering and non-engineering students alike with a valuable understanding of all the possible renewable energy conversion processes available to them and utilize them while taking care of its economics and sustainability.

Reynaldo M. Lantin
Retired Professor and Former Dean
College of Engineering and Agro-Industrial Technology (CEAT)
University of the Philippines at Los Baños (UPLB)
College, Laguna, Philippines

Preface

My main goal in writing this textbook is to provide engineers and practitioners interested in renewable energy with a solid and fundamental technical background in renewable energy conversion systems. The book has 14 chapters that cover about 15 renewable energy technologies, enough for a 15-week class on introduction to renewable energy conversions.

I have loaded up the book with a number of worked sample problems within each chapter. There are additional problems provided at the end of each chapter that instructors can use for teaching. I have added numerous equations, tables, and figures to clearly present each renewable energy technology.

Chapter 1 provides the basic concepts and conversion efficiency premises of most of the renewable energy technologies covered in the book.

Chapters 2 through 13 present and discuss aspects of all the renewable energy technologies one may find useful, including a couple of theories that are perhaps unconventional: cold fusion and gravitational field energy. I would encourage the reader to keep an open mind regarding these.

Each chapter begins primarily with resource estimates and calculations. I believe that when embarking on a renewable energy project, the first and essential step would be determining how much of this renewable energy is received in a given place, time, period, or season and on a long-term or yearly basis. Engineers can then use this resource analyses to develop feasibility studies.

Chapter 2 describes the technology for solar energy, the most abundant renewable energy resource available. My goal with this this chapter is to provide the general information that would allow an engineer to determine the available solar energy in a given locality anywhere around the world. As this inevitably involves numerous equations, calculations, and terminology regarding the movement of the earth around the sun, a more detailed discussion of each of the solar energy conversion technologies is beyond the scope of this book.

Chapter 3 is on wind energy. In this chapter, I have also focused on providing the engineer with a technical tool to effectively make an estimate of available wind energy in a given location using the Weibull distribution function for wind resource estimate.

Chapter 4 is on biomass energy conversion. As this was the focus of my previous textbook, this chapter presents mainly the major biomass conversion pathways now being commercialized worldwide. For a more in-depth discussion on biomass energy conversions, check out S. C. Capareda, *Introduction to Biomass Energy Conversions* Boca Raton, FL: CRC Press, Taylor and Francis Group, 2014.

The remaining chapters introduce the technology for other renewables—hydro power, geothermal energy, salinity gradient, and fuel cells—with a focus on calculations and estimates of available renewable energy potential for each. This is followed by chapters on tidal, wave, and ocean thermal energy conversion (OTEC) systems. I have added a chapter on human and animal power, a very important renewable resource for third world countries, and included an emerging, similarly lower energy–level renewable: piezoelectrics. A chapter on cold fusion and gravitational field energy, generally unpopular renewables, is also included.

The last chapter discusses the environmental, economic, and social costs of renewables. This chapter deals with energy sustainability issues, the levelized cost of renewable energy, and life cycle analyses of some of the renewable energy technologies.

This textbook comes with a solutions manual for all the problems given at the end of each chapter as well as a set of PowerPoint slides for use in a comprehensive renewable energy class. I welcome any suggestions and feedback for the improvement of this textbook.

Acknowledgments

I would like to express my gratitude to the following individuals for contributing their valuable time and expertise to complete this project:

Dr. Reynaldo M. Lantin served as my technical editor and provided valuable comments on the contents of the book and shared valuable and seemingly endless new and progressive ideas about renewable energy systems. He hired me as fresh graduate from college to lead a project on bioethanol production from cassava. This project became a nationwide program in the Philippines. This started my quest to learn about other renewable energy technologies, on which I built a career.

My daughter Alexa Jean Capareda put her B.A. English from UT Austin to good use as the English language and copy editor for this book. Her thorough and thoughtful editing certainly mended a lot of my writing faults and flaws. She also conceptualized the cover theme with the help of my supportive wife.

Ms. Rhonda Brinkmann (https://wordsmiths4u.wordpress.com/) provided many additional editing comments and also helped with the tedious task of requesting permissions for the use of numerous tables and figures used throughout the textbook.

I wish also to thank these individuals for the inspiration and support:

My former teachers at the Asian Institute of Technology in Thailand, Drs. Bhattacharya and R. H. B. Exell, are the main sources of my basic renewable energy teaching materials. Their lecture notes and their technical papers provided inspiration to complete this project for my renewable energy class. Dr. Bhattacharya guided me through my first textbook project. Unfortunately, he passed away a couple of years ago before seeing my second project.

Drs. Wayne A Lepori and Calvin B. Parnell, Jr., now both retired, have been very supportive mentors and colleagues at Texas A&M University. They taught me how to operate the fluidized bed gasifier for heat and power generation and how to design the gas clean up systems properly. I owe much of my knowledge of energy and air quality from these folks, but more importantly I am grateful for their friendship and unfailing support throughout my career.

Finally, I wish to thank everyone who has helped me in one way or another towards the goal of developing useful teaching materials and tools in advancing knowledge on emerging renewable energy technologies.

Sergio C. Capareda, PhD, PE
Professor and Faculty Fellow
Biological and Agricultural Engineering Department
Texas A&M University

Author

 **Sergio C. Capareda** is a Professor and Faculty Fellow at Texas A&M University (TAMU), USA. He holds a Bachelor of Science degree in agricultural engineering from the University of the Philippines at Los Baños (UPLB); a Master of Engineering degree in energy technology from the Asian Institute of Technology (AIT), Thailand; and a PhD in agricultural engineering from TAMU.

Capareda began his academic career in the field of renewable energy at UPLB. He developed the UPLB Biomass Energy Laboratory with funding from the Philippine Department of Energy (PDOE) and was Program Director for two World Bank–funded projects implemented by the PDOE on rural electrification and market assessment of renewable energy in the Philippines. He also developed the Biomass Energy Resource Atlas for the Philippines with funding from the U.S. Department of Energy and the U.S. Agency for International Development (USAID).

Upon joining the faculty at the TAMU Department of Biological and Agricultural Engineering in 2005, Capareda was tasked to redevelop the alternative energy program of the department. He developed and established the BioEnergy Testing and Analysis (BETA) Laboratory (https://betalab.tamu.edu) that year. The BETA Lab is currently being expanded to cover research and development for other major renewable energy technologies, such as solar, wind, and biomass power.

Dr. Capareda has authored or co-authored more than 100 refereed journal publications since 2003, two book chapters in the field of renewable energy and air quality, and a textbook titled *Introduction to Biomass Energy Conversions*. He holds a patent on Integrated Biofuel Production System and a patent for a Pyrolysis and Gasification System for Biomass Feedstock, which has now been licensed to private companies with various heat and power generation projects from various biomass resources, such as wood chips, poultry litter, municipal sludge, and municipal solid wastes (MSW).

A four-time recipient of the Returning Scientist Awardee from the government of the Philippines and a two-time recipient of the USAID-Stride Visiting Professorship Award, Dr. Capareda has been providing continuous support to various universities in the Philippines as a key consultant on their renewable and air quality teaching, research, and development initiatives.

Dr. Capareda is a licensed Professional Engineer in Texas and an active member of the American Society of Agricultural and Biological Engineers (ASABE).

1

Introduction to Renewable Energy

Learning Objectives

Upon completion of this chapter, one should be able to:

1. Define renewable energy and recognize when an energy resource becomes unsustainable.
2. Enumerate the advantages and disadvantages of renewable energy technologies.
3. Enumerate and describe the various renewable energy technologies.
4. Define energy conversion efficiencies and understand conversion efficiencies for each renewable energy technology.
5. Relate the overall environmental and economic issues concerning renewable energy conversion systems.

1.1 Introduction

Renewable energy is defined as energy from resources that are replenished at the same rate that they are used (Bhattacharya, 1983). A resource will cease to qualify as renewable if it is used in an unsustainable manner. Solar energy is a prime example of a renewable energy resource. The energy received by the earth from the sun each day is simply reradiated back into the atmosphere. Solar energy is utilized by simply delaying the return of this energy and converting it into some useful temporal applications. Most other renewable energies, such as biomass, wind, hydro power, and wave energy, would directly and indirectly be coming from or influenced by solar energy. By the above definition, geothermal energy does not fall into the category of renewable energy. This is because the process of harnessing geothermal energy removes heat from the earth's core and releases the energy to the atmosphere. There is no way the energy may be fully returned to the earth's core. However, for practical purposes, since the amount of energy currently removed from geothermal resources is still a small fraction of the overall available energy from earth's heat at its core, we can consider geothermal energy as a renewable resource. Widespread extraction of this heat would cause this resource to become unsustainable in the future and may have to be done at a reduced rate.

Most other renewable energy resources are results of the presence of the sun. Wind movements are created by uneven wind densities caused by uneven heating of the earth. Biomass energy is a result of photosynthetic activities from the sun, water, and nutrients from land. Hydro power is also a consequence of the hydrologic cycles of water from evaporation in water bodies to precipitation in some land masses, resulting in river flows. Of these discussed, geothermal energy is the only one not dependent on the sun.

This book will cover all the possible renewable energy resources that could become significant to mankind in the near future. These include solar, wind, biomass, and water power; geothermal energy; salinity gradient power; fuel cells; tidal and wave energy; ocean thermal energy; and the controversial cold fusion and gravitational energy. A chapter is also devoted to small-scale renewable energy conversion processes from piezoelectric systems as well as animal and human power. When (or even before) fossil fuel energy resources are depleted, the earth may solely depend on the above-mentioned renewable energy resources. The human race will have to learn how to harness these renewable energy resources in a sustainable manner. For example, the widespread utilization of biomass energy would cause deforestation in some areas and must be prevented. Biomass energy needs could also compete with food requirements of the growing population and must also be harnessed effectively.

1.2 Advantages and Disadvantages of the Use of Renewable Energy Resources

The world's energy consumption has increased at a steady rate over several hundred years. Man has shifted from the use of fuel wood to coal and, currently, the massive use of fossil fuels. It is believed that these resources (fossil fuels) will ultimately be replaced by renewable energy resources. The following section enumerates the advantages and disadvantages of these renewable energy resources.

1.2.1 Advantages

Perhaps the main advantage of using a renewable energy resource is its ability to sustain itself when used properly. There are other numerous additional advantages:

a. They are readily available in any part of the world.

b. Renewable energy could be extracted with limited resources.

c. The use of renewable resources will not cause serious harm to the environment.

d. The use of renewables provides diversity, balancing the harmful effects of using just a single energy or power source.

e. Wide ranges of products can now be derived from renewables.

f. Some renewable energy technologies do not require sophisticated components, parts, and materials.

1.2.2 Disadvantages

a. Renewables usually require large tracks of land.

b. Most renewable energy resources are not consistently available, and timeliness of production and use may be a challenge.

c. Storage of energy and power will also be a challenge for most renewables—some are available in large amounts for short periods, and this necessitates expenses on energy storage facilities.

d. Some renewables require special, costly materials (e.g., solar photovoltaic [PV] cells) with substantial carbon footprints during manufacturing.

e. Economics—for example, while the price of the solar PV cell has gone down below $1/watt peak (Wp) [$746/hp], the balance-of-systems costs (i.e., wirings, transmission, battery banks, etc.) are still relatively high.

The continued population growth in many countries will inevitably lead to increased energy use. If these countries do not have indigenous energy resources, they will be adversely dependent on foreign countries that control most of commercial fuel. These developing countries will have to rely on renewables to become energy independent and sustainable. There is an urgent need for these countries to expand their energy mix to include a larger percentage of renewables. Unfortunately, some renewable conversion systems will have to be made more economical than they are presently and would definitely require more research input from their governments.

Advantageously, many renewables have low carbon dioxide emissions as well as smaller carbon footprints. Some renewable processes, such as biomass conversion or photosynthesis, will surely help in delaying global warming events because they use up tonnes of carbon dioxide. Example Problem 1.1 shows that, theoretically, it would take about one and a half tonnes [1.65 tons] of carbon dioxide to produce a tonne [1.1 ton] of glucose sugar.

1.3 Renewable Energy Resources

1.3.1 Solar Energy

Solar energy is considered the prime renewable energy resource because most other renewables depend on the sun. There are eight thermodynamic pathways for solar energy conversion. These are depicted in Figure 1.1 and are explained simply in the succeeding sections.

a. *Solar Thermal Conversion:* Solar thermal conversion is the use of solar energy directly as heat. The two most common technologies in solar thermal conversion are solar drying and cooling. Drying of crops and agricultural products using the sun is an age-old process and is still in widespread use in many developing countries. Solar cooling, on the other hand, has not gained popularity. There had been simple solar refrigerators designed in the past, but those have not gained popularity. A revival of such technologies may be appropriate at this time.

b. *Solar Thermochemical Conversion:* Solar thermochemical conversion includes the use of chemicals that absorb heat and release heat for future use. One would also simply call this conversion pathway thermal to chemical conversion. The chemicals are sometimes simply called absorbents, whose phase may or may not change during absorption of solar energy. The classic example is that of potassium oxide conversion shown in Equation 1.1:

$$4KO_2 + Solar\ Energy \rightarrow 2K_2O + 3O_2 \tag{1.1}$$

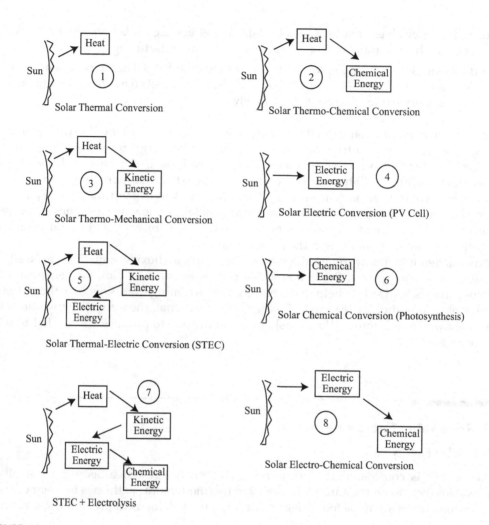

FIGURE 1.1
The eight thermodynamic pathways for solar energy conversion (adapted from Bhattacharya, 1983).

Equation 1.1 occurs over a temperature range of 300°C to 800°C [572°F to 1,472°F] (Duffie and Beckman, 2006) with a heat of decomposition of 2.1 MJ/kg [904.5 Btu/lb]. A similar reaction occurs for the decomposition of lead oxide (PbO_2) but the heat of decomposition is substantially less at 0.26 MJ/kg [112 Btu/lb]. Thus, other metal oxides may be future candidates for this reaction, provided they have higher heat of decomposition rates. The above phenomena are chemical changes in the compounds due to the action of heat (which may come from solar energy). Other chemicals only undergo phase changes, such those of Glauber's salt (Na_2SO_4). Upon heating Glauber's salt using solar energy at around 34°C [93.2°F], a solution is given off—the moisture with solid Na_2SO_4. Upon cooling, it gives off sensible heat say, 79.2°C [97°F] and heat of fusion for useful future applications.

c. *Solar Thermo-Mechanical Conversion:* The solar thermo-mechanical conversion process simply makes use of solar energy to heat up a fluid medium, usually water, to generate steam. The generated steam is used to run steam engines or steam turbines, thereby producing mechanical work. Any fluid medium may be used to generate mechanical power if it is able to turn the turbine blades. During the peak popularity of micro-turbines (shafting running at about 10,000 rpm), the coupling of solar energy could have been an ideal option, except that it is quite difficult to produce that much speed to turn the blades without concentrating the solar energy. This option could be revisited in the future when micro-turbine costs decrease and solar technologies that generate hot gases to run turbine shafts at high speeds become available.

d. *Solar Electric Conversion:* Solar electric conversion is simply the conversion of solar radiation directly into electrical power. The development of solar PV cells from various materials has made this conversion pathway one of the most dominant solar energy applications. Commercial PV systems are now ubiquitous in many countries worldwide with the cost per watt peak (Wp) decreasing every year. It will be demonstrated later that the cost is still the primary barrier against widespread adoption especially for large-scale applications.

e. *Solar Thermal Electric Conversion:* Solar thermal electric conversion (STEC) is the extension of item (c) above with the engine coupled to a generator. A steam turbine or a mechanical internal combustion engine may be used for this purpose. The key to this process is the use of heat to increase the kinetic energy of a fluid medium such that a mechanical device may be actuated to run a generator.

f. *Solar Chemical Conversion (Photosynthesis):* Solar chemical conversion is nothing more than the use of solar energy for the direct conversion into chemical energy. Photosynthesis or biomass production would fall under this category. The energy from the sun is used to combine carbon dioxide in the air and water from the soil to generate plant biomass. The classical photosynthesis chemical equation for biomass conversion is illustrated in Equation 1.2:

$$6CO_2 + 6H_2O + Solar\ Energy \rightarrow C_6H_{12}O_6 + 6O_2 \qquad (1.2)$$

The representative main product of Equation 1.2 is glucose. In most biomass plants, water is supplied from the roots, with the leaves collecting carbon dioxide via the stomata and sunlight captured by the chloroplasts in the leaves (Jones and Jones, 1997). Six moles of carbon dioxide are needed to produce a mole of glucose, with the release of 6 moles of oxygen.

g. *STEC plus Electrolysis:* STEC generates electricity, and predictably there will be an endless application of this process technology. An obvious application is the conversion of water into its hydrogen and oxygen components using electricity. The hydrogen produced may be used as fuel for many other power or fuel conversion systems. Electrolysis, by definition, is the process in which a chemical is broken down into its components using electricity. Thus, any representative chemical that would produce a useful product of fuel may be used. Water is the easiest choice since it is perhaps the cheapest chemical fluid to use.

h. *Solar Electrochemical Conversion:* Solar electrochemical conversion is the use of solar PV cells to cause a wide range of chemical transformation, generating fuel or energy along the way. It is very similar to item (g) above except that the electricity comes from the PV cells and not via thermal or mechanical means. Likewise, in this category, one may be able to enumerate several representative chemicals that may undergo transformations to generate additional heat, energy, or power.

Example 1.1: Carbon Dioxide Production from Glucose Conversion

Determine the amount of carbon dioxide needed (in tonnes) to produce a tonne of glucose using Equation 1.2.

SOLUTION:

a. Equation 1.2 may be converted into units of weight using the molecular weights of the compounds involved. The molecular weight of carbon is 12 g/mol, hydrogen is 1 g/mol, and oxygen is 16 g/mol:

$$6CO_2 + 6H_2O + Solar\ Energy \rightarrow C_6H_{12}O_6 + 6O_2$$

b. The amount of CO_2 required per tonne of glucose ($C_6H_{12}O_6$) will be calculated as follows:

$$\frac{6CO_2}{C_6H_{12}O_6} = \left[\frac{6\times(12+2\times(16))}{1\times[(6\times12)+(12\times1)+(6\times16)]}\right] = \frac{264\ g}{180\ g} = 1.5\frac{tonne\ CO_2}{tonne\ C_6H_{12}O_6}$$

c. Theoretically, it would require about one and a half tonnes [1.65 tons] of CO_2 to generate 1 tonne [1.1 tons] of glucose sugar.

The increased carbon dioxide concentration in the earth's atmosphere, due to widespread use of fossil fuels, is widely accepted to have caused global warming. To reduce the concentration of carbon dioxide, the best solution is to plant more trees to generate biomass. These biomass resources may then be used to produce renewable fuels to widely displace fossil fuels from non-renewable crude oils.

1.3.2 Wind Energy

Wind energy results from the uneven heating of the earth by the sun. Part of the earth is being warmed by the sun during the day, causing the rise of air mass, being displaced by colder air masses, creating air movements. The simplest way to estimate the energy that may be derived from wind is to understand the wind kinetic energy equation shown in Equation 1.3. The energy from the wind is proportional to the product of its mass and the square of its velocity. In the metric system, the mass is in units of kg and the wind velocity is in units of meters per second:

$$KE = \frac{1}{2}mv^2 \tag{1.3}$$

where
  m = mass of air, kg
  v = wind velocity, m/s

The units are simplified below such that the product using the above units will result to units of energy, Joule [Btu], as shown below:

$$KE(Joule) = \frac{1}{2} \times (kg) \times \left(\frac{m}{s}\right)^2 = \frac{1}{2} \times \left(\frac{kg \times m}{s^2}\right) \times m = \frac{1}{2} N \times m = \frac{1}{2} Joule$$

$$KE(Btu) = \frac{1}{2} Joule \times \frac{Btu}{1,055 \ Joules} = \frac{Btu}{2,110} = 0.000474 \ Btu$$

When the units are substituted, the following resulting calculations are made to arrive at the final unit of kinetic energy in Joules. Note that by Newton's law of motion, $F = m \times a$, the unit of force in metric units, is in Newtons; the mass is in kg; while the acceleration due to gravity is in units of m/s². Thus, 1 N is equal to 1 kg-m/s². If this relationship is used to simplify the units in Equation 1.3, one will come up with kinetic energy with a unit of N-m, which is equivalent to a Joule. Example Problem 1.2 shows how this equation is used with corresponding units of conversion.

Example 1.2: Kinetic Energy Available from Moving Wind

Determine the kinetic energy available from 10 kg of air moving at a speed of 10 m/s. Report the answer in units of Joules and in Btu.

SOLUTION:

a. Using Equation 1.3, the given values are simply plugged into the equation:

$$KE = \frac{1}{2} \times 10 \ kg \times \left(\frac{10 \ m}{sec}\right)^2 \times \frac{N - sec^2}{kg - m} = 500 \ N - m = 500 \ J$$

b. Thus, the equivalent unit in the English system is shown below:

$$KE = 500 \ J \times \frac{Btu}{1,055 \ J} = 0.47 \ Btu$$

c. Note the difference in the order of magnitude of Joules and Btu. One should always remember the conversion of Joules to Btu using the relationship 1,055 Joules/Btu.

Wind power is a growing market and the yearly production output has increased significantly over the last 10 years. Between 2008 and 2012, wind power has provided about 36.5% of all new generating capacity in the United States (EIA, 2018). It is currently the fastest-growing source of electricity production in the world (Wind Energy Foundation, 2016). The total installed wind capacity in the United States at the end of 2017 was about 89,077 MW. The state of Texas had the highest installed capacity in the United States with a reported generating capacity of more than 22,637 MW at of end of 2017. This was followed by the state of Oklahoma (7,495 MW) and the state of Iowa (7,308 MW) (AWEA, 2018). Wind power production is very dynamic with one state leading the others interchangeably besides Texas.

1.3.3 Biomass Energy

Biomass energy is one thermodynamic pathway for solar energy conversion through photosynthesis. Every year, plant biomass is produced from solar energy, water, and nutrients from the soil or water bodies. The product is not a simple glucose compound, as shown in Equation 1.2, but a wide range of compounds that would still have to be converted into useful heat or fuel. The plant biomass is a complex mixture of chemical compounds, such as cellulose, hemi-cellulose, sugars or carbohydrates, proteins, and fats, together with numerous inorganic compounds. Consequently, biomass comes with variable heating values or energy contents. These biomass resources are used primarily as food for the increasing population worldwide, but they are now used to generate fuels for transport and chemicals for industrial uses. There is always the question of food versus fuel needs, and this friction makes the biofuel developed from biomass rather expensive.

The three most popular products from biomass that contribute to commercial fuel requirements are bio-ethanol, biodiesel, and biogas. Bio-ethanol comes from fermentation of simple sugars or the conversion of starch into ethanol—via conversion of starch into soluble sugars (saccharification) and fermentation of sugars into ethanol. Both sugars and starch are food products and may be more important for human consumption than for use in vehicles for transport. Ultimately, bio-ethanol must come from the ligno-cellulosic portion of the biomass so that it will not compete with food sources of the human population. Unfortunately, this conversion process is still quite expensive due to the high costs of enzymes that are able to break down the ligno-cellulose into simple sugars. Ruminants are humans' only competitors for ligno-cellulosic biomass resources—their primary food source.

The conversion of glucose sugar into ethanol is depicted ideally by Equation 1.4. Example Problem 1.3 shows that more than 1.5 kg [3.3 lbs] of sugar is needed to generate a liter [0.2642 gallons] of ethanol.

$$C_6H_{12}O_6 \; + \; yeast \; \rightarrow \; 2C_2H_5OH \; + \; 2CO_2 \; + \; Heat$$
$$\left(180 \text{ kg}\right) \qquad\qquad \left(92 \text{ kg}\right) \quad \left(88 \text{ kg}\right) \qquad\qquad (1.4)$$

Example 1.3: Amount of Sugar Needed per Volume of Ethanol Produced

Determine the amount of pure sugar (in kg) needed to generate a liter of pure ethanol following Equation 1.4 above. Convert these units into number of pounds of sugar needed per gallon of ethanol. Assume the density of ethanol to be 0.789 kg/L. For other conversion factors used in this example, refer to the Appendix at the end of this textbook.

SOLUTION:

a. Amount of pure glucose needed to generate a liter of ethanol:

$$\frac{kg \; glucose}{Liter \; ethanol} = \frac{180 \; kg \; glucose}{92 \; kg \; ethanol} \times \frac{0.789 \; kg \; ethanol}{Liter \; ethanol} = 1.54 \frac{kg}{L}$$

b. Amount of sugar needed to produce a gallon of ethanol:

$$\frac{lbs \; sugar}{gallon \; ethanol} = \frac{1.54 \; kg \; glucose}{Liter \; ethanol} \times \frac{2.2 \; lbs}{1 \; kg} \times \frac{3.785 \; L}{gallon} = 12.8 \frac{lbs}{gallon}$$

Ethanol is not the only fuel being commercialized from biomass. The second-most popular biofuel is biodiesel that has been converted from various vegetable oil resources. In the United States, soybean oil is the prime source of commercial biodiesel. However, due to the rising cost of soybean oil as feedstock, there came changes in the primary source of material for making biodiesel, and more became sourced from cheaper oil such as animal fats and recycled/used oil. Aside from biodiesel, numerous other commercial hydrocarbon fuels are currently being produced from biomass. The science behind the production of commercial gasoline, aviation fuel, and diesel fuel from biomass has been demonstrated (Capareda, 2013). Unfortunately, it is the economics of production, coupled with the high cost of the feedstock itself, that will delay the widespread use of these products from biomass. More research projects are now directed toward various other feedstock variants, such as microalgae, to bring the overall cost of production to the lowest possible level. Still, there are many issues confronting the biomass-to-fuel route, including logistics of transport, cost of biomass, as well as processing cost (in particular, enzymes and other catalysts used) to make biomass fuel competitive with hydrocarbon fuels from crude oil. At this point, there is not one biofuel that is the predominant replacement for conventional hydrocarbon fuel. However, gasoline in the United States will continue to contain ethanol (at least 10%–15%), while diesel engines will contain biodiesel (at least 5%–10%) because of the implementation of the U.S. Renewable Feel Standards (USEPA, 2015). The standards call for blending bio-ethanol in gasoline fuel and biodiesel in diesel engines. Each year, the target amount of biofuels produced will increase, and this would guarantee the sale of such fuels. The government will issue renewable energy identification numbers (RINS). For example, the proposed 2016 target for cellulosic biofuels is 206 million gallons [779.71 million liters] (mainly bio-ethanol), 1,800 million gallons [6,813 million liters] of biomass-based diesel, and 3,400 million gallons [12,869 million liters] of advanced biofuels (usually bio-butanol or some aviation fuel-related products) (USEPA, 2015). With this standard, biorefineries in the United States will continue to produce biofuels despite the low price of crude oil. At the beginning of each year, all biofuel production facilities would be issued RINs and would be guaranteed sales of their biofuel product.

1.3.4 Hydro Power

Hydro power produces 21% of the world's electricity supply (EIA, 2017). Over 1 million MW of combined power output was produced in the year 2012 with a theoretical potential of around 2,800 GW. There are reportedly 2,000 hydro power plants in the United States with a total capacity of 92,000 MW (USDOE, 2013). The largest hydro power plant in the United States is the 7,600 MW Grand Coulee Power Station in Columbia River, Washington State.

Worldwide, it is debatable as to which country has the biggest hydro power plant. Pentland (2013) reported that the Three Gorges Dam in China is the largest with a generating capacity of 22,500 MW, and in second place is the Itaipu Dam of Brazil and Paraguay. Conca (2017) pointed out that this ranking done by generating capacity is misleading since the installed capacity is the maximum power a plant could produce only when everything is running perfectly. However, all hydro power plants do not run continuously, and production is dependent on their power factor. The power factor is the ratio of actual power delivered versus the installed capacity. Conca (2017) reported that in 2016, the power factor of the Three Gorges Dam is 48% (or producing only 93 billion kWh versus the 193 billion kWh maximum). The power factor of Itaipu Dam is 84%, and Itaipu actually produces 103 billion kWh of energy and not its nameplate-installed capacity of 14,000 MW. In this scenario, the Itaipu Dam is still considered the largest producer of hydro power in the world.

The Asian continent has contributed 24% of the world's power supply from hydro power plants, followed by North America (23%) and Europe (17.1%). The countries with the most power production from hydro power plants include Canada (315 GWh), China (309 GWh), Brazil (282 GWh), and the United States (255 GWh). Developing countries should likewise take advantage of this renewable energy resource and continue to develop micro–hydro power units for rural villages.

Bodies of water were used commonly in ancient times to generate mechanical and electrical power. Numerous water wheels were designed in the past and installed along river banks to convert kinetic energy into mechanical power to drive sawmills and grinders. Perhaps the most common application of falling water bodies nowadays is for electrical power production by taking advantage of its dynamic head. Equation 1.5 illustrates the power derived from a water body with a given volumetric flow rate, Q (in units of kg/s), and a dynamic head, H (in meters). Power is simply calculated from the product of these two parameters. Example Problem 1.4 shows how to generate about 200 W of electrical power from a water body with a volumetric flow rate of about 35 L/m and a dynamic head of just 1.5 m:

$$Power\,(Watts) = Q \times H \qquad\qquad (1.5)$$

where
Power = theoretical power from water, kW
Q = volumetric flow, kg/s
H = dynamic head, m

Example 1.4: Theoretical and Actual Power Derived from Water Stream

Determine the theoretical and actual power that can be derived from a water stream with a volumetric flow rate of 35 L/s [554.82 gpm] and with a dynamic head of 1.5 m [4.92 ft]. Assume an overall conversion efficiency of 40%.

SOLUTION:

a. Equation 1.5 will be used with the density of water value of 1 kg/L [8.327 lbs/gal] and other conversion factors to arrive at the correct units of power in watts. This will require the use of the gravitational constant 9.8 m/s² [32.2 ft/s²] as well as Newton's Law relationship as shown below:

$$Theo.Power\,(W) = \frac{35\,L}{s} \times \frac{1\,kg}{L} \times \frac{1.5\,m}{1} \times \frac{N-s^2}{1\,kg-m} \times \frac{9.8\,m}{s^2} \times \frac{W-s}{N-m} = 515\,W$$

$$Theoretical\,Power\,(hp) = \frac{515\,Watts}{1} \times \frac{hp}{746\,Watts} = 0.69\,hp$$

b. The actual power that could be derived will be less due to conversion efficiencies as shown below:

$$Actual\,Power\,(W) = Theo.Power \times Efficiency = 515\,W \times 0.40 = 206\,W$$

$$Actual\,Power\,(hp) = 206\,W \times \frac{hp}{746\,Watts} = 0.276\,hp$$

The example shows that over 200 W (0.268 hp) of continuous power may be generated even if the dynamic head is only one and a half meters (~5 ft) with a flow of 35 L/s. Unfortunately, 35 L/s is equivalent to about 555 gpm of water, which is quite a lot of water flow. If the flow is not enough, the dynamic head should be high enough to generate the

FIGURE 1.2
Low-cost micro–hydro power units for rural villages (courtesy of Asian Phoenix Resources, Ltd, Canada with the trademark name PowerPal™, used with permission).

same magnitude of electrical power. Developing countries such as Vietnam have developed these micro–hydro power units ranging from 200, 500, to 1,000 W of electrical power output. Asian Phoenix Resources Ltd of Canada is currently distributing this commercial micro-hydroelectric generator unit, as depicted by the flyer cover shown in Figure 1.2.

1.3.5 Geothermal Energy

The world's geothermal energy output in 2015 exceeded 12.8 GW. Figure 1.3 shows the world's geothermal provinces that are mostly concentrated along the Pacific Rim and

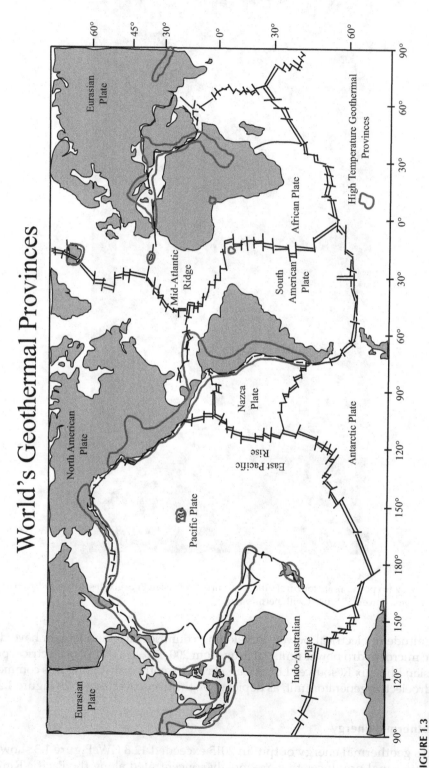

FIGURE 1.3
Map showing the world's geothermal provinces.

some parts of Europe. The regions with the highest geothermal resource bases are the North American continent (21%) and Asia (21%), followed by Eastern Europe (17%), South America (14%), and Africa (13%) (GEA, 2015). Most other regions have below 10% of the total contribution, with the Pacific Islands, including Hawaii, comprising 9% of the total. In the year 2015, the top three countries with the highest geothermal output included the United States (3,386 MW), the Philippines (1,904 MW), and Indonesia (1,191 MW). Note that these countries are situated ideally along the Pacific Coast. In the United States, about 80% of the total electrical power output comes from the California geysers. The total U.S. output is around 31.6% of the world's output (GEA, 2015).

Harnessing energy from geothermal resources is nothing more than making use of high temperature and superheated steam to run steam turbines via the thermodynamic steam cycle or the Rankine Cycle. The working fluid in this cycle is alternately vaporized and condensed. Ideally, dry saturated steam enters the steam turbine and expands to higher pressure, moving the turbine blades, which are coupled to a generator to produce electrical power. The steam is then condensed to a saturated liquid at constant pressure and temperature. The liquid is then pumped back to the reservoir under ideal conditions. Of course, the ideal cycle is never achieved in most conventional power plants, and numerous losses are incurred throughout the power production cycle. The technology for the production of electrical power has been quite established, and minimal breakthroughs or advanced research projects are being developed. Perhaps attention should be focused on taking advantage of low-temperature heat for common applications of heat rather than electrical power.

A most interesting success story concerning the use of geothermal resources for electrical power is that of a Reykjavik community in Iceland. Prior to utilizing geothermal energy, 90% of the town's heat was provided by fossil fuels, causing massive pollution. Today, more than 95% of the country's urban population uses geothermal energy to heat their homes, and almost 100% of the country's energy comes from renewable sources (Mims, 2008). Reykjavik now remains one of the cleanest and most sustainable cities in the world primarily due to its fortunate geographical location. Reykjavik has a geothermal resource base underneath its land. In fact, most sidewalks there are heated using geothermal energy.

Engineers interested in electrical power generation should know the basic thermodynamic cycles for the conversion of steam from a geothermal well into useful energy. The majority of electricity produced in the United States comes from steam power plants. These plants operate by the same principle regardless of whether the steam is coming from combustion of coal or natural gas or from fission processes in a nuclear power plant. The ideal thermodynamic cycle used is the Rankine cycle depicted in Figure 1.4.

The thermal efficiency (η_{th}) of the ideal Rankine Cycle is given by Equation 1.6. The numerator in this equation is the work done on the fluid (i.e., the difference between the work output [W_{out}] in the turbine and the work input [W_{in}] by the pump). The denominator is the energy input or the energy available from the steam (Q_{in}). Example Problem 1.5 shows how Equation 1.6 is to be used, while the chapter on geothermal energy will describe how these given values were estimated:

$$\eta_{th} = \frac{W_{out} - W_{in}}{Q_{in}} \times 100\% \qquad (1.6)$$

The Ideal Rankine Cycle

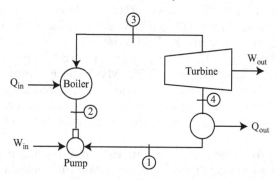

FIGURE 1.4
The ideal Rankine Cycle used in geothermal power systems.

Example 1.5: Ranking Cycle Theoretical Conversion Efficiency

Determine the theoretical thermal conversion efficiency of a Rankine Cycle with the following data: (a) W_{out} = 560 kJ/kg [0.2413 Btu/lb], (b) W_{in} = 3 kJ/kg [0.001293 Btu/lb], and (c) Q_{in} = 2,400 kJ/kg [1.034 Btu/lb]. The mass flow rate of the steam is around 22,500 kg/hr [49,500 lbs/hr].

SOLUTION:

a. It would be intuitive to convert all values to energy units rather than energy per unit of weight, even though the energy per unit of weight may also be used for efficiency calculations. Thus, the work output may be converted into kW:

$$W_{out} = 560 \frac{kJ}{kg} \times \frac{22,500\ kg}{hr} \times \frac{1\ hr}{3,600\ s} \times \frac{kW-s}{kJ} = 3,500\ kW$$

$$W_{out} = 3,500\ kW \times \frac{1,000\ Watts}{1\ kW} \times \frac{hp}{746\ Watts} = 4,691.7\ hp$$

b. Likewise, the W_{in} and Q_{in} will be converted the same way:

$$W_{in} = 3 \frac{kJ}{kg} \times \frac{22,500\ kg}{hr} \times \frac{1\ hr}{3,600\ s} \times \frac{kW-s}{kJ} = 18.75\ kW$$

$$W_{in} = 18.75\ kW \times \frac{1,000\ Watts}{1\ kW} \times \frac{1\ hp}{746\ Watts} = 25.13\ hp$$

$$Q_{in} = 2,400 \frac{kJ}{kg} \times \frac{22,500\ kg}{hr} \times \frac{1\ hr}{3,600\ s} \times \frac{kW-s}{kJ} = 15,000\ kW$$

$$Q_{in} = 15,000\ kW \times \frac{1,000\ Watts}{1\ kW} \times \frac{1\ hp}{746\ Watts} = 20,107.24\ hp$$

c. Thus, the efficiency may be calculated:

$$\eta_{th} = \frac{W_{out} - W_{in}}{Q_{in}} \times 100\% = \frac{3,500\ kW - 18.75\ kW}{15,000\ kW} \times 100\% = 23.2\%$$

1.3.6 Salinity Gradient

Salinity gradient technology takes advantage of the osmotic pressure differences between salt water and fresh water. If a semi-permeable membrane is placed between sealed bodies of salt water and fresh water, the fresh water will gradually permeate through the filter by a process called osmosis. This occurrence will create pressure differences between these two bodies of water and energy may be extracted. This is one of the renewable technologies with no emissions of CO_2 or other effluents that would have global environmental effects. Unlike solar and wind energy, this process is non-periodic and may be suitable for small-scale or large-scale applications. Unfortunately, no large-scale plants have been established as yet, perhaps due to the expected high capital cost of plant construction, including the material cost of membranes.

Salinity gradient technology is still believed to be one of the largest sources of renewable energy that has yet to be fully exploited. Salt water is one of the earth's most abundant resources and should be utilized for renewable energy or power. The osmotic pressure of seawater is about 20 atmospheres, and when this salt water mixes with fresh water, the free energy estimate is equivalent to about 200 m of energy (hydraulic head) lost and believed to be the source of energy to drive turbines and electrical generators (Emami, et al., 2013). Another study shows that the energy density is about 22,300 kW [29,892.76 hp] for every 3,000,000 cubic meters per day [79.26 million gallons/day] of salt water (Loeb, 2002) or about 7.43 W/m³ [0.000282 hp/ft³] of seawater. Because of the vast volume of seawater available throughout the earth, this potential energy is quite significant—an estimated total global potential of around 2.6 TW (terrawatts, or 10^{12} watts) was reported (Wick and Schmitt, 1977), which is much more than the global energy consumption in 2008 (Veerman, et al., 2011). The two most common salinity gradient systems are pressure retarded osmosis (PRO) and reversed electro-dialysis systems (RED). These systems will be discussed extensively in the Salinity Gradient chapter.

1.3.7 Fuel Cells

Quite a few technical materials (books and articles) have been written about fuel cells. This is also a technology that is not very intuitive to work with from a non-technical standpoint. The concept of an electrochemical conversion device is not uncommon to many engineering processes. If one is able to generate renewable fuel such as hydrogen gas and would want to convert this into electrical power without combusting this potentially explosive fuel, one would simply need two electrodes that would convert the chemical energy of this fuel directly into electrical energy. Usually on the anode side of the electrode, hydrogen is dissipated into its ionic component (or will ionize) and would release some of its electrons (Equation 1.7a). On the cathode side, one would need an oxidant that would take in this ionized hydrogen component and react with the oxidant to produce water (Equation 1.7b). This event will create a voltage potential through the continuous use of the hydrogen ions, and the electrons will bring about electric current:

$$\text{Anode Side} \quad 2H_2 \rightarrow 4H^+ + 4e^- \tag{1.7a}$$

$$\text{Cathode Side} \quad 4e^- + 4H^+ + O_2 \rightarrow 2H_2O \tag{1.7b}$$

In the combustion of hydrogen and oxygen, heat is released. In the electrochemical reaction between hydrogen and oxygen in a fuel cell, electricity and heat are produced. The heat produced by combustion is not the same as the electricity produced in a fuel cell. This difference is thermodynamically referred to as Gibbs free energy. The high heating value for the combustion of hydrogen is around 285.8 kJ. The theoretical efficiency of fuel cells is around 83% as shown in Example Problem 1.6.

Example 1.6: Theoretical Efficiency of Fuel Cells

Calculate the theoretical electrical energy efficiency of converting hydrogen and oxygen into electrical energy via fuel cell using the following data: (a) the electricity produced was 237.2 kJ/mol, and (b) the heating value of hydrogen was 285.8 kJ/mol. Compare this efficiency with the practical electricity production from a hydrogen fuel cell with an electrical output of only 154 kJ/mol.

SOLUTION:

a. The electrical energy efficiency (*EEE*%) is simply the ratio of output over energy input as follows:

$$EEE(\%) = \frac{electricity\ produced}{HHV\ of\ fuel\ used} \times 100\% = \frac{237.2\ kJ}{285.8\ kJ} \times 100\% = 83\%$$

b. Thus, the theoretical efficiency is higher than when hydrogen is combusted in an internal combustion engine, with overall combustion efficiencies of around 30% to 35%.

c. The practical efficiency is calculated as follows:

$$Practical\ EEE(\%) = \frac{154\ kJ/mol}{285.8\ kJ/mol} \times 100\% = 54\%$$

d. The practical efficiency is likewise higher than in internal combustion engines.

One interesting area in the field of fuel cells is the possibility of utilizing microbes for the production of electrical energy, called microbial fuel cells. The extensive work of some pioneers in this field, such as Professor Logan of Penn State University and of some European researchers, has been quite remarkable (McAnulty, et al., 2017). These groups were able to identify certain groups of microbes available in wastewater that have the potential to generate an appreciable amount of electrical power. Their goal is to treat the wastewater while producing electrical power that is more than the energy required to bring down the biological oxygen demand or chemical oxygen demand of wastewater. If these systems can be proven to be economically feasible, it could create a sustainable energy source while reducing environmental pollution.

1.3.8 Tidal Energy

Tidal energy comes as a result of attractive forces between the earth, the sun, and the moon. For people along shorelines, the variation in ocean tides over the course of the day may seem insignificant. However, because of the vastness of shorelines as well as the magnitude of seawater that may be used for energy production, tidal energy is a renewable energy resource that may be of great potential for replacing fossil fuels. The rise and fall of water levels as a result of gravitational attraction between the moon and the earth and between the sun and the earth are called semi-diurnal tides. These tides last

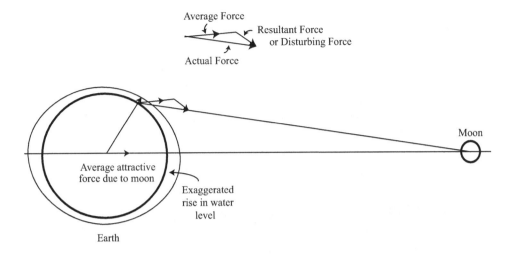

Attractive Forces Between the Earth and the Moon

FIGURE 1.5
Attractive forces between the earth and the moon that cause the variations in water levels along the shores.

around 12 hours and 25 minutes. Spring tides are those variations observed during the full moon and new moon as the earth, sun, and moon line up. During this period, the tides are higher than normal. If the earth–moon and the earth–sun relationships are at right angles, the gravitational forces are subtractive, causing neap tides that are subdued or low (Bhattacharya, 1983). Figure 1.5 shows the typical attractive forces between the earth and the moon causing differences in water levels across the globe. There are two "humps" shown, and these occur every 24 hours and 50 minutes, the time of the moon's rotation of the earth. Similar tides are produced by the sun, except the cycle periods are different and spread out over the course of a year, likewise due to the earth's rotation around the sun. Thus, the peak variations in sea tides actually happen every 12 hours and 25 minutes, and the maximum difference in height (between low tide and high tide) is called the tidal range. On average, these tidal ranges could vary between 3 meters to as high as 5 meters in many parts of the world (Encyclopedia Britannica, 2018). However, in some prime locations, the highest tides could have a range between 12 and 16 meters. The discussed location is found in the Bay of Fundy on the Atlantic Coast of Canada (Encyclopedia Britannica, 2018). This place has been a tourist spot that people visit just to observe these great tidal variations.

The amount of energy that could be derived from tides is calculated the same way as hydro power plants using Equation 1.5. Example Problem 1.7 shows the magnitude of energy that may be generated from tides.

Example 1.7: Actual Power and Energy from Tides

Residents of a household living near the shore would like to build a reservoir that could generate tidal power for their needs. The yearly average tide is about 1 meter [3.28 ft]. The owner built a reservoir that can contain 5,000 cubic meters [1.321 million gallons] of sea water and plans to drop about 55 L/s [871.86 gpm] of this water when the tide is lowest with an average head of 0.5 meters [1.64 ft] for about 12 hours. Estimate the amount of actual power generated during the given period if the overall conversion efficiency is 60%. Assume the density of seawater to be about 1.03 kg/L [8.577 lbs/gal]. How much energy (in kWh) was produced during this period?

SOLUTION:

a. Equation 1.5 will be used with the density of sea water at 1.03 kg/L and (other conversion factors) to arrive at the correct units of power in watts. This will require the use of gravitational constant 9.8 m/s² as well as the Newton's Law relationship as shown below:

$$Actual\ Power\ (W) = \frac{55\ L}{s} \times \frac{1.03\ kg}{L} \times 0.5\ m \times \frac{N-s^2}{1\ kg-m} \times \frac{9.8\ m}{s^2} \times \frac{W-s}{N-m} \times 0.6 = 166.6\ W$$

$$Actual\ Power\ (hp) = 166.6\ Watts \times \frac{hp}{746\ Watts} = 0.2233\ hp$$

b. The energy produced is calculated as follows:

$$Energy\ (kWh) = 166.6\ Watts \times 12\ hrs \times \frac{kW}{1,000\ W} = 2\ kWh$$

$$Energy\ (hp-hr) = 2\ kWh \times \frac{1,000\ Watts}{1\ kW} \times \frac{hp}{746\ Watts} = 2.68\ hp-hr$$

1.3.9 Wave Energy

Wave energy is also a consequence of solar and wind energy. This renewable energy resource is available only from the oceans, and extraction equipment must be able to operate in marine environments. The implications include maintenance and construction costs as well as lifetime reliability. The power produced must be transported to land via long-range power distribution systems. The energy converter component parts must be capable of withstanding severe peak stresses in storms and ocean currents. A machine that is used to capture energy from waves is called a wave energy converter (WEC).

The basic governing equation for wave power is given in Equation 1.8. In this equation, the density (ρ), acceleration due to gravity (g), wave height (H), and wave period (T) must be known:

$$Power\left(\frac{kW}{m}\right) = \frac{\rho g^2}{64\pi} H^2 T \qquad (1.8)$$

The power and energy that can be harnessed from waves is substantial. Huge waves over longer distances can generate power up to several megawatts (MW). Equation 1.9 may be simplified further as shown in Equation 1.8. The constant that replaced other constant parameters is equivalent to around 0.5:

$$Power\left(\frac{kW}{m}\right) = \left(0.5\frac{kW}{m^3 s}\right) H^2 T \qquad (1.9)$$

The wave energy chapter will discuss various wave energy converters developed over the years. There are also companies that have started to commercialize these WECs and have slowly overcome technical issues and demonstrated their viability. Example

Problem 1.8 shows how the equivalent constant was estimated from the given wave power equation of constant acceleration due to gravity and water density.

Example 1.8: Estimated Value of Constant in Wave Equation

Determine the estimated value of constant such that g and ρ do not appear in the equation (i.e., value of $\rho g^2/64\pi$, or units of kW/m³/s). Assume that the acceleration due to gravity is 9.8 m/s² and that the density of water is 1,000 kg/m³.

SOLUTION:

1. The value of constant is approximately 0.5 as shown below:

$$\frac{\rho g^2}{64\pi} = \frac{1{,}000\ kg}{m^3} \times \left(\frac{9.8\ m}{s^2}\right)^2 \times \frac{1}{64 \times 3.1416} \times \frac{kW}{1{,}000\ W} = 0.478\frac{kW}{m^3 s} = 0.5\frac{kW}{m^3 s}$$

$$\frac{\rho g^2}{64\pi} = 0.5\frac{kW}{m^3 s} \times \frac{1{,}000\ W}{1\ kW} \times \frac{hp}{746\ W} \times \frac{1\ m^3}{\left(3.28\ ft\right)^3} \times \frac{60\ s}{min} = 1.140\frac{hp}{ft^3 min}$$

2. For a quick estimate of wave power, Equation 1.8 may be used. However, the density of salt water may be different from the density of clean water, and this must be taken into account for the actual wave power estimate.

1.3.10 Ocean Thermal Energy Conversion Systems

Ocean thermal energy conversion systems (OTECs) use the heat energy stored in the earth's ocean systems to generate electrical power. Ideally, OTECs work best if the temperature difference between the colder ocean water and the warm surface water is at its maximum or at least 20°C [36°F] (Finney, 2008). Conditions where the temperature differences are at their maximum are solely found in the tropical areas of the earth. These areas are between the Tropic of Cancer (23.45° north latitude) and the Tropic of Capricorn (–23.45° south latitude). To take advantage of these temperature differences, the cold water from beneath the oceans must be transported to the surface level to create a massive temperature gradient. This would require large-diameter intake pipes submerged several kilometers into the depth of the ocean. While this is an insurmountable task, the potential of generating billions of watts of electrical power is promising. While numerous scientists proposed tapping thermal energy from the oceans as early as 1881 (French physicist Jacques Arsene d'Arsonval), it was only in the early 19th century (1930s) that engineers started building prototypes for ocean thermal energy conversion systems (Georges Claude, d'Arsonval's student) (Finney, 2008). These prototypes produced 22 kW [29.5 hp] of electricity using a low-pressure turbine. In the chapter devoted to OTEC, we will cover the various ways of generating electrical power from OTECs. Because the temperature difference is quite small, one would therefore opt to use a chemical that has a low boiling point, thus being able to generate and produce a significant power differential and overall resulting power output. These systems are called closed cycles, whereby a refrigerant (such as ammonia) is used to rotate the turbine and generate electricity. For example, the ideal energy conversion efficiency for about a 20°C [68°F] temperature difference is only about 8%, while the actual efficiencies may be between only 3% and 4%. This is illustrated in a simple example shown in Example Problem 1.9. In a technical report published by Vega (1995), it was estimated that approximately 4 m³/s [141.15 ft³/s] of warm seawater and 2 m³/s (70.575 ft³/s) of cold water (with a ratio of 2:1) with a nominal temperature difference of 20°C [68°F] are required per MW of exploitable or net electricity.

Example 1.9: Actual Conversion Efficiency of OTEC Systems

Ammonia was used as a working fluid in an OTEC system. Actual energy conversion efficiency is defined as the net power (w_{net}) divided by the heat supplied to the system (Q_h). The net power was found to be 49,000 kW [65,683.65 hp], while the heat supplied to the ocean water was found to be 1,235,000 kW [1,655,495.98 hp]. Determine the actual energy conversion efficiency from this application.

SOLUTION:

a. Since the units are already the same, the thermodynamic efficiency is calculated simply:

$$\eta_{cycle}\left(\%\right) = \frac{w_{net}}{Q_h} \times 100\% = \frac{49,000}{1,235,000} \times 100\% = 3.97\%$$

b. Therefore, actual power may only be in the single digit range for OTECs.

1.3.11 Human, Animal, and Piezoelectric Power

Many developing countries harness the power of animals to do work. However, one should put into perspective the small magnitude of these energy resources with respect to the world's requirements for energy and power. Human power is only about one-tenth that of an animal, and most animals employed as power sources will contribute only around 746 watts [or 1 hp] (Too and Williams, 2001). It would no doubt take an unrealistic number of animals or human beings to satisfy even a meager amount of energy and power needs for useful applications. However, as humans increasingly require electronic gadgets in their daily lives, tapping small amounts of power from renewable energy resources toward phones and small computers may be worthwhile.

As the world progresses toward making electrical components in the nano range, there is an opportunity to accommodate power sources from renewables over fossil fuels. A great example is the proliferation of devices such as piezoelectric conversion systems through which humans and animals can generate energy and power by simply walking.

Piezoelectricity is the electric charge that accumulates in certain solid materials, such as ceramics or crystals, and even biological matter, such as bones, DNA, and proteins, in response to applied mechanical stresses. In humans and animals, these mechanical stresses can be triggered by the simple act of walking. A piezoelectric effect may also be generated in response to temperature change. A very simple example is the use of lead zirconate titanate crystal. This material generates measurable piezoelectricity when it is deformed even by about 0.1% of its original dimension (Howells, 2009).

In recent years, there have been several projects that realized the ability to generate larger amounts of power through harvesting kinetic energy from the simple act of walking. For example, if one were to attach such a device to one's shoes, small amounts of energy could be generated and could power a small electronic gadget. There are floors in some train stations in Japan that are designed to generate continuous power as pedestrians pass. These energy resources may be used to maintain lighting systems in hallways and walkways—literally, small steps toward less reliance on fossil fuels (Mail Foreign Service, 2008).

These amounts of energy are obviously quite small compared to the total energy requirement of a nation. However, simulation studies have shown that with proper piezoelectric design, a building may be able to meet about 0.5% of its own annual energy needs (Elhalwagy, et al., 2017). If this were implemented in thousands of buildings in key cities of the country, the positive impact would be considerable. High-traffic areas should be the first application sites of these types of energy. The compounded energy expenditures contribute to satisfying the greater energy requirements of offices and train or bus stations. To apply this principle to animal energy expenditure, one would have to wait until an equivalent system were designed in the walkways of a dairy farm, perhaps while the cows are being milked, such that the energy required for the milking process may also be harnessed through piezoelectric means (Paripati, 2013).

1.3.12 Cold Fusion and Gravitational Field Energy

Cold fusion and gravitational field energy are two controversial topics in the scientific field. Cold fusion is a hypothetical type of nuclear reaction that would occur at or near room temperature. The sun's energy comes from fusion reaction, in which hydrogen nuclei combine to form helium nuclei. Two nuclei of deuterium (D), which is an isotope of hydrogen (H), combine in many competing reactions, resulting in a release of energy through fusion. The liberated energy is radiated as solar energy (Bhattacharya, 1983). If these reactions could occur at "normal" temperatures, then we would have an endless supply of energy and power since seawater rich in deuterium may be a viable source of enormous amounts of energy.

In 1989, Martin Fleischman (University of Southampton), then one of the world's leading electrochemists, and Stanley Pons (University of Utah) reported that they developed an apparatus that produced anomalous heat (excess heat) of a magnitude they asserted could occur only through nuclear processes. Of course, many scientists tried to replicate these experiments to no avail (Platt, 1998). The two scientists have since lost their reputations, and the study conducted by the U.S. Department of Energy (USDOE) concluded that there was no conclusive evidence of cold fusion. To date, a small number of scientists continue to investigate cold fusion phenomena that are now often referred to as low-energy nuclear reactions (Krivit, 2018). Studies of cold fusion should be evaluated with a level of skepticism. At this stage, results are highly theoretical, but the imagined potential is substantial.

Gravitational field energy is also another potential resource of note. Gravitational energy is the potential energy associated with the gravitational field. This is also associated with tachyon field energy. Tachyons, defined as particles that move faster than light, are also purely hypothetical at this time but have begun to be within the scope of research in the field of physics (Nieper, 1983). So far, the best book that provides illuminating discussions on potential energy from gravitational fields is *Revolution in Technology, Medicine and Society: Conversion of Gravity Field Energy*, originally written in 1983 by a German physician, Hans Nieper. Dr. Nieper was looking for an alternative cure for cancer and found himself looking into the potential of gravitational field energy to provide the world with an endless supply of energy. His book discussed the works of Nikola Tesla and the development of an engine that would be powered solely by gravitational field energy. It also included discussions on of the development of the N-machine, a machine envisioned and invented by Bruce DePalma. DePalma's machine eliminated the use of fossil fuels with an efficiency of more than 100%, defying our understanding of the conservation of energy. DePalma collaborated with Dr. Paramahamsa Tewari of India on the genesis of free power generation.

Such discussions provide lofty though appealing goals surrounding potential free energy for mankind. This particular group of scientists believed that Tesla had indeed developed an engine that could run solely on gravitational field energy. An electrical device that could capture this energy would provide the world with an endless supply of energy and power—as long as the earth has its gravitational pull. DePalma did not live to actualize his invention, and fellow scientists from India, Germany, and the United States can only continue to ponder this ambitious theory of free energy from gravity (Nieper, 1983).

1.4 Renewable Energy Conversion Efficiencies

Many of the renewable energy resources described above may need to be converted into some useful form in order to be of practical benefit. The sun's energy must be converted into some form of heat of electricity for its potential to be harnessed. Thus, one should understand the definition or concept of conversion efficiencies. Conversion efficiencies come in different forms, and the basic efficiency is a simple ratio of output product over a given input material as shown in Equation 1.10. The input could be the amount of solar energy received in units of W/m², and the output could be the electrical power received over a given time period and must be converted into the same unit of W/m². Example Problem 1.10 provides an example of solar energy conversion efficiency for electrical power, which is around 15%:

$$\eta(\%) = \frac{Output}{Input} \times 100\% \tag{1.10}$$

Example 1.10: Conversion Efficiency of Solar PV Systems

Determine the efficiency of converting solar energy received on a solar PV panel if the incoming solar radiation was measured to be 970 W/m² and the electrical output of the solar panel with an area of 0.6975 m² was measured to be 35 volts and 3 amps.

SOLUTION:

a. From the basic electrical power equation, Volt × Ampere = Watts, and the electrical power output could be easily calculated as follows:

Power (Watts) = 35 Volts × 3 Amperes = 105 Watts (0.141 hp)

b. The input energy may be calculated as follows and converted into the same unit as the output energy (in watts):

$$Input\ Energy = \frac{970\ W}{m^2} \times 0.6975\ m^2 = 676.6\ W\left[0.907\ hp\right]$$

c. Thus, the conversion efficiency is calculated as follows:

$$\eta(\%) = \frac{Output}{Input} \times 100\% = \frac{105\ W}{676.6\ W} \times 100\% = 15.5\%$$

d. Only 15.5% of the incoming solar radiation is converted into electrical power.

Other conversion efficiencies are a little more complicated, as we will demonstrate in the coming chapters. For example, the conversion of biomass into liquid fuels would take a complex route if the conversion process took several stages of conversion from solid materials into gaseous material and finally into liquid form. The conversion efficiency will likewise be based on the energy content of the final fuel versus the energy of the biomass that was used to make the liquid fuel. The overall conversion efficiencies will be lower in some cases. In fact, the conversion of solar energy into biomass is quite low, on average between 2% and 3% for most species, and in some species (e.g., C3 photosynthesis plants), it is around 4.6% maximum. For C4 photosynthesis plants, it is around 6% (Zhu, et al., 2008). Hence, if we calculate the overall conversion efficiency of making fuel from biomass and ultimately use the solar energy as input, the overall conversion efficiencies will be in single digits or a fraction. Note that these conversion efficiencies were based on the current global CO_2 concentration of around 380 ppm. If the atmospheric CO_2 concentration goes up to 700 ppm, the 6% conversion efficiencies for C4 photosynthesis plants will significantly diminish (Zhu, et al., 2008). The field of renewable energy and biomass energy conversion for food, fiber, or fuels is much affected by global climate condition—primarily the concentration of CO_2 in the atmosphere. The final chapter of this book will deal with the sustainability of renewables and how global warming comes into play.

1.5 Renewable Energy Resources—Why?

The primary reason that renewable energy is very important for mankind is the issue of the sustainability of our energy resources as well as environmental pollution caused by our massive use of fossil fuels. While all the renewable energy resources described in this chapter may be considered sustainable energy resources, all must have a certain "renewable mode" of usage; that is, the rate of their consumption must be less than the rate of generation. A very simple example is the extensive use of biomass that leads to unsustainability. Biomass energy may be utilized only in conjunction with a plan for massive reforestation as biomass is harvested. The other important consideration is the depletion of nutrients from land where biomass is grown. Therefore, a good program to recycle nutrients back to land must also be initiated. The final chapter of this book is devoted to discussions of steps and measures to ensure that negative environmental effects are minimized. A popular issue is the greenhouse effect caused by increased levels of CO_2 emissions brought on by the massive use of fossil fuels, further leading to unwanted climate change. The current global average atmospheric carbon dioxide in 2016 was 402.9 ppm (Lindsey, 2017). Atmospheric CO_2 emissions were never higher than 300 ppm starting 800,000 years ago. It was reported that the last time CO_2 emissions were this high was more than 3 million years ago, when environmental temperature was 2°C to 3°C (3.6°F–5.4°F) higher than during the preindustrial era and the sea level was 15 to 25 meters (50–80 ft) higher than today (Lindsey, 2017). We have been observing massive changes in the weather conditions the last few years, and many scientists attribute this to the greenhouse effect due to increased CO_2 concentration in the atmosphere. This phenomenon is perhaps the single most important reason that the world population should decrease the use of fossil fuels or develop processes to reduce CO_2 emissions from the atmosphere, such as massive reforestation or the planting of more biomass resources that utilize this atmospheric carbon. We may also see technologies that utilize the carbon dioxide in air and convert this into other forms suitable for the beneficial use of the human population.

There are now genetically engineered bacteria that can inhale CO_2 and produce energy (Javelosa, 2016). Harvard University professor of energy Daniel G. Nocera was able to genetically engineer a type of bacteria that is able to ingest hydrogen gas and carbon dioxide and convert them into ethanol. He believed that this particular bacteria (*Ralston eutropha*) could convert sunlight 10 times more efficiently than plants (Javelosa, 2016). This would indeed be a breakthrough if proven on wide-scale systems. A study by Schlager, et al. (2017) showed that another type of methanogenic microorganism from an anaerobic digestate was able to ingest CO_2 to improve their growth, thereby reducing atmospheric carbon concentrations. Products of this process include acids and alcohols such as methanol. We are now seeing similar research studies with the same hope of reducing greenhouse gas emissions via biological microbial growths.

1.6 Summary and Conclusion

Renewable energy resources will play a more urgent role as fossil fuels become scarce or approach depletion. Perhaps the last obstacle to overcome for widespread adoption is the economic viability of converting a renewable energy resource into its useful applications of heat, fuel, or electrical power—the basic necessities of the commercial world. Cost will always be an issue in the installation of a renewable energy technology. For example, the payback period for solar energy conversion equipment for power generation is usually around 30 years. Nowadays, with the decreasing cost of solar panels to less than $1/Wp [$746/hp], the return to investment may be lowered if the electricity cost in an area is quite high. An extensive economic projection and analysis must be made before embarking upon long-term renewable energy source adoption. Perhaps the best way to harness renewable energy technologies is to create a variable mix and balance their utilization. Thus, one must be aware of the economics and technical limitations of all types of renewable energy resources. The goal of this book is to present all possible renewable energy technologies such that engineers may be able to make intelligent and carefully evaluated decisions around the economics and technical limitations of a particular renewable.

The economics of renewables are to be compared with the economics of fossil fuels, and historical projections of these fossil fuels must also be evaluated accordingly. The U.S. Energy Information Administration forecasts renewable energy to be the fastest-growing power source through 2040. New investments in renewable energy have risen to about $50 billion in 2015 from just $9 billion in 2004 (Hulac, 2015), mainly through installation of millions of new PV systems. The key to solar energy's future, especially PV systems, is battery storage. To make the technology sustainable, each solar-harnessing household must have affordable deep-cycle batteries in order to save solar power for evening use. Such developments in the solar field will be crucial to pursue.

1.7 Problems

1.7.1 Carbon Dioxide Required to Make Carbohydrates

P1.1 Determine the amount of carbohydrates or glucose produced (in tonnes) for every tonne of carbon dioxide used using Equation 1.2.

1.7.2 Kinetic Energy of a Mass of Wind

P1.2 Determine the kinetic energy available from 100 kg of air moving at a speed of 12 m/s. Report the answer in units of Joules and in Btu.

1.7.3 Carbon Dioxide Production during Ethanol Fermentation

P1.3 Determine the amount of carbon dioxide produced (kg) for every kg of glucose consumed in its conversion into ethanol. Use Equation 1.4 shown below in your calculations. Convert these units into English units of lbs CO_2 per lb glucose.

$$C_6H_{12}O_6 + \text{yeast} \rightarrow 2C_2H_5OH + 2CO_2 + \text{heat}$$

1.7.4 Theoretical and Actual Power from Water Stream

P1.4 Determine the theoretical and actual power that can be derived from a water stream with a volumetric flow rate of 55 L/s and with a dynamic head of 2.0 m. Assume an overall conversion efficiency of 45%.

1.7.5 Theoretical Thermal Conversion Efficiency of Rankine Cycle

P1.5 Determine the theoretical thermal conversion efficiency of a Rankine Cycle with the following data: (a) W_{out} = 650 kJ/kg, (b) W_{in} = 3.5, and (c) the Q_{in} = 2,500 kJ/kg. The mass flow rate of the steam is around 22,000 kg/hr.

1.7.6 Fuel Cell Efficiencies

P1.6 Calculate the actual electrical energy efficiency of converting hydrogen and oxygen into electrical energy via fuel cell technology using the following data: (a) the electricity produced was 180 kJ/mol, and (b) the heating value of hydrogen was 285.8 kJ/mol. Compare this efficiency with the practical electricity production from a hydrogen fuel cell having an electrical output of only 130 kJ/mol.

1.7.7 Tidal Power Calculations

P1.7 Residents of a household living near the shore would like to build a reservoir that could generate tidal power for their needs. The yearly average tide height is about 1.5 meters. The owner built a reservoir that can contain 8,000 cubic meters of sea water and plans to drop about 35 L/s of this water when the tide is lowest, with an average head of 1 meter for about 12 hours. Estimate the amount of actual power generated during that period if the overall conversion efficiency is 60%. Assume the density of seawater to be about 1.03 kg/L. How much energy was produced during this period?

1.7.8 Solar Water Heater Conversion Efficiency

P1.8 Determine the efficiency of converting solar energy received through a solar water heater if the average incoming solar radiation for an hour was measured to be 900 W/m^2. The water was heated from 25°C to 75°C for this period with an average mass flow rate of 4 kg/s on a 1 m^2 collector area.

1.7.9 OTEC Energy Conversion

P1.9 Ammonia was used as a working fluid in an OTEC system. Find the actual energy conversion efficiency, defined as the net power (w_{net}) divided by the heat supplied to the system (Q_h). The net power was found to be 35,000 kW, while the heat supplied to the ocean water was found to be 1,300,000 kW.

1.7.10 Solar PV Conversion Efficiency

P1.10 Determine the efficiency of converting solar energy received on a solar PV panel if the incoming solar radiation was measured to be 1,100 W/m² and the electrical output of the 1 m² solar panel was measured to be 40 volts and 3.5 amps.

References

AWEA. 2018. US Wind Energy Facts at a Glance. American Wind Energy Association Fact Sheets. Washington, DC. Available at www.awea.org. Accessed June 27, 2019.

Bhattacharya, S. C. 1983. Lecture Notes in "Renewable Energy Conversion" Class. Asian Institute of Technology, Bangkok, Thailand.

Capareda, S. C. 2013. Introduction to Biomass Energy Conversions. CRC Press, Taylor and Francis Group, Boca Raton, FL.

Conca, J. 2017. The biggest power plants in the world—hydro and nuclear. Forbes Magazine Online, August 10. Available at: https://www.forbes.com/sites/jamesconca/2017/08/10/the-biggest-power-plants-in-the-world-hydro-and-nuclear/#3f19e88e2c88. Accessed April 5, 2018.

Duffie, J. A. and W. A. Beckman. 2006. Solar Engineering of Thermal Processes. 3rd Edition. John Wiley and Sons, New York, NY.

EIA. 2017. Electric Power Annual 2016. December 2017. US Energy Information Administration, Independent Statistics and Analysis, US Department of Energy, Washington, DC. Available at: https://www.eia.gov/electricity/annual/pdf/epa.pdf. Accessed June 27, 2019.

EIA. 2018. Annual Energy Outlook 2018. US Department of Energy, Washington, DC. Available at: https://www.eia.gov/outlooks/aeo/. Accessed June 27, 2019.

Elhalwagy, A. M., M. Y. M. Ghoneem and M. Elhadidi. 2017. Feasibility study for using piezoelectric energy harvesting floor in buildings' interior spaces. Energy Procedia 115: 114–126.

Emami, Y., S. Mehrangiz, A. Etermadi, A. Mostafazadeh and S. Darvishi. 2013. A brief review about salinity gradient energy. International Journal of Smart and Clean Energy 2, (2): 295–300.

Encyclopedia Britannica. 2018. Bay of Fundy, Bay Canada. Written by Editors of Encyclopedia Britannica and Coastal Landforms written by Richard A. Davis, last updated March 16, 2018. Available at: https://www.britannica.com/science/coastal-landform#ref500208. Accessed April 13, 2018.

Finney, K. A. 2008. Ocean thermal energy conversion. Guelph Engineering Journal 1: 17–23.

Geothermal Energy Association. 2015. 2015 Annual US and Global Geothermal Power Production Report. February. Available at: http://geo-energy.org/reports/2015/2015%20Annual%20US%20%20Global%20Geothermal%20Power%20Production%20Report%20Draft%20final.pdf. Accessed April 12, 2018.

Howells, C. A. 2009. Piezoelectric energy harvesting. Energy Conversion and Management 50: 1847–1850.

Hulac, B. 2015. Strong future forecast for renewable energy. E&E News Sustainability of Scientific American Magazine. April 27. Available at: https://www.scientificamerican.com/article/strong-future-forecast-for-renewable-energy/. Accessed April 16, 2018.

Javelosa, J. 2016. This genetically-engineered bacteria can inhale CO_2 and produce energy. Futurism. May 30. Available at: https://futurism.com/this-new-bacteria-can-inhale-co2-and-produce -energy/. Accessed April 16, 2018.

Jones, M. and G. Jones. 1997. Advanced Biology. Cambridge University Press, Cambridge, UK.

Krivit, S. B. 2018. Experts testify before Congress on future of US fusion energy research and ITER. New Energy Times. March 28. Available at: http://news.newenergytimes.net/. Accessed April 13, 2018.

Lindsey, R. 2017. Climate change: atmospheric carbon dioxide. October 17. of Climate Watch Magazine. Available at: https://www.climate.gov/news-features/understanding-climate/ climate-change-atmospheric-carbon-dioxide. Accessed April 16, 2018.

Loeb, S., 2002. Large-scale power production by pressure-retarded osmosis, using river water and seawater passing through spiral modules. Desalination 143 (2): 115–122.

Mail Foreign Service. 2008. The power of the commuter: Japan uses energy-generating floor to help power subway. December 12. Available at: http://www.dailymail.co.uk/news/article-1094248/ The-power-commuter--Japan-uses-energy-generating-floor-help-power-subway.html. Accessed April 13, 2018.

McAnulty, M. J., V. G. Poosarla, K. Y. Kim, R. Jasso-Chavez, B. E. Logan and T. K. Wood. 2017. Electricity from methane by reversing methanogenesis. Nature Communications 8 (15419). DOI: 10.1038/ncomms15419.

Mims, C. 2008. One hot island: Iceland's renewable geothermal power. Scientific American. October 20.

Nieper, H. A. 1983. Revolution in Technology, Medicine and Society: Conversion of Gravity Field Energy. MIT Verlag, Oldenburg, Germany.

Paripati, P. 2013. Animal weight monitoring system. US Patent Number 9,226,481 B1. Granted January 15, 2016. US Patent Office, Department of Commerce, Washington, DC.

Pentland, 2013. World's largest electric power plant. Forbes Magazine Online, August 26. Available at: https://www.forbes.com/sites/williampentland/2013/08/26/worlds-39-largest-electric-power -plants/#3fe65b9858da. Accessed April 5, 2018.

Platt, C. 1998. What if cold fusion is real? Wired Online Magazine. November 1. Available at: https:// www.wired.com/1998/11/coldfusion/. Accessed April 13, 2018.

Schlager, S., M. Haberbauer, A. Fuchsbauer, C. Hemmelnair, L. M. Dumitru, G. Hinterberger, H. Neugebauer and N. S. Sariciftci. 2017. Bio-electrocatalytis application of microorganisms for carbon dioxide reduction to methane. ChemSusChem 10: 226–233. DOI: 10.1002/ cssc.201600963.

Too, D. and C. Williams. 2001. Determination of the crank-arm length to maximize power production in recumbent-cycle ergometry. Human Power 51: 1–8.

USDOE. 2013. 2012 Year End Wind Power Capacity in the US by States. US Department of Energy, Golden, CO. Available at www.usdoe/gov.

USEPA. 2015. Renewable Fuel Standard Program: Standards for 2014, 2015 and 2016 and Biomass-Based Diesel Volume for 2017 Final RUle. 40 CFR Part 80. Federal Register 80 (239). December 14. Environmental Protection Agency, Washington, DC.

Veerman, J., M. Saakes, S. J. Metz and G. J. Harmsen. 2011. Reverse electrodialysis: a validated process model for design and optimization. Chemical Engineering Journal 166 (1): 256–268.

WEF. 2016. US Wind Industry Fast Facts. Wind Energy Foundation. Available at: http:// windenergyfoundation.org/about-wind-energy/us-wind-industry-fast-facts/. Accessed April 5 2018.

Wick, G. L. and W. R. Schmitt. 1997. Prospects for renewable energy from the sea. Marine Technology Society Journal 11: 16–21.

Zhu, X.-G., S. P. Long and D. R. Ort. 2008. What is the maximum efficiency with which photosynthesis can convert solar energy into biomass? Current Opinion in Biotechnology 19: 153–159.

2

Solar Energy

Learning Objectives

Upon completion of this chapter, one should be able to:

1. Estimate the available solar energy received at any place on the earth.
2. Enumerate the nomenclatures for determining the solar radiation received on a solar conversion device.
3. Enumerate the various solar energy conversion technologies.
4. Define energy conversion efficiencies in some solar energy devices.
5. Describe the instruments that measure solar energy.
6. Size appropriate solar conversion devices.
7. Relate basic economic issues concerning solar energy conversion systems.

2.1 Introduction

Solar energy is considered the prime renewable energy resource. It is perhaps one of the most reliable energy resources; however, it is still beset with high energy costs of conversion in many of its useful applications. The primary reason for this is that the energy requirement of the human race has increased dramatically such that the low-density energy contributed by the sun is not enough for our daily needs. Solar energy will have to be concentrated or magnified in order to satisfy our energy needs, requiring higher capital costs for overall systems. The economic return for most solar energy technologies is still quite high, and the manufacturing costs have not yet diminished significantly enough to be of widespread use. Many countries, especially the United States, have some subsidies to encourage the use of this technology. As unit prices decrease, we are seeing an increase in use. One would have to know the ways of measuring the sun's available resources as well as the total unit cost to be able to make an informed estimate of solar energy's economic return.

This chapter is devoted to describing how energy from the sun may be measured at any given place on the earth, identifying solar energy applications, making estimates of energy conversion efficiencies, and evaluating the cost-effectiveness of a chosen application. In this chapter, only the major solar conversion pathways—solar thermal conversion, photovoltaic (PV) conversion, and other solar power devices —will be discussed in depth.

Other solar conversion devices will also be briefly reviewed to provide the reader with a complete perspective of the benefits of all solar energy conversion systems.

2.2 The Solar Constant and Extraterrestrial Solar Radiation

The amount of energy received on a unit area of a surface perpendicular to the ray of the sun, outside of the earth's atmosphere, on an average distance from the sun, is termed solar constant (G_{sc}). This value was originally set at 1,353 W/m² (Duffie and Beckman, 1981), but through improved data and more accurate measuring instruments, the accepted value of the solar constant is now around 1,367 W/m² (Duffie and Beckman, 2006). This is the value that will be used for this book.

However, the amount of solar energy received in the earth's atmosphere, called extraterrestrial solar radiation, varies at any given season, time of day, and time of year. The equation to estimate this parameter is given in Equation 2.1:

$$G_{on} = G_{sc} \times \left(1 + 0.033 \times cos\frac{360n}{365} \right)$$

(2.1)

where

G_{sc} = the solar constant, 1,367 W/m²
n = day of the year, with January 1 equal to $n = 1$.

Table 2.1 makes it easier to calculate the variable n. Example Problem 2.1 shows how Equation 2.1 may be used to estimate extraterrestrial solar radiation on January 1 or on any given day.

Example 2.1: Determination of Extraterrestrial Solar Radiation on Any Given Day

Determine the extraterrestrial solar radiation received on earth in January 1. Plot the variation of extraterrestrial solar radiation received with the time of year on a normal plane. Use G_{sc} = 1,367 W/m².

TABLE 2.1

Table to Easily Estimate the Value of n for a Given (Non–Leap-Year) Month

Month	n for ith Day of Month	Month	n for ith Day of Month
January	i	July	$181 + i$
February	$31 + i$	August	$212 + i$
March	$59 + i$	September	$243 + i$
April	$90 + i$	October	$273 + i$
May	$120 + i$	November	$304 + i$
June	$151 + i$	December	$334 + i$

Note: On a computer, the angles are in radians. In this case, you must convert degree angles into radians by multiplying the degree angle with $\pi/180$. The computer syntax for the value of pi is as follows: "pi()" (i.e., the letter pi followed by open and closed parentheses).

SOLUTION:

a. The equation for extraterrestrial radiation is given as follows:

$$G_{on} = G_{sc} \times \left(1 + 0.033 \times \cos \frac{360n}{365} \right)$$

b. $n = 1$ and substituting the values for n results in the following:

$$G_{on} = 1,367 \times \left(1 + 0.033 \times \cos \frac{360 \times 1}{365} \right) = 1,412 \frac{W}{m^2}$$

c. If the value of n varies from 1 to 365 and the resulting values are plotted, the graph shown here will illustrate the variations in extraterrestrial solar radiation throughout the year.

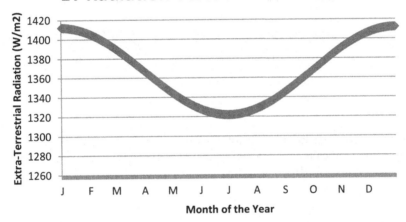

ET Radiation Versus Time of Year

d. Perhaps surprisingly, the values of extraterrestrial solar radiation are high during winter and low during summer.

2.3 Actual Solar Energy Received on the Earth's Surface

The solar energy received on the earth's surface are of two types: diffuse radiation and direct or beam radiation. The latter reaches the earth's surface without having been blocked by clouds or other obstructions. Diffuse solar radiations received on the earth's surface are those that are intercepted by the clouds. Some solar devices, such as PV systems, rely mostly on direct or beam radiation and hardly have any significant output during very cloudy days.

Figure 2.1 shows the simplified sketch of the distribution of solar energy on an average day in the tropics (Bhattacharya, 1983). Of about 1,370 W/m² of total solar radiation radiated to the earth, around 80 W/m² is scattered back to space, about 350 W/m² is reflected back to space, and about 190 W/m² is absorbed by the atmosphere. Thus, only about 500 W/m² is direct or beam solar radiation, and about 250 W/m² is diffuse radiation, as

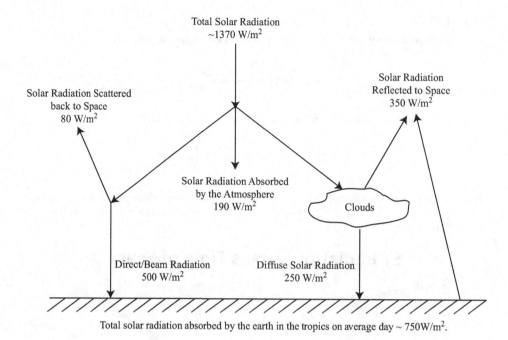

FIGURE 2.1
A simplified sketch of the distribution of solar radiation absorbed by the earth on a typical day in the tropics.

these are blocked by the clouds. The total solar radiation absorbed by the earth on an average day in the tropics is around 750 W/m². This value goes up to around 1,100 W/m² during very clear days and drops to as low as 200 W/m² on cloudy days.

2.4 Solar Energy Measuring Instruments

In the past, to measure the amount of solar energy received on the earth's surface, one would have used a crystal ball—literally, a transparent solid glass spherical ball. The crystal ball acts as a prism and directs a concentrated solar beam to its back end, where a piece of paper or cardboard is placed along the direct path of sunshine for that given day. This would burn the paper as clear sunshine is received. At the end of the day, a technician would measure the actual length of the burnt path and take the ratio to the theoretical sunshine hours possible. If there is an established correlation between the amounts of the solar energy in units of W/m² to sunshine duration in hours, then the actual amount of solar energy as well as sunshine duration is recorded for that day. This unit is called the Campbell-Stokes sunshine recorder. This unit is still in use in many developing countries. The unit requires daily readings and may not be able to measure the sunshine duration during early morning (dawn) or during late afternoon (dusk), when the sun's energy is not enough to burn paper, even though the sun is definitely out. This chapter will present ways to estimate theoretical sunshine duration (N) as well as the theoretical amount of solar energy received at any given location on the earth on any given day (H_o) or time of the day (I_o) in the absence of physical obstruction and absorbance.

FIGURE 2.2
A photo of a simple research-grade solar pyranometer.

As described in the previous section, solar radiation comes in two forms—diffuse and direct solar radiation. Instruments to measure the two are readily available nowadays. The most common is the solar pyranometer. A pyranometer is a device that measures total or global solar radiation or the solar radiation flux density in units of W/m². It has a sensor that relates the amount of solar energy received from all directions, regardless of form. If one is interested in measuring direct solar radiation, a pyrheliometer is used. This device has a long tube that allows only the direct portion of the solar radiation to be measured. The device tracks the sun perpendicularly through any given time of day. The pyrheliometer is quite expensive to own. One may use two pyranometers, which are relatively cheaper, to measure both the direct and the diffuse portions of the solar radiation. With two pyranometers, one is provided with a shaded ring that blocks the direct solar radiation from hitting the sensor, while the other measures global or total solar radiation. The difference in readings between the two pyranometers is the direct solar radiation received in a given unit of time. PV devices respond very well to direct solar radiation and not to the diffuse solar radiation. Hence, direct radiation data are important for these types of devices. A photo of a simple pyranometer is shown in Figure 2.2.

2.5 Solar Time

In calculating the amount of solar energy received at a given location at any given day of the year and time of day, one must keep in mind the concept of solar time. Solar time is defined here as the time based on the apparent angular motion of the sun across the sky with solar noon, the time the sun crosses the meridian of the observer (Duffie and

Beckman, 2006). Solar time is not the same as our standard or clock time. The difference is given by Equation 2.2:

$$\text{Solar Time} = \text{Standard Time} + 4 \times (L_{std} - L_{loc}) + E \tag{2.2}$$

where

    standard time = our clock time

    L_{std} = meridian or standard longitude on which the local time is based (e.g., College Station, Texas, is based on 90° meridian)

    L_{loc} = meridian of the observer (e.g., College Station, Texas, is exactly 96.32° West)

    E = the equation of time, in minutes, given by Equation 2.3:

$$E = 229.2 \times (0.000075 + 0.00186 \; \cos B - 0.032077 \; \sin B - 0.014615 \; \cos 2B - 0.04089 \; \sin 2B) \tag{2.3}$$

B is given by Equation 2.4, where n is the day of the year:

$$B = (n-1) \times \frac{360}{365}, \; in \; degrees \tag{2.4}$$

Example Problem 2.2 shows how solar time is calculated based on a given location and clock time.

Example 2.2: Determination of Solar Time on a Given Date and Location

Determine the solar time corresponding to 10:30 a.m. Central Standard Time on October 25, 2014, in College Station, Texas. The longitude of College Station, Texas, is 96.32 West, and this town's time was based on a 90° meridian.

SOLUTION:

    a. For October 25, 2014, $n = 298$

    b. B is then calculated using Equation 2.4:

$$B = (n-1) \times \frac{360}{365} = (298-1) \times \frac{360}{365} = 292.9°$$

    c. The equation of time E is then calculated using Equation 2.3:

$$E = 229.2 \times (0.000075 + 0.00186 \cos(292.9) - 0.032077 \sin(292.9)$$
$$- 0.014615 \cos(2 \times (292.9)) - 0.04089 \sin(2 \times (292.9))) = 16.01$$

    d. The longitude correction is then calculated:

$$\text{Longitude Correction} = 4 \times (90 - 96.32) = -25.28 \text{ minutes}$$

    e. The solar time is calculated as follows:

$$\text{Solar Time} = 10.50 + (-25.28/60) + (16.01/60) = 10.35$$

f. Note that the longitude correction and the equation of time, *E* is converted into units of hours to be consistent with the calculation of solar time. The fraction may be brought back into units of minutes by multiplying this by 60 minutes to an hour.

g. So, the solar time is actually 10:21 a.m.

One may be able to shift back and forth between using solar time and the actual (clock) time using the above procedures. When on a computer performing these calculations, one must remember that the computer uses radians for the angle as opposed to degrees, as in *B* in Equation 2.4.

2.6 Geometric Nomenclatures for Solar Resource Calculations

There are various nomenclatures that must be known in order to make an estimate of how much solar energy is received from a solar collector on any given day of the year or time of day. The movement of the sun in the sky is predictable enough that one may be able to estimate the solar energy received on the surface of the earth at any given place, day, and time. However, one must be familiar with the terminology or nomenclature. Refer to Figure 9.2 for the geometric relationships of some important solar angles relative to a solar collector. One rule of thumb in orienting a fixed solar collector (if in the northern hemisphere) is to orient it due south (north if you are in the southern hemisphere) with a slope, β, equal to the latitude of the place. For example, in College Station, Texas, the fixed solar collector must be oriented due south at a slope of about around 30°. This is because the latitude of College Station, Texas, is 30.61° North. If the solar collector is directly due south, it will have a surface azimuth angle of 0°. Otherwise, a solar azimuth angle will be formed. As a convention and recommended in the book by Duffie and Beckman (2006), the solar azimuth angle that is not due south will be negative if the displacement is east of south and positive if this is west of south. In Figure 9.2, the surface azimuth angle is positive at an angle γ.

The zenith is the vertical line, perpendicular to the horizontal surface. Hence, the zenith angle θ_z is the angle between the zenith and the line to the sun's rays. Another important geometric term is the solar altitude angle α_s—the angle between the horizontal line and the line to the sun. This is the complement of the zenith angle, as shown in the Figure 2.3. The solar azimuth angle in Figure 2.3 is the same as the surface azimuth angle γ_s. If one were to project the ray of the sun vertically on a horizontal surface and measure the angle that it makes with the south, this will provide the solar azimuth angle.

Not shown in the figure is one of the most important angles, the declination angle, used in many solar resource calculations. The declination angle is the angular position of the sun at solar noon with respect to the plane of the equator. Solar noon is defined here as the time the sun crosses the meridian of the observer. There is also a convention for the solar declination angle δ; that is, north is positive, and south is negative. Note that the solar declination angle only varies between −23.45° to +23.45°. If one were ask how many times the sun would be directly above one person as it crosses his meridian, the answer is two—one during the summer solstice and another during the winter solstice. But this is true only if his latitude is below 23.45° North or 23.45° South. This is because when one plots the

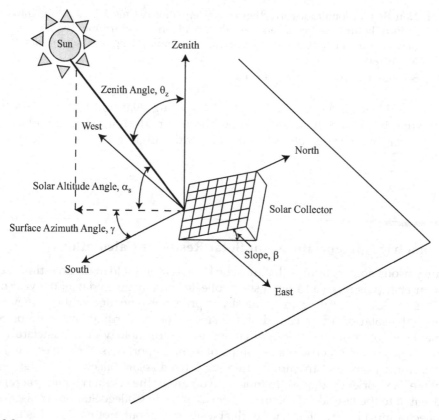

FIGURE 2.3

Nomenclatures in the estimate of solar radiation received on a given surface, oriented at various directions on the earth; some important geometric angles relating a solar collector with sun's angular position.

vertical projection of the sun on the earth's surface, called the sun path diagram, the path will follow a regular sinusoidal shape that covers only those latitudes.

In solar resource calculations, latitude is assigned the symbol φ (phi), and the value varies between +90 using the convention of north as positive and south as negative. Other nomenclature in tracking the movement of the sun is the hour angle ω. This is the angular displacement of the sun east or west of the local meridian due to the rotation of the earth on its axis at 15° per hour. The convention to be followed is negative for the morning and positive for the afternoon, with zero being solar noon. Accordingly, the hour angle ω at 11:00 a.m. (solar time) will be −15° and −90° at 6:00 a.m. solar time.

The solar declination δ may be calculated using Equation 2.5:

$$\delta = 23.45 \times sin\left(360 \times \frac{284 + n}{365} \right)$$ (2.5)

where

n = the day of the year

There is a more recent and accurate equation presented by Iqbal (1983), as cited by Duffie and Beckman (2006), but this new, more complicated equation does not warrant use for a quick and simple calculation of declination angle δ. The error using the older equation is less than 0.035° and therefore does not have a significant effect on overall solar resource calculations. Example Problem 2.3 shows how this value is used. Use the Iqbal equation for a more accurate estimate of solar declination. For the purposes of this book, it is enough to use the simple declination Equation 2.5.

Example 2.3: Calculation of Declination Angle on a Given Day

Calculate the declination angle on January 1, 2014.

SOLUTION:

1. Using the declination Equation 2.3 and with a value of $n = 1$, the resulting answer is shown here:

$$\delta = 23.45 \times sin\left(360 \times \frac{284 + n}{365} \right) = 23.45 \times sin\,281.1° = -23.01°$$

2. Thus, on January 1, 2014, the sun's variation from the zenith line is almost at its maximum as it crosses someone's meridian.

Perhaps the single most important equation for solar energy resource calculation is the incidence angle, θ. This is the angle between the beam radiation on a surface and the normal to that surface. The cosine of this angle will resolve the component perpendicular to the solar collector in question. If one knows the amount of direct or beam solar radiation received on a solar collector, the cosine of this angle will determine the percentage of the beam radiation that the solar collector actually received. The other component will not be absorbed by the collector. The incidence angle is given by Equation 2.6:

$$Cos\theta = sin\delta sin\phi cos\beta - sin\delta cos\phi sin\beta cos\gamma + cos\delta cos\phi cos\beta cos\omega + cos\delta sin\phi sin\beta cos\gamma cos\omega \\ + cos\delta sin\beta sin\gamma sin\omega \tag{2.6}$$

The solar incidence angle is dependent upon all the other nomenclatures discussed in this section. A practitioner would always relate the declination angle to a certain day of the year since it is only dependent on the value n. The latitude of the place is defined by the parameter φ and the slope of the collector is given by the parameter β. If the solar collector is oriented due south, then the azimuth angle will be zero. Using these variables, one would be able to estimate the incidence angle at any given time of the day. Remember that the time of day will specify the parameter ω. Example Problem 2.4 illustrates how the incidence angle is calculated for a given solar collector with given orientations.

Example 2.4: Calculation of Angle of Incidence of Beam Radiation on Collector Surface

Calculate the angle of incidence of beam radiation on a solar collector surface located in College Station, Texas, at 10:30 a.m. (solar time) on October 25, 2014, if the surface is tilted at an angle of 30° and pointed 15° west of south. The latitude of College Station, Texas, is 30.61° North.

SOLUTION:

a. First calculate the value of $n = 298$.

b. Then the other values are as follows: $\omega = -22.5°$, $\gamma = 15°$, $\beta = 15°$, and $\varphi = 30.61°$.

c. The solar declination angle is then calculated:

$$\delta = 23.45 \times sin\left(360 \times \frac{284 + 298}{365}\right) = -13.12°$$

d. With all the parameters known, Equation 2.6 may be used to estimate the incidence angle as follows:

$$Cos\theta = sin(-13.12)sin(30.61)cos(30) - sin(-13.12)cos(30.61)sin(30)cos(15)$$
$$+ cos(-13.12)cos(30.61)cos(30)cos(-22.5)$$
$$+ cos(-13.12)sin(30.61)sin(30)cos(15)cos(-22.5)$$
$$+ cos(-13.12)sin(30)sin(15)sin(-22.5)$$

$$Cos\theta = \left[(-0.227)\times(0.509)\times(0.866)\right] - \left[(-0.227)\times(0.861)\times(0.50)\times(0.966)\right]$$
$$+ \left[(0.974)\times(0.861)\times(0.866)\times(0.924)\right]$$
$$+ \left[(0.974)\times(0.509)\times(0.50)\times(0.966)\times(0.924)\right]$$
$$+ \left[(0.974)\times(0.50)\times(0.259)\times(-0.383)\right] = 0.838$$

e. $Cos\ \theta = 0.838$ and $\theta = 33.04°$.

f. If the magnitude of solar energy received on the surface is known, about 83.8% of this is actually absorbed by the collector at this time of the day.

Equation 2.6 may be simplified for other special cases or for special orientation of a flat collector. For example, if a flat plate collector is sloped toward the south (or north) where $\gamma = 0$ degrees, the Equation 2.6 is simplified and the last term is dropped since cosine of 0 is 1, and the simplified equation is shown in Equation 2.7:

$$Cos\theta = sin\delta sin\phi cos\beta - sin\delta cos\phi sin\beta cos\gamma + cos\delta cos\phi cos\beta cos\omega + cos\delta sin\phi sin\beta cos\gamma cos\omega \quad (2.7)$$

If one is interested in calculating the solar energy received on a vertical wall of a building, the angle β becomes $90°$, and Equation 2.6 becomes simplified to Equation 2.8:

$$Cos\theta = -sin\delta cos\phi cos\gamma + cos\delta sin\phi cos\gamma cos\omega + cos\delta sin\gamma sin\omega \quad (2.8)$$

If the collector is laid flat on the ground (or on a horizontal surface), Equation 2.6 is simplified to Equation 2.9:

$$Cos\theta_z = cos\delta cos\phi cos\omega + sin\delta sin\phi \quad (2.9)$$

Equation 2.9 is a familiar equation since the angle of incidence becomes the zenith angle of the sun, hence the subscript θ_z. In fact, the equation for horizontal surfaces (Equation 2.9) leads to the calculation of the sunset (or sunrise) hour angle or the estimate of the sunrise and sunset for a given location and day of the year. Equation 2.9 may be rearranged to calculate for the value of ω, which we now call the sunset (or sunrise) hour angle ($ω_s$), as shown in Equation 2.10, when $θ_z$ is equal to 90°:

$$cosω_s = -\frac{sinδsinϕ}{cosδcosϕ} = -tanδtanϕ \tag{2.10}$$

Note that the angle φ is associated with a location (the latitude) and that the declination angle δ is associated with a given day of the year. If one is perhaps working for a television or radio station and would like to calculate and report the sunrise and sunset times, Equation 2.10 is simply used. One other use of Equation 2.10 is to determine the possible total number of daylight hours in a given day and location. This is shown in Equation 2.11. In this equation, the sunset (or sunrise) hour angle is simply converted into number of hours, noting that the sun moves at 15° per hour over the horizon:

$$N = \frac{2}{15} cos^{-1}(-tanδtanϕ) \tag{2.11}$$

Example Problem 2.5 shows how the hour angle equation is used to determine the sunrise time on a given day and location, including estimating the total possible daylight hours.

Example 2.5: Calculation of Hour Angle, Time of Sunrise, and Number of Daylight Hours on a Given Day and Location

Calculate the hour angle, time of sunrise, and number of daylight hours on October 15, 2014, in College Station, Texas. The latitude of College Station, Texas, is 30.61°.

SOLUTION:

a. The hour angle is first determined using Equation 2.10:

$$cosω_s = -\frac{sinδsinϕ}{cosδcosϕ} = -tanδtanϕ$$

$$ω_s = cos^{-1}(-tanδtanϕ) = cos^{-1}(-tan30.61 × tan - 13.12)$$

$$ω_s = cos^{-1}(-tan30.61 × tan - 13.12) = cos^{-1}(0.1379) = 82.07°$$

b. This hour angle is then recalculated to correspond to the solar time by assuming that at a sunrise hour angle of 90°, the sunrise would be 6:00 a.m. (or the sunset at 6:00 p.m. solar time). Since the angle is less than 90°, one would already speculate that the real sunrise time should be after 6:00 a.m.

c. The fraction of angle between 90° and 82.07° (90 – 82.07 = 7.93°) is simply converted by ratio and proportion to the time value to give the number of minutes shown below:

$$Minutes = \frac{7.93°}{} × \frac{hr}{15°} × \frac{60\ minutes}{hr} = 31.72\ minutes$$

d. Therefore, the sunrise should be at 6:31.72 a.m. (solar time).

e. The number of possible daylight hours is calculated from Equation 2.11, and the results are shown below:

$$N = \frac{2}{15}cos^{-1}\left(-tan\delta tan\phi\right) = \frac{2}{15}cos^{-1}\left(-tan30.61 \times tan - 13.12\right) = 10.94 \; hrs$$

f. As a result, for this day, one would expect close to 11 hours of sunshine unless the sun is blocked by clouds at some point during the day.

Equation 2.11 is also a useful equation in determining and simulating the possible daylight hours for the whole year such that the planting and flowering behavior of some photoperiods (or plants that respond to changes in daylight hours) may be timed properly. For example, the chrysanthemum (*Crysanthemum morifolium*) is a plant whose vegetative growth and flowering is affected by the number of daylight hours. To initiate flowering for this plant, careful timing that corresponds with a number of daylight hours is necessary. Chrysanthemums are short-day plants that require longer daylight hours to trigger flowering. Some nurseries put a shade over this type of plant if flowering is not wanted. When it is an ideal time for flowering, timed to the peak sales period, they remove the shade (Kahar, 2008).

2.7 Extraterrestrial Solar Radiation on a Horizontal Surface

It is possible to estimate the daily theoretical or maximum solar radiation received on the earth's surface using knowledge of the parameters introduced in Section 2.5 together with that of extraterrestrial solar radiation introduced in Section 2.2. If one were to integrate Equation 2.1 from sunshine to sunset by multiplying this with the cosine of θ_z (i.e., for a horizontal surface), the daily extraterrestrial solar radiation H_o may be estimated. This is given in Equation 2.12. However, the calculated values are highly theoretical and assume that there are no obstructions or that there is absolutely no atmosphere to block sunshine:

$$H_o = \frac{24 \times 3600 G_{sc}}{\pi}\left(1 + 0.033cos\frac{360n}{365}\right) \times \left(cos\phi cos\delta sin\omega_s + \frac{\pi\omega_s}{180}sin\delta sin\phi\right) \qquad (2.12)$$

Equation 2.12 is consistent with all the parameters used in earlier sections. Example 2.6 illustrates the way to estimate the day's solar radiation on a horizontal surface in the absence of atmosphere.

Example 2.6: Calculation of Theoretical Solar Radiation Received on a Given Location at a Given Day

Determine the H_o—the day's solar radiation on a horizontal surface in the absence of atmosphere in College Station, Texas, on October 25, 2014. The latitude of College Station, Texas, is 30.61°. Use the results from previous example problems.

SOLUTION:

a. The following have already been given or calculated: solar constant = 1,367 W/m², $n = 298$, $\varphi = 30.61°$ and $\delta = -13.12°$, and $\omega_s = 82.07°$. Substituting these values in Equation 2.12 leads to the following results for H_o:

$$H_o = \frac{24 \times 3600 G_{sc}}{\pi}\left(1 + 0.033\cos\frac{360 \times 298}{365}\right)$$

$$\times \left(cos30.61cos - 13.12\,sin82.07 + \frac{\pi \times 82.07}{180}\,sin30.61sin - 13.12\right)$$

$$H_o = 37,595,111 \times (1.0134)$$

$$\times\left[\left((0.861)\times(0.974)\times(0.99) + (1.4324)\times(0.509)\times(-0.227)\right)\right]$$

$$H_o = 25,350,463\frac{J}{m^2} = 25.35\frac{MJ}{m^2}$$

b. As such, in the absence of atmosphere, one would expect to receive solar radiation of approximately 25.35 MJ/m².

Note that if one were to encode Example Problems 2.1 to 2.6 in spreadsheet software and run this simulation from day 1 to day 365, a complete plot of possible daily maximum solar radiation may be estimated, assuming that there is no atmosphere.

To make an estimate of the hourly solar radiation I_o in a given time frame, Equation 2.13 may be used:

$$I_o = \frac{12 \times 3600 G_{sc}}{\pi} \times \left(1 + 0.033\cos\frac{360 \times n}{365}\right)$$

$$\times \left(cos\delta cos\phi \times (sin\omega_2 - sin\omega_1) + \frac{\pi(\omega_2 - \omega_1)}{180}\,sin\delta sin\phi\right) \tag{2.13}$$

Most parameters used in this equation have already been defined in earlier sections. The values of ω_1 and ω_2 are the times in question and are previously defined. Example 2.7 illustrates how this equation may be used.

Example 2.7: Calculation of Theoretical Hourly Solar Radiation Received on a Given Location at a Given Time of Day

Determine the I_o, the day's solar radiation, on a horizontal surface in the absence of atmosphere in College Station, Texas, on October 25, 2014, between the hours of 10 and 11 a.m. (solar time). The latitude of College Station, Texas, is 30.61°. Use the results from previous example problems.

SOLUTION:

a. The following have already been given or calculated: solar constant = 1,367 W/m², $n = 298$, $\varphi = 30.61°$ and $\delta = -13.12°$, and $\omega_1 = -30°$ and $\omega_1 = -15°$. Substituting these values in Equation 2.13 leads to the following results for I_o:

$$I_o = \frac{12 \times 3600 \times 1,367}{3.1416} \times \left(1 + 0.033\cos\frac{360 \times 298}{365}\right)$$

$$\times \left(cos30.61cos - 13.12 \times (sin(-15) - sin(-30)) + \frac{3.1416 \times (-15 - (-30))}{180}\,sin30.61sin - 13.12\right)$$

b. The individual terms have values as follows:

$$I_o = 18,797,555 \times (1.0134)$$
$$\times \left[(0.861 \times 0.974 \times (-0.259 + 0.5) + 0.2618 \times 0.509 \times -0.227) \right]$$
$$I_o = 18,797,555 \times (1.0134) \times \left[(0.2021) + (-0.03025) \right] = 3,273,647 \frac{J}{m^2}$$

c. The hourly solar radiation received during this time frame will be around 3.27 MJ/m² in the absence of atmosphere.

In both of the above calculations for daily and hourly solar radiation, it is assumed that the atmosphere is free of obstructions, clouds, or other interference. The use of Equations 2.12 and 2.13 allows one to make a projection of daily solar radiation throughout the year as well as variations within the day. The actual data will vary according to the given location and its unique characteristics. The one parameter that will vary is the clearness index, K_t. This will be discussed in the next section.

2.8 Available Solar Radiation on a Particular Location

The theoretical maximum solar radiation received on the surface of the earth will vary according to location, season, and time of the day. The previous sections have covered how to make these estimates. The actual solar radiation received in a given place may be determined by setting up solar energy measuring instruments. Solar radiation received on the ground comes in two types: direct or beam radiation (H_b) and diffuse radiation (H_d). The sum of these values is the global solar radiation received on a surface. The most common instrument for measuring global solar radiation is the pyranometer, shown in Figure 2.4.

The solar pyranometer may also be used to measure diffuse solar radiation if it is equipped with a shading ring. This ring will block off any direct or beam solar radiation. If one has two solar pyranometer units and a shaded ring, the amount of direct solar radiation may be calculated by taking the difference between the two readings. There is an instrument that is specifically used to measure direct or beam solar radiation. This is called a pyrheliometer. The unit has a very small aperture and only receives the direct component of the solar radiation. It tracks the sun at every minute and thus allows only the direct beam radiation to be measured. The unit is quite expensive with all the control systems that must be in place. Still, the cheapest way to measure direct and diffuse solar radiation is with the use of two pyranometers. The shaded ring would have to be adjusted almost daily to ensure that only diffuse radiation is received in one unit.

There are empirical relationships that may be used to estimate the H_b and H_d. The most popular is the Collares-Pereira and Rabl (1979) correlation shown in Equations 2.14 and 2.15:

$$\frac{H_d}{H} = \begin{cases} 0.99 \; for \; K_t \leq 0.17 \\ 1.188 - 2.272K_t + 9.473K_t^2 - 21.865K_t^3 + 14.648K_t^4 \; for \; 0.17 < K_t < 0.75 \\ -0.54K_t + 0.632 \; for \; 0.75 < K_t < 0.80 \\ 0.2 \; for \; K_t \geq 0.80 \end{cases} \qquad (2.14)$$

FIGURE 2.4
Photo of a research-grade solar pyranometer mounted on a solar panel.

$$K_t = \frac{H}{H_o}$$

(2.15)

where
H_d = diffuse solar radiation
H_o = daily solar radiation in absence of atmosphere
H = actual solar radiation received
K_t = clearness index

Since H_o may be readily estimated at any given location on the earth following examples shown in Section 2.6, the clearness index K_t may be established for a given location if year-long data are available. The clearness index is the ratio of actual solar radiation received on a surface to the extraterrestrial solar radiation in the absence of atmosphere. For example, the 30-year running data published in College Station, Texas, has established the clearness indices for every month throughout the year (shown in Table 2.2). The clearness index is highest during the month of August (usually the hottest month in this location in Texas), with a value of 54% of the maximum solar radiation, and lowest in January (28%). Example Problem 2.8 shows how this equation is used.

Example 2.8: Determination of Clearness Index on a Given Location

Determine the clearness index K_t and the fraction of solar energy that is beam and the fraction that is diffuse for the month of October if the solar radiation actually received was 13 MJ/m².

TABLE 2.2

Solar Radiation Data in College Station, Texas (NSRDB, 2006)

Month	H (MJ/m²)	H₀ (MJ/m²)	Kₜ
January	5.93	20.91	0.28
February	7.60	25.86	0.29
March	10033	31.56	0.33
April	11.99	36.74	0.33
May	13.89	39.97	0.35
June	12.62	41.18	0.31
July	18.21	40.56	0.45
August	20.72	38.10	0.54
September	17.31	33.77	0.51
October	12.99	28.07	0.46
November	8.49	22.53	0.38
December	10.05	19.67	0.51
Yearly Mean	12.51	31.58	0.40

SOLUTION:

a. The clearness index K_t is simply calculated using Equation 2.15 by first determining the extraterrestrial solar radiation in the absence of atmosphere as illustrated in Example Problem 2.6. The value of $H_o = 25.35$ MJ/m²:

$$K_t = \frac{H}{H_o} = \frac{13MJ/m^2}{25.35MJ/m^2} = 0.513$$

b. The second version of Equation 2.14 will be used:

$$\frac{H_d}{H} = 1.188 - 2.272K_t + 9.473K_t^2 - 21.865K_t^3 + 14.648K_t^4$$

$$\frac{H_d}{H} = 1.188 - 2.272(0.513) + 9.473(0.513)^2 - 21.865(0.513)^3 + 14.648(0.513)^4$$

$$\frac{H_d}{H} = 1.188 - 1.166 + 2.493 - 2.952 + 1.0145 = 0.5775$$

c. Thus, about 57.75% of the solar radiation received was diffuse radiation, or 7.5 MJ/m².

d. The amount of beam or direct solar radiation is the difference, or about 5.5 MJ/m².

The plot of solar radiation received in College Station, Texas, throughout the year in units of MJ/m² is shown in Figure 2.5, showing highest amounts during the summer months of June to August.

The previous sections have demonstrated how to make an estimate of the solar radiation received on a surface at any given season, month, day, or even time of day. To make a year-round estimate of solar energy, the clearness index of a given location is necessary. If a solar collector is installed at a given angle or orientation, the amount of solar energy received on this surface may also be estimated. The basic rule of thumb in orienting a flat plate solar collector is to incline it at an angle equal to the latitude of the place. It should be facing south if the location

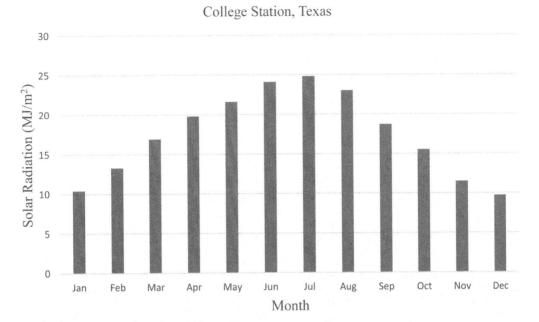

FIGURE 2.5
Average solar radiation received in College Station, Texas, for a year (NSRDB, 2006).

in question is in the northern hemisphere and the reverse if it is located in the southern hemisphere. The previous sets of equations may also be used to face the collector against the sun, with minute-by-minute tracking throughout the day. The next section will discuss the various solar energy conversion devices that may be used and installed in a given location.

2.9 Solar Energy Conversion Devices

Of the eight thermodynamic pathways for solar energy conversion discussed in Chapter 1, this chapter will only discuss the most common systems. These include solar thermal conversion devices (solar heating and cooling systems), solar PV systems, and other general solar power generation devices.

2.9.1 Solar Thermal Conversion Devices

2.9.1.1 Solar Refrigerators

Solar refrigerators are not particularly new. These systems were developed several decades ago. The precursors to solar refrigerators are kerosene absorption refrigerators (e.g., the old Electrolux kerosene refrigerators). The principles are the same, and this is depicted in Figure 2.6 (Exell, 1983).

The solar refrigeration concept is fairly simple. At the initial stage, an ammonia-water system is prepared on a vessel (on the left of the above figure) at a concentration of around 45%. The vessel on the right is empty and kept at low pressure, usually less than 3 atmospheres.

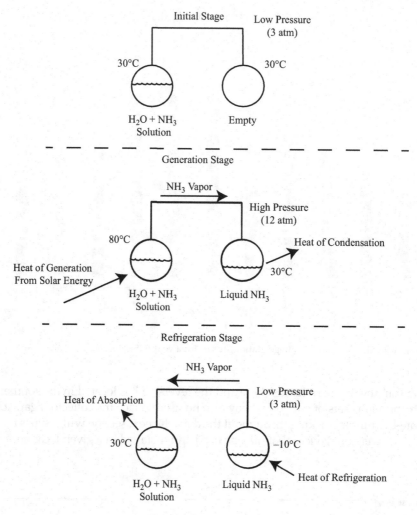

FIGURE 2.6
Schematic diagram for a solar refrigeration system (Exell, 1983).

During the generation stage, whereby solar energy is used to heat the vessel on the left, ammonia vapor is allowed to evaporate and condense on the vessel on the right. The key to the design is a contraption between the vessels that allows only the ammonia vapor to go through the second vessel while keeping the moisture or water from the first vessel. This event, occurring during a hot, sunny day, would create high pressure of around 12 atmospheres on the second vessel while keeping almost pure liquid ammonia in the second receptacle. In this scenario, the heat of generation is solely coming from the sun. The temperature in the solar collector is around 80°C and to be prevented from reaching boiling temperature such that the moisture will not travel to the second vessel. At the end of the day, most liquid ammonia has been condensed on the second vessel. During the evening, the liquid ammonia is allowed to vaporize and returned to the solar collector. This is the refrigeration step. Note that the pure liquid ammonia vaporizes at around 10°C, and if this evaporation step occurs in coils that are enclosed in a vessel with water, the water will freeze or turn into ice. The solar collector back insulation is simply removed to allow the tubes

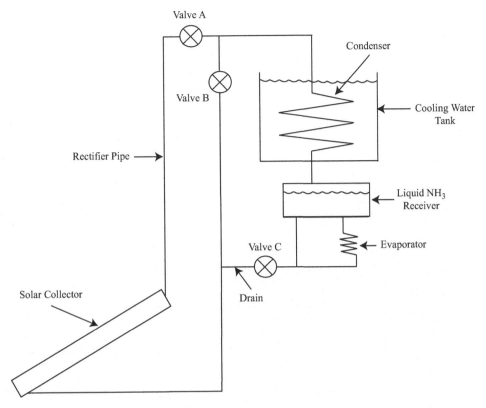

FIGURE 2.7
Schematic of a simple solar refrigerator (Exell, 1983).

that contain the water-ammonia mixture to reabsorb the liquid ammonia, and this process completes the cycle. The schematic of a simple solar refrigerator is shown in Figure 2.7. The solar collector has tubes that contain the ammonia-water mixture and includes removable insulation that may be taken away (or lowered) in the evening. The rectifier pipe allows the ammonia to evaporate while directing the moisture back to the solar collector. Valve A is usually open during the generation step, and the ammonia vapor goes through a huge condenser with cooling water. The liquid ammonia is collected in a receiver and throttled to the evaporator during the refrigeration step. This is where the cooling effect is exhibited.

The practical pressure and temperature profile and relationship in a simple solar refrigerator is shown in Figure 2.8. The ammonia concentrations could vary between 34% and 45%, and the temperature ranges from –10°C to close to 100°C. The pressure varies between atmospheric to around 1,160 kPa.

The amount of ammonia distilled varies linearly with daily global solar radiation; a typical trend for this solar-assisted distillation is shown in Figure 2.9. About 10 kg of ammonia may be distilled when the global solar radiation is around 25 MJ/m². On a clear day in the tropics with a solar collector area of around 20 square meters, over 30 kg of ice may be produced.

2.9.1.2 Solar Dryers

Solar dryers are also very common solar thermal conversion devices. The principle is to heat cold air and pass it though materials that are being dried. A simple solar dryer made

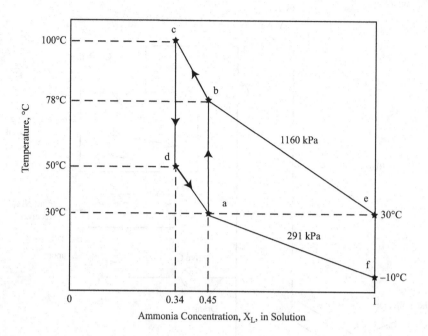

FIGURE 2.8
Practical pressure-temperature profile in a solar refrigerator (Exell, 1983).

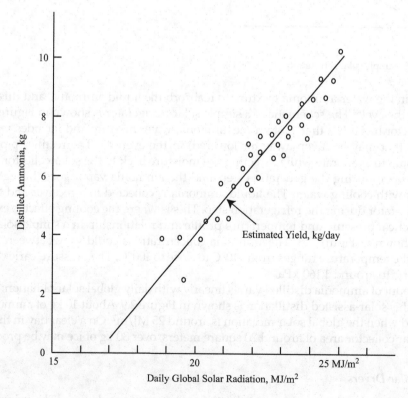

FIGURE 2.9
Typical performance of a simple solar refrigerator (Exell, 1983).

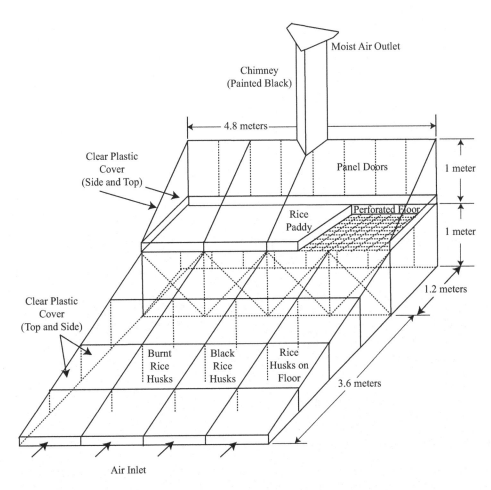

FIGURE 2.10
Simple solar dryer for agricultural products (Exell, 1980).

for half a tonne of rice paddy is shown in Figure 2.10. This dryer is made up of plastic sheets and wood for structural support. Burnt rice husks or coconut husks are laid down on the floor absorber area such that solar radiation is absorbed. Cold air enters through the inlet on the front end of the system and is heated as it travels through the solar collector. A chimney must be designed and installed above the perforated floor for drying agricultural products. This chimney should be painted black to effectively bring hot and moist air out of the drying floor and, in effect, encourage the movement of air.

2.9.1.3 Solar Water Heaters

Solar water heaters are used to heat cold water in the winter such that it may be used for household water use. There are two types of solar water heaters: natural circulation and forced circulation systems. Shown in Figure 2.11 are various types and configurations of the systems. Natural circulation systems have tanks mounted above the solar collector, while forced circulation system tanks may be placed at the same level as the collector. These systems would require a pump to move the water from one section to another. Some sensors may be needed to direct water flow.

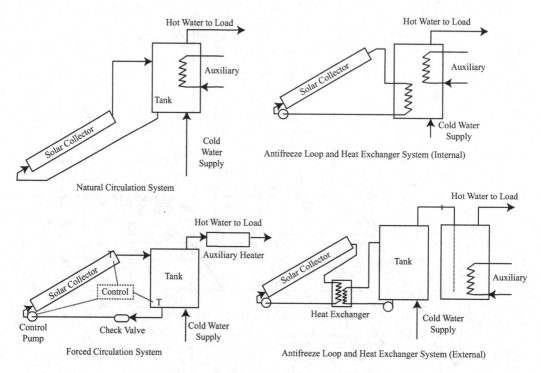

FIGURE 2.11
Various types of solar water heaters (Bhattacharya, 1983).

2.9.2 Solar Photovoltaic (PV) Systems

Solar energy may be converted directly into electrical energy through the use of PV systems. The amount of electrical power produced is proportional to the amount of solar radiation received. Currently, the initial capital cost of a solar PV system is still its main drawback. The solar module is the heart of the solar electric system. There are various materials used to manufacture solar cells, the individual devices that convert solar energy into electricity. The solar PV types are as follows:

a. Mono-crystalline silicon. This material has a conversion efficiency of about 14% and has the best efficiencies among the most common types. However, this is more expensive than multi-crystalline silicon. It is characterized by its plain dark color.

b. Multi-crystalline silicon. This type has a conversion efficiency of around 11% and is less expensive than the mono-crystalline silicon. It is composed of several silicon crystals and is easier to manufacture than mono-crystalline types.

c. Amorphous silicon. This type is commonly used in small appliances, such as watches and calculators. Its efficiency and long-term stability are significantly low. This material is rarely used in power applications.

Figure 2.12 shows the schematic of a solar PV home system. The system components include the solar PV panel, the battery control unit, a deep-cycle battery, inverters to convert the direct current (DC) output to alternating current, as well as some electrical connections and loads. Figure 2.13 illustrates how the solar array is used to supply a village

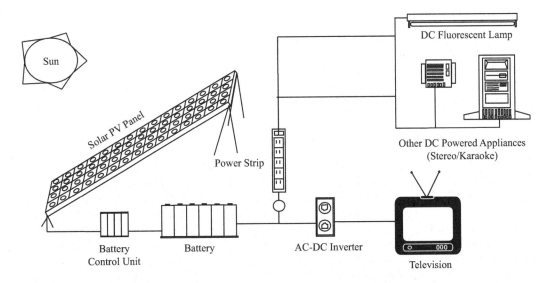

FIGURE 2.12
Schematic of a simple solar PV home system.

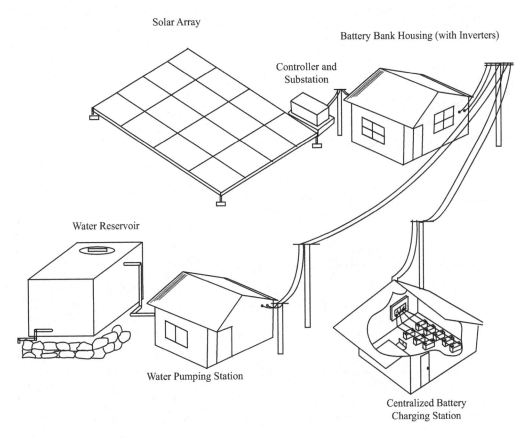

FIGURE 2.13
Schematic of a solar village power system.

with continuous electrical power. Usually, the solar array has a controller and substation that supplies electrical power to a battery bank. This is usually housed in another enclosed building. Inverters are also in place to convert DC power to AC power as required by some appliances. There may also be separate housing for a centralized battery charging station. The electrical power output may also be directed to other applications, such as water pumps that supply the village with potable water.

2.9.3 Solar Thermal Electric Power Systems

Solar thermal electric power systems take advantage of the sun's energy and convert it first into thermal energy. However, flat plate collectors do not provide a high temperature, and as a result, the solar energy is usually concentrated and directed to a single absorber. As shown in Figure 2.14, solar energy is directed by numerous heliostats (or reflectors) to reflect the solar energy into a central receiver system (usually a boiler). This boiler will then absorb enormous amounts of solar energy, thereby producing a very high-temperature steam, normally around 950°F [510°C] at around 1,400 psi (95 atmospheres). This superheated steam is then directed to a steam turbine to move the turbine blades, and the turbine shafting is connected to a generator to produce electrical power. This is then connected to the grid. The spent steam goes through a cooling process where it is pumped to a feed water heater before being brought back to the central receiver system. The capacity of these systems may range from a low of 10 MW to as high as 200 MW of electrical power, depending upon the available area for solar energy collection (Bhattacharya, 1983).

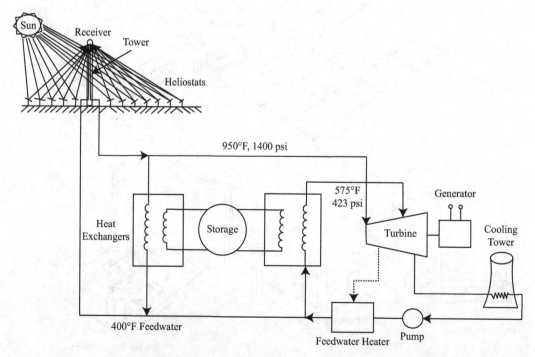

Central Receiver Solar Thermal Electric Power Plant (STEC), 10 - 200 MWe

FIGURE 2.14
Schematic of a solar thermal electric power generation system (Bhattacharya, 1983).

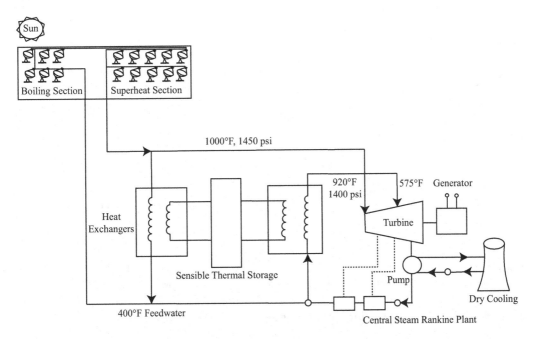

Parabolic Dish-Steam Transport and Conversion System

FIGURE 2.15
Solar thermal power systems with distributed collectors (Bhattacharya, 1983).

2.9.4 Solar Thermal Power Systems with Distributed Collectors

Some solar thermal power generation systems separate the boiler and superheat sections before superheated steam goes to the turbine. This is depicted in Figure 2.15. The boiling and superheat sections use parabolic dishes instead of heliostats to concentrate the sun's energy. Each parabolic dish would have its own absorber to raise the temperature of water significantly and turn it into steam. The output from the boiling section is then directed to the superheat section. The superheat temperature may be above 1,000°F [538°C], and the pressure could also be as high as 1,450 psi (98.6 atmospheres) (Bhattacharya, 1983).

2.9.5 Solar Thermal Power Systems with Distributed Collectors and Generators

Another version of a solar thermal conversion system is an independent power generation system shown in Figure 2.16. In this system, the parabolic dishes are also equipped with a small heat engine generator. Thus, the output of each collector is already electrical power. The advantage of this design is that the transport of steam is minimized and only electrical power is transmitted over longer distances. The electric collection system may have a battery bank as a standby storage system to ensure that the load to the grid is made uniform and constant.

The average cost of a solar thermal collector system varies from a low of around $25/m² to as high as $60/m² over the years (EIA, 2012). From 1998 to 2004, there had been a decline in this average cost. However, due to fluctuations in energy costs—notably natural gas costs in 2005-2007—the thermal cost has increased significantly. The average cost should be around $30-$40/m² as shown in Figure 2.17.

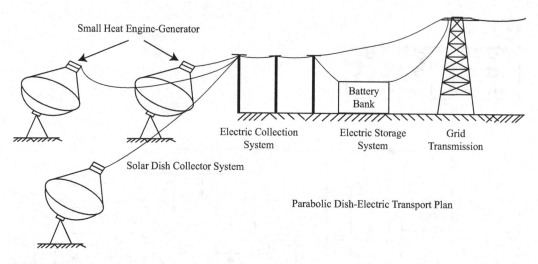

FIGURE 2.16
Solar thermal power system with distributed collector and engine (Bhattacharya, 1983).

2.9.6 High-Temperature Solar Heat Engines

Solar concentrators are also used to generate high-temperature steam that can drive external combustion engines, such as Stirling or Rankine engines. The Southern California Edison Company has numerous facilities (10 MW) installed in the Mojave Desert (Viera da Rosa, 2013). The system now uses intermediate working fluid ($NaNO_3/KNO_3$ 60:40 mix) instead of water. Due to the salt's corrosive nature, the use of much more corrosive-resistant material, such as better-quality stainless steel, is necessary. The configuration of

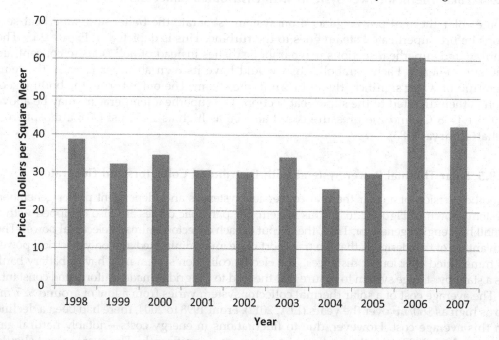

FIGURE 2.17
Average prices of solar thermal collector systems (EIA, 2012).

the system is similar to Figure 2.15. The salt leaves the tower at 565°C [1,049°F], and the return steam temperature is at 288°C [550°F].

Another company, Bright Source Energy, Inc., also a leading solar thermal company in the United States, has constructed the world's largest solar project in California's Mojave Desert called the Ivanpah Project, located at Ivanpah Dry Lake. Its gross power is 392 MW, and its net power is 377 MW (fully operational in 2013). The facility spans 3,500 acres [14.2 km²] and is reported to serve 140,000 households (Bright Source Energy, Inc., 2014). The project is considered the largest solar thermal power plant in the world. Ivanpah was constructed by Bechtel and is operated by NRG Energy.

2.10 Solar Collector System Sizing

Sizing a solar collector system, specifically a PV system, is quite simple and straight-forward. For example, the major components of a solar PV system are the following:

a. PV panel

b. Battery

c. Controller

d. Wires

e. Load in the form of appliances, lights, or motors

The solar cell module or the PV system is the heart of the solar electric system. It converts solar radiation directly into electricity. The power output of the solar cell increases as solar radiation is increased, more or less in direct proportion. Note that a solar PV system is more sensitive to direct solar radiation than diffuse radiation. The balance of system (BOS) is comprised of the other components listed above besides the solar PV system.

There are numerous other PV systems based on other materials, such as cadmium telluride (CdTe), gallium arsenide (GaAs), or copper indium gallium selenide (CuInGaSe), and numerous other thin films.

The battery stores the energy delivered by the solar collector and provides power for the load or appliances. Batteries for solar applications are usually of the deep-cycle kinds and are more expensive than the regular automotive batteries. Batteries for solar applications last about five years, an improvement compared to the typical three-year lifespan of automotive batteries. Deep-cycle batteries are designed to survive up to 80% discharge (NAW&S, 2014).

The cables or wires used for PV systems must be selected such that unnecessary losses can be avoided—proper sizing is important. Cables should always be as short as possible and larger in size to carry the required load. For solar PV applications, stranded wires are usually recommended.

Solar PV sizing begins with load calculations. The daily load requirement (DLR) must be first calculated by adding the individual load-in watts and their corresponding hours of operation (in hours). Note that energy can be added but not power. DLR is simply the product of load-in watts and time in hours and has the units of watt-hours. On larger systems, the kWh load is usually the unit used. This is shown in Equation 2.16:

$$DLR(kWh) = L(kW) \times t(h) \qquad (2.16)$$

Once the DLR is estimated, the next step is to estimate battery size. Battery capacity is usually rated in either watt-hours or ampere-hours. The latter is most common. The equations to use are shown in Equation 2.17 and Equation 2.18. In Equation 2.17, BC_{wh} is the battery capacity requirement in watt-hours. The constant factor that is commonly used for calculating battery capacity is 10, meaning that there will be a 10% daily depth discharge or that it will take 10 days before the battery will totally be out of power. The battery capacity in ampere-hours is given in Equation 2.18, where the battery capacity in units of watt-hours is divided by the system voltage (V). If one were to go to a battery store and purchase a battery, the salesperson would typically ask for the ampere-hour (AH) capacity needed:

$$BC_{wh} = DLR \times 10 \tag{2.17}$$

$$BC_{AH} = \frac{BC_{wh}}{V} \tag{2.18}$$

The PV panel watt-peak (Wp) requirement is then calculated using two factors—the climate generation factor (CGF) and the overall system efficiency (%). This is given in Equation 2.19. The CGF varies depending upon the climate of the site location and also upon global geographic location. For example, the CGF in Thailand is 3.43, whereas in European countries (the European Union), the value used is 2.93 (El-Shimy, 2007). The overall system efficiency (not the PV efficiency) is a function of PV design and usually in the range close to 85%:

$$W_p\left(Watts\right) = \frac{DLR}{CFG \times \xi_s} \tag{2.19}$$

The wire size is calculated using Equation 2.20. The wire size (Aw) is usually reported in units of mm, while W_L is the load in watts, V is the system voltage, and L is the length of wire in meters. The standard wire gauge in the United States is called the American wire gauge (AWG), also known as the Brown and Sharpe wire gauge. The dimensions of the wires are given in ASTM Standard B 258. The cross-sectional area of each gauge is important for determining current carrying capacity. Increasing gauge numbers mean decreasing wire diameters. For example, AWG#16 has a diameter of 1.291 mm [0.0508 in], AWG#14 has a diameter of 1.628 mm [0.0641 in], AWG#12 has a diameter of 2.053 mm [0.0808 in], and AWG#10 has a diameter of 2.588 mm [0.1019 in]:

$$A_w\left(mm\right) = 0.04 \times \left(\frac{W_L}{V}\right) \times L \tag{2.20}$$

Example 2.9 illustrates the procedure for sizing PV systems as well as getting an estimate of load, battery capacity, Wp calculations, and wire sizing.

Example 2.9: Sizing Solar PV Systems

A typical barn lights up its entrance from 6 p.m. to 9 p.m., the feeding stall lamp for an hour, and the feed water pump for two hours. The entrance lamp wattage is 10 watts, and the feeding stall lamp requires 10 watts. The pump consumes 1.4 amperes at 12 volts. (a) What is the energy requirement of the barn? (b) Compute the size of the battery and (c) the PV panel needed for the system in watt-peak (Wp), assuming a CGF of 3 and overall system efficiency of 85%. (d) What size of wire is recommended for the PV battery and battery, assuming 10 meters length?

SOLUTION:

a. The total energy may be easily calculated by making a tabulation of loads and their wattage and hours of use to estimate energy in watt-hour as shown in the table below.

Load	Wattage	Hours	Watt-Hour
Entrance Lamp 1	10	3 hrs	30
Feeding Stall Lamp 2	10	1 hr	10
Pump	$1.4 \times 12\ V = 16.8$	2 hrs	33.6
Total	36.8 watts		73.6 watt-hour

Therefore, the energy requirement for this barn each day is 73.6 watt-hour.

b. Battery capacity is estimated using Equation 2.17 and Equation 2.18 as shown:

$$BC_{wh} = DLR \times 10 = 73.6 \times 10 = 736\ Watt - hrs$$

$$BC_{AH} = \frac{BC_{wh}}{V} = \frac{736\ Watt - hrs}{12\ Volts} = 61.3\ Ampere - hours$$

The battery capacity that is available is usually given in nominal values, for example, a 70 ampere-hour battery, and this will be selected, which is a higher rating.

c. The PV panel Wp calculation is shown using Equation 2.19 and the assumptions given:

$$W_p\ (Watts) = \frac{DLR}{CFG \times \xi_s} = \frac{73.6\ Watt - hours}{3 \times 0.85} = 28.86\ Watts$$

The PV panel to be selected must also be higher and of nominal size. In this case, a 30-Wp unit may be selected. Some stores also sell 45-Wp units.

d. Finally, the wire size may be calculated from the PV to the battery and from the battery to the load as shown below and using Equation 2.20:

$$A_w\ (mm) = 0.04 \times \left(\frac{W_L}{V}\right) \times L = 0.04 \times \left(\frac{28.86\ Watts}{12\ Volts}\right) \times 10 = 0.96\ mm\ (AWG\#16)$$

$$A_w\ (mm) = 0.04 \times \left(\frac{W_L}{V}\right) \times L = 0.04 \times \left(\frac{36.8\ Watts}{12\ Volts}\right) \times 10 = 1.23\ mm\ (AWG\#16)$$

One should be able to find corresponding wire sizes very easily from electrical wiring stores as well as from electricity textbooks.

2.11 Economics of Solar Conversion Devices

The economics of solar conversion devices, particularly PV panels, depends much on the capital cost of the panels. The cost of balance of these systems, including operation and maintenance, could account for about 68% of PV system pricing (GTM, 2012). The payback period decreases as the price of PV panels and balance of system costs decrease over time. These panels could last between 25 and 40 years. Several years ago, the price per watt-peak was in the range of $6-$8; nowadays, the typical cost ranges from $1 to $3. China's production cost is near $1/Wp, and some reports even quote the price as below $1/Wp.

As the price of PV units goes down, one will also see a decline in the cost of electricity produced by these PV panels. The point at which the cost of PV electricity is equal to or cheaper than the price of grid power is defined as "grid parity." This condition is achieved more easily in areas that receive abundant sunshine (such as in the tropics) as well as in areas with high costs of electricity (such as California, the Philippines, and Japan). We can infer that in these areas, there will be an in increase in the installation of PV systems.

Example 2.10: Sizing Household PV Systems

A household owner would like to net meter his residential electrical power load using solar energy. The average solar radiation in his area is around 5 kWh/m²/day. He uses about 10,000 kWh of electricity per year. He pays for electricity at $0.10/kWh. (a) What is the minimum area of solar PV system his household would need if the overall conversion efficiency is 11%? (b) At a price of $2/Wp, what is the initial investment and payback period required to recover the initial capital cost only and based on the payment for electricity alone? Assume a CGF of 3 and efficiency of 89% for the calculation of peak wattage.

SOLUTION:

a. The first step is to calculate the DLR for this household. This can simply be calculated using the yearly requirement divided by the number of days in a year, assuming 365 days/yr for a non–leap year as shown:

$$DLR\left(\frac{kWh}{day}\right) = \frac{10,000\ kWh}{yr} \times \frac{1\ yr}{365\ days} = 27.4 \frac{kWh}{day}$$

b. Then, using the efficiency equation, we find the amount of solar energy needed to satisfy this DLR:

$$\eta = \frac{Output}{Input} \times 100\%$$

$$0.11 = \frac{27.4\ kWh/day}{Input} \times 100\%$$

$$Input\left(\frac{kWh}{day}\right) = \frac{27.4\ kWh/day}{0.11} \times 100\% = 249 \frac{kWh}{day}$$

c. Then the minimum area needed to satisfy this output energy is calculated as follows:

$$Area\left(m^2\right) = \frac{249\ kWh}{day} \times \frac{m^2 - day}{5\ kWh} = 49.8\ m^2$$

d. Next, the watt-peak calculation is found with the following equation, assuming a CGF of 3 and efficiency of 85%:

$$W_p\left(Watts\right) = \frac{DLR}{CGF \times \xi_s} = \frac{27,400\ Watt - hours}{3 \times 0.89} = 10,262\ Watts$$

e. At \$2/Wp, the initial investment is about \$20,524. If he uses the grid power at \$0.10/kWh, he would be paying around \$1,000/yr for electricity. Consequently, recovering the initial capital investment alone and paying for net-metered electricity would require 20.52 years.

The US Department of Energy, through the National Renewable Energy Laboratory (NREL), has developed a computer model called SAM (Systems Advisor Model). This model calculates performance and financial metrics of renewable energy systems. Students and other technical experts in solar energy may use the results from SAM to evaluate the economic and technical performance of numerous solar energy devices and technologies. The software has recently been expanded to cover other renewables, such as wind, geothermal, biomass, and conventional power systems (Blair, et al., 2014). This software is open for use by any person technically knowledgeable in renewables, including those with advanced knowledge in simulations. SAM's advanced simulation options include sensitivity analyses and statistical analyses that are used in Monte Carlo–type simulations and to observe the effect of weather variability. One would typically prepare all variables in a Microsoft Excel spreadsheet and use that to run the full program with ease. The software can be downloaded for free at https://www.nrel.gov/analysis/sam/.

2.12 Summary and Conclusions

This chapter introduces the step-by-step procedure for making a sound and accurate estimate of the amount of solar radiation that may be received in any given location across the world through any given time of day or day of the year. The amount of energy received by the earth from the sun is predictable across all seasons if the atmosphere is free of obstructions. This magnitude will vary slightly above or below the value of solar constant, assumed to be 1,367 W/m^2 using the best data from the National Aeronautics and Space Administration (NASA) (Duffie and Beckman, 2006). However, the amount of solar energy received on the earth's surface would be reduced due to reflectance, absorbance, or losses through the earth's dusty or moist atmosphere and may be as low as 100-200 W/m^2 during cloudy days or as high as 1,100 W/m^2 during clear days. For any location on the earth, one has to establish the clearness index, K_t, defined as the actual solar energy received over its maximum value, either monthly or for a defined season. With the clearness index known, the total solar radiation (both diffuse and direct) may be estimated more accurately and simulation studies made. In the absence of accurate data to establish K_t for a given locality, one may use the nearest city for estimates or perform simple interpolations between city locations. Numerous cities in the world have already established these values, and most solar energy textbooks have published these values (e.g., Duffie and Beckman, 2006).

Once the solar energy resource calculations are made, an engineer may be able to design solar conversion devices and estimate efficiencies of conversions. These calculations vary according to the choice of particular conversion system utilized. The solar PV system is perhaps the easiest one to estimate since this device has fairly established solar energy conversion efficiencies into electrical power. However, the installer of PV devices must be familiar with the proper orientation of solar collection in terms of its azimuth angle, tilt, and other parameters presented in this chapter. The simple rule of thumb in orienting fixed solar collectors for people in the northern hemisphere is to orient the collector facing due south with an angle equal to the latitude of the place. Keeping in mind the concept of solar time versus actual clock time, one can simulate the performance of solar collectors at any given time of the day for a year. Similarly, one may be able to estimate the theoretical duration of sunshine each day of the year, including times of sunrise and sunset, through the equations presented in this chapter. There are photoperiod crops, such as chrysanthemums, which are very sensitive to sunshine duration (for the crop to initiate flowering or continue its vegetative stage). Equation 2.11 may be used in this case.

2.13 Problems

2.13.1 Extraterrestrial Solar Radiation

P2.1 Determine the extraterrestrial solar radiation received on the earth on June 15, 2014. Plot the variation of extraterrestrial solar radiation received with the time of year on a normal plane. Use $G_{sc} = 1,367$ W/m^2.

2.13.2 Solar Time

P2.2 Determine the solar time corresponding to 8:30 a.m. Central Standard Time on June 15, 2014, in College Station, Texas. The longitude of College Station, Texas, is 96.32 West, and this town's time was based on a 90° meridian.

2.13.3 Solar Declination Angle

P2.3 Calculate the solar declination angle on June 15, 2014.

2.13.4 Angle of Incidence

P2.4 Calculate the angle of incidence of beam radiation on a solar collector surface located in College Station, Texas, at 8:30 a.m. (solar time) on June 15, 2014, if the surface is tilted at an angle of 30° and pointed 25° west of south. The latitude of College Station, Texas, is 30.61° North.

2.13.5 Hour Angle, Time of Sunrise, and Number of Daylight Hours

P2.5 Calculate the hour angle, time of sunrise and number of daylight hours on June 15, 2014, in College Station, Texas. The latitude in College Station, Texas, is 30.61° North.

2.13.6 Theoretical Daily Solar Radiation, H$_o$

P2.6 Determine the H$_o$—the day's solar radiation on a horizontal surface in the absence of atmosphere in College Station, Texas, on June 15, 2014. The latitude of College Station, Texas, is 30.61° North. Use the results from previous problems if possible.

2.13.7 Theoretical Hourly Solar Radiation

P2.7 Determine the I_o—the day's solar radiation on a horizontal surface in the absence of atmosphere in College Station, Texas, on June 15, 2014 between the hours of 8 a.m. and 9 a.m. (solar time). The latitude of College Station, Texas, is 30.61° North. Use the results from previous example problems.

2.13.8 Clearness Index to Estimate Beam and Diffuse Radiation

P2.8 Determine the clearness index, K_t, and the fraction of solar energy that is beam radiation and the fraction that is diffuse radiation for the month of June 15, 2014, in College Station, Texas, if the solar radiation actually received is 15 MJ/m².

2.13.9 Sizing Solar PV Panels

P2.9 A typical barn lights up its entrance from 6 p.m. to 6 a.m., its feeding stall for two hours (6 p.m.-8 p.m.), and its feed water pump for two hours. The wattage of the entrance lamp is 10 watts, and the feeding stall lamp is 10 watts, whereas the pump consumes 2 amperes at 12 volts. (a) What is the energy requirement of the barn? (b) Compute the size of the battery and (c) the PV panel needed for the system in watt-peak (Wp) assuming a CGF of 3 and overall system efficiency of 85%. (d) What size of wire is recommended for the PV battery and battery, assuming 15 meters length?

2.13.10 Economics of Solar Energy

P2.10 A homeowner would like to net meter his residential electrical power load using solar energy. The average solar radiation in his area is around 4 kWh/m²/day. He uses about 12,000 kWh of electricity per year. He pays for electricity at $0.12/kWh. (a) What is the minimum area of solar PV system this household would need if the overall conversion efficiency is 15%? (b) At a price of $1.5/Wp, what is the payback period and initial investment required to recover only the initial capital cost and based on the payment for electricity alone? Assume a CGF of 3 and efficiency of 85% for the calculation of peak wattage.

References

Bhattacharya, S. C. 1983. Lecture Notes in "Renewable Energy Conversion" Class. Asian Institute of Technology, Bangkok, Thailand.

Blair, N., A. P. Dobos, J. Freeman, T. Neises and M. Wagner. 2014. System Advisor Model, SAM 2014.1.14: General description. National Renewable Energy Laboratory Technical Report NREL/TP-6A20-61019, February. Contract No. DE-AC36-08GO28308. US Department of Energy, Golden, CO.

Boyle, G. 2004. Renewable Energy: Power for a Sustainable Future. Oxford University Press, Oxford, UK.

Bright Source Energy, Inc. 2014. Ivanpah Project Fact Sheet. Bright Source Energy, Inc., Oakland, CA.

Collares-Pereira, M. and A. Rabl. 1979. The average distribution of solar radiation—correlations between diffuse and hemispherical and between daily and hourly insolation values. Solar Energy 22: 155–163.

Duffie, J. A. and W. A. Beckman. 2006. Solar Engineering of Thermal Processes. 3rd Edition. John Wiley and Sons, New York, NY.

Duffie, J. A. and W. A. Beckman. 1981. Solar Engineering of Thermal Processes. 1st Edition. John Wiley and Sons, New York, NY.

EIA. 2012. Renewable Energy Annual 2009. US Energy Information Administration, Independent Statistics and Analysis, US Department of Energy, Washington, DC. 20585. January. Available at: https://www.eia.gov/renewable/annual/pdf/rea_report.pdf. Accessed April 17, 2018.

El-Shimy, M. 2017. Approximate sizing of photovoltaic arrays: the panel generation factor (PGF)-proof. November 10. Chapter 3: Operational Characteristics of Renewable Sources. Challenges and Future Prospectives. In: El-Shimy, M. (Editor). Economics of Variable Renewable Sources for Electric Power Production. Germany: Lambert Academic Publishing, Germany.

Exell, R. H. B. 1980. Basic design theory for a simple solar dryer. Renewable Energy Review Journal 1 (2).

Exell, R. H. B. 1983. The theory of a simple solar refrigerator. Renewable Energy Review Journal 5, (2).

GTM. 2012. Solar balance-of-system costs account for 68% of PV system pricing: new GTM Report. GTM Research Spotlight Issue. November 15. GTM: A Wood Mackenzie Business. Available at: https://www.greentechmedia.com/articles/read/Solar-Balance-of-System-Accounts-for-68-of-PV-System-Pricing-New-GTM-Repo#gs.gu36T=E. Accessed April 18, 2018.

Iqbal, M. 1983. An Introduction to Solar Radiation. Academic Press, Toronto, Canada.

Kahar, S. Ab. 2008. Effects of photoperiod on growth and flowering of Crysanthemum moridolium Ramat cv. Reagan Sunny. Journal of Tropical Agriculture and Food Science 36 (2): 01–08.

Kreith, F. and J. F. Kreider. 2011. Principles of Sustainable Energy. CRC Press, Taylor and Francis Group, Boca Raton, FL.

NSRDB. 2006. National Solar Radiation Data Base: 1991-2005 update: typical meteorological Year 3. Available at: http://rredc.nrel.gov/solar/old_data/nsrdb/1991-2005/tmy3/by_state_and_city.html. Accessed April 17, 2018. US Department of Energy, National Renewable Energy Laboratory (NREL), Golden, CO.

Nelson, J. 2003. The Physics of Solar Cells. Imperial College Press, London.

Northern Arizona Wind & Sun (NAW&S). 2014. Deep cycle battery FAQ. Copyright 1998-2014. Available online at: https://www.solar-electric.com/learning-center/batteries-and-charging/deep-cycle-battery-faq.html. Accessed April 18, 2018.

Vieira da Rosa, A. 2013. Fundamentals of Renewable Energy Processes. 3rd Edition. Academic Press, Elsevier Publications, Walthan, MA.

3

Wind Energy

Learning Objectives

Upon completion of this chapter, one should be able to:

1. Estimate the available wind energy from actual wind speed data.
2. Enumerate the various wind energy applications.
3. Define maximum and practical conversion efficiencies.
4. Describe the instruments used to measure wind energy.
5. Design the size and dimensions of a windmill appropriate for a given site or application.
6. Relate the overall environmental and economic issues concerning wind energy conversion systems.

3.1 Introduction

Wind energy is a consequence of solar energy. The earth receives heat from the sun on one side (the side facing the sun) while its opposite side cools. This uneven heating of the earth gives rise to wind movement. In the heated regions, air rises as a result of a decrease in density while being replaced by cooler, slightly dense wind, creating movement. The rising of hot air is a frequent and basic occurrence. This event is depicted in Figure 3.1. In addition, since the earth moves on its own axis, a tangential component of wind movement exists, giving way to easterly or westerly wind directions, called trade winds. There are also local variations in wind movements due to the presence of mountains and barriers, such as vegetation or forest cover.

Extracting energy from the wind is achieved by designing a device that converts the kinetic energy from the wind into mechanical energy by way of windmills and wind aero-generators.

In the past, water pumping in most parts of the United States relied on multi-bladed windmills. Such windmills move at quite slow speeds, producing high torque as necessary for pumping water from deep wells. As we have harnessed mechanical power into electrical power using electromagnets, engineers have now also designed very efficient wind aero-generators that convert wind energy into electrical energy. However, these generators require quite high speeds and necessitate the use of windmills that move at a faster rate. These are usually wind generators with a few blades each. One might assume

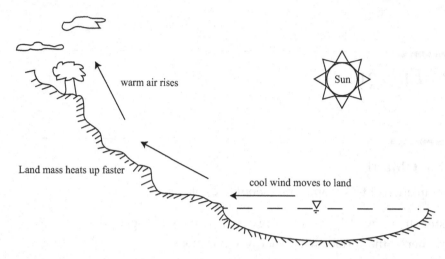

FIGURE 3.1
Typical generation of wind movement.

that the more blades a windmill has, the faster it goes, but in fact, the fastest wind aero-generators are those with fewer blades. It is indeed possible to design a windmill with only a single blade. In this design, there must be a counterweight on the opposite end of a single-bladed windmill so that there will be no imbalance in weight distribution. The most popular wind aero-generators for generating electrical power are those with two, three, or four blades.

The variations of wind speed due to height are also known—at higher elevations, wind speed is higher. To take advantage of this, windmills are usually raised to higher elevations to avail of higher wind speeds. Windmills with towers greater than 60 meters [200 ft] must have clearance from aviation authorities, especially in the United States, since aviation authorities have jurisdiction of airspace above that height. Details of wind speed variations due to elevation and topography will be discussed in later sections. Wind speed data are now available in most cities and counties from weather stations that report basic meteorological data or from airports that have wind-measuring instruments from a standard 10-meter elevation [32.8 ft]. When these data are used, the wind speed at the tower height must be adjusted using empirical relationships to correct wind variations with different heights. Air density is also a function in the calculations and must be known. It will also be demonstrated in later sections that wind power varies with the cube of wind speed, and wind energy calculations based on average wind speed can underestimate the actual wind power. One must also have long-term running data to make projections of wind potential in an area. If one plots the wind speed data over time for a given location, the distribution of these data from a given histogram is neither a bell-shape curve nor normally distributed. Consequently, one should rely on a different statistical distribution besides the normal curve, that is, those that are skewed to one side either as the binomial distribution in common statistics or in the Weibull distribution. About 1%–2% of the sun's energy is converted into wind energy, and this is reported to be about 50-100 times more energy than all the biomass plants in the world that capture their energy from the sun (Busby, 2012).

3.2 Basic Energy and Power Calculation from the Wind

The main basis for the calculation of wind energy potential comes from the kinetic energy equation shown in Equation 3.1. In this equation, the kinetic energy from the wind is a function of its mass and the square of the velocity. The most common unit for this energy is Joules (1,055 Joules = 1 Btu):

$$KE = \frac{1}{2}mv^2 \qquad\qquad (3.1)$$

where
  m = mass of wind, in kg
  v = wind speed, in m/s

Example 3.1 shows how the kinetic energy from a given mass of wind is estimated. Note that a simple conversion is needed to eliminate the unit where the unit second is squared. This would come from Newton's Law of force as shown in Equation 3.2. In this equation, the force (in units of Newton) is the product of mass and acceleration. The unit of mass is in kg, while the unit of acceleration is in m/s². One may use this relationship to convert the complicated unit combination of kg-m²/s² into Joules by simply multiplying this with 1 or unity, having a unit of N-s²/kg-m from Newton's Law:

$$Force\,(Newton) = m \times a = kg \times \frac{m}{s^2} = Newton\,(N) \qquad\qquad (3.2)$$

where
  F = force in units of Newton (N)
  m = mass (kg)
  a = acceleration due to gravity (9.8 m/s²)

Example 3.1: Kinetic Energy from Wind

Estimate the kinetic energy from wind moving at a speed of 10 m/s having a mass of 100 kg. Convert this into English units assuming a conversion of 1,055 Joules per Btu.

SOLUTION:

  a. The kinetic energy is simply calculated using Equation 3.1 and the Newton's Law relationship shown in Equation 3.2:

$$KE = \frac{1}{2}mv^2 = \frac{1}{2} \times 100\,kg \times \left(\frac{10\,m}{s}\right)^2 \times \frac{N-sec^2}{kg-m} \times \frac{Joules}{N-m} = 5,000\,Joules$$

  b. In English units, the conversion is quite simple:

$$KE = 5,000\,Joules \times \frac{Btu}{1,055\,Joules} = 4.7\,Btu$$

There is a significant difference between energy and power. Energy is a fixed value, for example, the energy contained in a 100 kg [220 lbs] mass of wind and moving at a speed of 10 m/s [32.8 ft/s]. This is like holding a glass of milk with an energy content of, for example, 165 kcal (1 kcal = 4,184 J = 3.966 Btu) and carrying it from the refrigerator to the table. That energy is fixed and contained in the glass no matter where it is taken. Energy is also synonymous with work, or force multiplied by a given distance, as shown in Equation 3.3. Work has the same unit as energy. Thus, if this milk is being carried from the refrigerator to the table at a distance of 10 meters [32.8 ft] while expending a force of 10 N, the energy or work done is equal to 100 Joules (i.e., the product of force and distance) [0.095 Btu]:

$$Work\,(Joules) = Force \times Distance = N - m = Joules \tag{3.3}$$

where
    work = units of Joules
    force = units of N
    distance = units of m

Power, on the other hand, is a rate. Think of this parameter as a rate of doing work or, specifically in wind calculation, the rate at which energy is being extracted from the wind. If the energy unit is in Joules, then the unit of power is in Joules/s, which is a power unit equivalent to watts (i.e., Joules/s = watts). Thus, the power from the wind is simply the kinetic energy divided by a unit of time. It may seem awkward to divide this energy by time, but one may consider it as extracting energy from a mass of wind per unit of time or use the mass flow rate (kg/s) of wind. The classical wind power equation is given by Equation 3.4 and how the unit of watt is derived:

$$Wind\ Power\ (Watts) = \frac{1}{2}\dot{m}v^2$$

$$Wind\ Power\ (Watts) = \frac{1}{2} \times \frac{kg}{sec} \times \left(\frac{m}{s}\right)^2 \times \frac{N-s^2}{kg-m} \times \frac{Joule}{N-m} \times \frac{Watt-s}{Joule} = \frac{1}{2}\,Watts \tag{3.4}$$

where
    wind power = watts
    $\dot{m}$ = mass rate (kg/s)
    v = wind speed (m/s)

The mass rate of wind flow is also related to density, the cross-sectional area perpendicular to wind flow, as well as the velocity, as shown in Equation 3.5. The product of this mass flow rate term is also shown following Equation 3.5 to illustrate how the mass flow rate unit was derived:

$$\dot{m}\left(\frac{kg}{s}\right) = \rho Av$$

$$\dot{m}\left(\frac{kg}{s}\right) = \rho \times A \times v = \frac{kg}{m^3} \times m^2 \times \frac{m}{s} = \frac{kg}{s} \tag{3.5}$$

where
  $\dot{m}$ = mass rate (kg/s)
  ρ = density of wind (kg/m³)
  A = cross-sectional area perpendicular to wind flow (m²)
  v = wind speed (m/s)

If one would now incorporate this mass flow rate relationship to the wind power equation, the resulting relationship is the crucial equation in wind power calculation as shown in Equation 3.6. Example 3.2 shows how this equation is used in calculations:

$$Wind\ Power\ (Watts) = \frac{1}{2}\dot{m}v^2$$

$$Wind\ Power\ (Watts) = \frac{1}{2}\rho \times A \times v \times v^2 \tag{3.6}$$

$$Wind\ Power\ (Watts) = \frac{1}{2}\rho A v^3$$

where
  wind power = watts
  ρ = air density, usually 1.2 kg/m³ at standard temperature and pressure (STP) [or a value of 0.075 lb/ft³ in English units]
  A = cross-sectional area perpendicular to flow (m²)
  v = wind speed (m/s)

Example 3.2: Theoretical Power Extracted from Wind

Determine the power being extracted from wind with a mass rate of flow of 100 kg/s [220 lbs/s] at a velocity of 10 m/s [32.8 ft/s]. Assume an air density of 1.2 kg/m³ [0.075 lbs/ft³] over a cross-sectional area of 10 m² [107.584 ft²]. Express the units in the English system.

SOLUTION:

  a. Equation 3.6 is used as shown with corresponding units:

$$Wind\ Power\ (Watts) = \frac{1}{2}\rho A v^3$$

$$Power\ (W) = \frac{1}{2} \times \frac{1.2\ kg}{m^3} \times 10\ m^2 \times \left(\frac{10\ m}{s}\right)^3 \times \frac{N - s^2}{kg - m} \times \frac{J}{N - m} \times \frac{W - s}{J} = 6,000\ W$$

  b. The wind power of 6,000 W is converted into English units using the simple conversion factor of 746 W per horsepower:

$$Power\ (hp) = 6,000\ W \times \frac{hp}{746\ W} = 8.04\ hp$$

Note that in Equation 3.6, the power from the wind is proportional to the cube of the wind speed and not to the average wind speed. The wind speed data from various weather stations around the country usually record wind speed as an average value over a given period of time and at a height of about 10 meters [32.8 ft].

It is important to note the difference in power calculations if the data are based on an average value versus the cube of that value. For example, one meteorological station only measures wind speed in the middle of the day and calls that the day's average wind speed—say, 3 m/s [9.84 ft/s]. Another station measures the wind speed in the morning, at noon, and in the afternoon—say, values of 2, 3, and 4 m/s, respectively [6.56, 9.84, and 13.12 ft/s]. Note that the average wind speed for the two stations are the same, that is, 3 m/s [9.84 m/s]. However, if one calculates the wind power based on Equation 3.6 (assuming the air density and area to be the same), the resulting power estimates are 81 (27 + 27 + 27) versus 91 (8 + 27 + 64). In this case, there is an underestimation of wind power if the data are based on average values as opposed to having the data of the cube of wind speed. This is illustrated in Example 3.3.

Example 3.3: Underestimation of Wind Power Using Single Average Wind Speed Data

Estimate the available theoretical power from wind in an area where the daily average wind speed was recorded as 5 m/s [16.4 ft/s] and compare this to an area where the readings are done in the morning, at noon, and in the afternoon, with readings of 4, 5, and 6 m/s, respectively [13.12, 16.4, and 19.68 ft/s]. Calculate the power at each measurement period and calculate the average power for the day assuming air density of 1.2 kg/m³ [0.075 lbs/ft³] and an area of 100 m² [1,075.84 ft²].

SOLUTION:

a. The calculation is done using Equation 3.6 successively, and the power is taken from the average of the readings as shown:

$$Wind\ Power\ (W) = \frac{1}{2} \times \frac{1.2\ kg}{m^3} \times 100\ m^2 \times \left(\frac{4\ m}{s}\right)^3 = 3{,}840\ W$$

$$Wind\ Power\ (hp) = 3{,}840\ Watts \times \frac{hp}{746\ Watts} = 5.15\ hp$$

$$Wind\ Power\ (W) = \frac{1}{2} \times \frac{1.2\ kg}{m^3} \times 100\ m^2 \times \left(\frac{5\ m}{s}\right)^3 = 7{,}500\ W$$

$$Wind\ Power\ (hp) = 7{,}500\ Watts \times \frac{hp}{746\ Watts} = 10.05\ hp$$

$$Wind\ Power\ (W) = \frac{1}{2} \times \frac{1.2\ kg}{m^3} \times 100\ m^2 \times \left(\frac{6\ m}{s}\right)^3 = 12{,}960\ W$$

$$Wind\ Power\ (hp) = 12{,}960\ Watts \times \frac{hp}{746\ Watts} = 17.37\ hp$$

b. The average power using compounded data was 8,100 W [10.86 hp], while that of just using a single piece of data for the day was 7,500 W [10.05 hp]. In this instance, there is a slight increase in power estimate when more data are available or when variations within the day are taken.

c. Note also that energy may be added, but power units cannot be added. The average power may be calculated.

3.3 The Worldwide Wind Energy Potential

It has been estimated that globally, the wind power from land areas with at least 6.9 m/s wind speed (15 mph) could produce about 72 terawatts (TW) of electricity, or about 72 million MW of power, as opposed to the current generating capacity of only 200,000 MW (Busby, 2012). There is large potential in wind power especially for small-scale applications in agriculture and on ranches. There is an atlas online that contains wind maps of more than 30 countries and wind data from about 80 more countries around the world. This is managed by the National Laboratory for Sustainable Energy in Denmark (Kuhn, 2012). The wind resource map for Europe shows that the highest wind potential is in the Scotland area, followed by areas in England as well as countries in the northern shores, such as the Netherlands, Belgium, Denmark, and those in the northern portion of Spain. High wind potentials are also identified near the shores of Monaco and Slovenia as well as surrounding the Mediterranean Sea.

In the United States, the U.S. Department of Energy's National Renewable Energy Laboratory (NREL) has mapped out the nation's wind energy potential, including individual maps for each state (NREL, USDOE, 2018). There are also other reports accessible online (Elliott, et al., 1986). The American Wind Energy Association (AWEA) also published U.S. wind power potential. The wind power resource potential in the United States is about 10.5 million MW. As of 2016, the installed wind power capacity in the United States is reported by the AWEA to be around 75,714 MW. According to most recent data, Texas leads all the states in installed capacity with about 18,531 MW, or about over 24% of all the United States, followed by Iowa at 6,356 MW and California at 5,662 MW. Most states, including Alaska (62 MW) and Hawaii (203 MW), have respectable installed capacity as well (AWEA, 2016). Surprisingly, the southeastern states, such as Florida, Louisiana, Arkansas, Mississippi, Alabama, Georgia, South Carolina, Kentucky, North Carolina, and Virginia, have not demonstrated any installed capacity on the most recent NREL maps (USDOE, 2018). This is primarily because the NREL maps have projected very little potential in these areas. However, wind power is site specific, especially in the rural and agricultural areas, and these data could change if smaller wind systems are installed. For example, some initiatives in Tennessee displayed about 29 MW of installed capacity as of 2016 (AWEA, 2016). In Texas, the areas with the highest potential are in the El Paso area (east of Texas) as well as in the Panhandle area, where most large-scale wind farms are already in operation. Figure 3.2 shows the most recent wind map of the United States.

3.4 The Actual Energy and Power from the Wind

The maximum power or energy that could be recovered from the wind is determined by momentum theory and reported to be only 59.3% of the theoretically available power (Manwell, et al., 2012). This is the classical Betz coefficient, published in 1919 by the famous German physicist Albert Betz. According to Betz, no wind turbine can capture more than 59.3% (or 16/27) of the kinetic energy available in the wind (Bhattacharya, 1983). The most common depiction of this theory is shown in Figure 3.3, including various types of windmills. Note that in some publications similar to this figure, the multi-bladed windmill is replaced by the Savonius windmill. The correct version is that which shows a multi-bladed

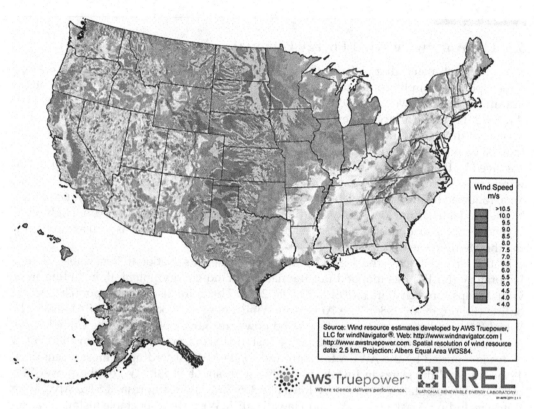

FIGURE 3.2
Utility-scale land-based 80-meter wind map of the United States (courtesy of NREL, USDOE, 2018).

power coefficient of 0.15 instead of 0.30. The Savonius windmill should instead have a power coefficient of 0.30. The X-axis of this graph is the tip-speed ratio or (TSR). This is the ratio of the speed of the tip of the blade divided by the wind speed. The speed of the tip portion of the blade is usually several times faster than the wind speed, especially

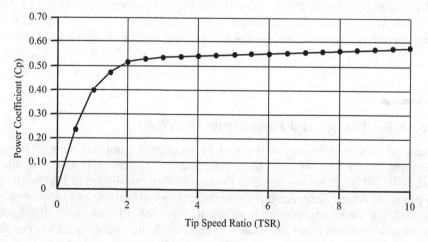

FIGURE 3.3
Power coefficient curve following the Betz coefficient.

for the two-bladed windmill or the Darrieus windmills, which have a TSR of around 6 (Manwell, et al., 2002). However, these windmills are difficult to start and will not move at lower wind speeds. Usually, a Savonius windmill is placed in the middle of a Darrieus windmill to initiate the early movement at a low wind speed. As the initial inertia from the Savonius is reached, it can be used to start the movement of the Darrieus windmill. The Savonius and multi-bladed windmills usually have lower cut-in wind speed (or the wind speed when the turbine starts to move). For a power-generating wind turbine that moves a lot faster, there is also a limit to rotation called the cut-out wind speed. This cut-out wind speed is the maximum speed allowed by the wind turbine for safety purposes. The wind turbines that achieve this speed are usually disengaged from the load or the generator so that there will be no excess in power and the generator is not damaged.

Commercial windmills can only achieve between 75% and 85% of this maximum value. The overall conversion efficiency for the wind turbine is, at best, close to only 50% (Aggeliki, 2011).

A typical power coefficient curve is shown in Figure 3.3. The Y-axis in this curve is the power coefficient, C_p, while the X-axis is the TSR. The TSR is defined as the ratio of the blade tip speed to the wind speed. As the tip of the windmill blade usually moves faster than the wind speed, positive values from above zero to greater than 10 may be possible (Manwell, et al., 2002). All windmill designs fall under this power coefficient curve. Figure 3.4 shows the performance

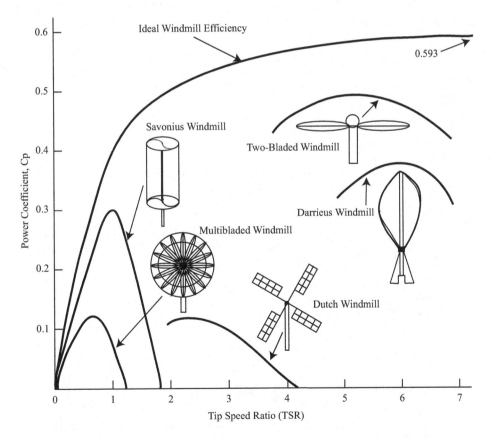

FIGURE 3.4
Power coefficient for some windmill designs (Manwell, et al., 2002).

curves for various windmill designs. The two-bladed windmill has an overall efficiency close to 50%. A multi-bladed windmill commonly used for water pumping has an overall efficiency of close to only 15%, while a Savonius windmill has an overall efficiency of close to 30% (Manwell, et al., 2002).

3.5 Actual Power from the Wind

The actual power from the windmill may be estimated by incorporating the product of power coefficient (C_p) and the overall efficiency coefficient (E_o) into Equation 3.6; this is shown in Equation 3.7. In this equation, the overall efficiency coefficient E_o is a function of both the mechanical efficiency and the electrical efficiency of conversion (in the case of electricity generation). Note that the value of the product of C_p and E_o should be less than the Betz coefficient (0.593). In some literature, the actual power is calculated with simply the use of Equation 3.6 with only the overall conversion coefficient E_o. This form of an equation assumes that the overall conversion efficiency included the power coefficient (C_p) and the reduction in efficiency due to mechanical friction and electrical efficiency from generators (in the case of electricity production). Take note that the power coefficient is very much dependent upon the TSR as well as the type of windmill, as shown in Figure 3.2. Example 3.4 shows how actual power from a windmill is estimated:

$$Actual\ Wind\ Power\ (Watts) = \frac{1}{2}\rho A v^3 E_o \qquad (3.7)$$

where
actual wind power = watts
ρ = air density, usually 1.2 kg/m³ [0.075 lbs/ft³] at STP
A = cross-sectional area perpendicular to flow (m²)
v = wind speed (m/s)
E_o = overall efficiency (includes mechanical and electrical efficiencies)

Example 3.4: Actual Power Generated from Windmills

Determine the actual power generated from a windmill with an effective diameter of 10 meters, air density of 1.225 kg/m³ [0.0764 lbs/ft³], average wind speed of 6 m/s [19.68 ft/s], and overall conversion efficiency of 45%.

SOLUTION:

a. Equation 3.7 is used as follows:

$$P_a(W) = \frac{1}{2}\rho A v^3 E_o = \frac{1}{2} \times 1.225\frac{kg}{m^3} \times \frac{\pi \times (10\ m)^2}{4} \times \left(\frac{6\ m}{s}\right)^3 \times 0.45 = 4,676\ W$$

$$P_a(hp) = 4,676\ Watts \times \frac{hp}{746\ Watts} = 6.27\ hp$$

b. Therefore, the actual power is significantly lower due to mechanical and electrical conversion losses.

3.6 Windmill Classification

Windmills are classified into various types according to different criteria. Thus, a windmill may be categorized according to its application, orientation of its axis, position of the blades, number of blades, and, sometimes, materials used. The most common classification is based on the way the blades are oriented; that is, wind turbines are classified as either vertical axis or horizontal axis wind turbines.

3.6.1 Classification according to Speed

3.6.1.1 High-Speed Windmills

High-speed wind turbines are usually those with fewer blades, usually fewer than four. Their speed could be more than 300 revolutions per minute (rpm). Electric power generators generally require these high-speed turbines. These wind turbines have lower torque output and are not suited to the high mechanical torque requirement of some applications, such as water pumping. High-speed wind turbines are also light in nature and do not carry large inertia from their blades. The blades should be made of light materials such as ceramic or plastic. Note that most electrical generators run at around 1,800 to 3,600 rpm, and in order for the electric generator to produce power, the shafting speed must be increased by 6 to 12 times the original speed of the windmill. This requires a device, usually a gear combination, that can increase the speed accordingly (Nelson, 2014).

3.6.1.2 Low-Speed Windmills

Low-speed wind turbines are those that move at a few hundred rpm. They are used for high-torque mechanical applications, such as water pumping or for wood sawmills. The materials for the blades are usually heavy to impart inertia on the turbines as the blades move. The gear system is made up of direct drives and does not require gearboxes to increase speed. For water pumping purposes, a circular plate is placed at the end of the windmill drive shaft to convert the rotary motion of the wind turbine to a reciprocal and linear motion as the simple motion for water pumping or sawing. There are also windmill types that are installed near the ground (usually horizontal axis types). Since the wind speed at lower elevations is usually low, we can expect windmills at such elevations to rotate at very low speeds. If an application requires higher rotational speed, some gearboxes are needed to magnify the speed coming from the primary shafting of the windmill. The main advantage of these types of windmills is that gearboxes and generators placed on the ground do not require posts of metal as aboveground supports.

3.6.2 Classification according to Position of Blades

3.6.2.1 Upwind Windmills

Upwind windmills have blades are placed in front, against the wind. There is usually a tail or wind vane that adjusts the direction of the blades perpendicular to the wind direction. Maximum wind power is achieved when the wind turbine blades are directly perpendicular to the direction of wind. As such, for upwind windmills, this guide vane is very necessary.

3.6.2.2 Downwind Windmills

Downwind windmills do not require a tail or a vane guide. The wind blades are on the downwind location and virtually "catch" the wind as the windmill moves against it. The hub or body will always be perpendicular to the wind direction. This body or hub of the machine acts as the wind vane on these upwind designs. Newer wind turbines do not have vanes to direct the wind blades against the wind. They typically have wind sensors that direct the orientation of the windmill perpendicular to the wind direction. In the event of reliable wind direction and wind speed sensors, the whole turbine assembly is oriented electronically to generate the maximum power at any given time.

3.6.3 Classification according to Orientation of Blade Axis

3.6.3.1 Vertical Axis Windmills

The most common vertical axis windmills are the following:

1. Savonius windmills
2. Darrieus windmills

The easiest way to fabricate a low-cost Savonius windmill is to use a 200 L metal drum. Standing on its circular base, one would cut this metal drum symmetrically from top to bottom. One would then slide the two half drums apart and place vertical shafting in the middle. The whole assembly could be raised to take advantage of higher wind speeds at higher elevations. The main advantage of a vertical axis windmill is that the drive shaft could be located at the lower end of the main shaft to provide easy access during lubrication and routine maintenance. The main disadvantage is that wind speeds at lower heights are low and would generate low energy output.

The Darrieus windmills are like extra-large egg beaters. Their blades are made of fiberglass, with a cross section similar to an airfoil or wind turbine blade and bent in an oval shape as depicted in Figure 3.2. A Darrieus windmill has a very high cut-in wind speed and would require an initial "push" from a windmill that has a lower cut-in wind speed, such as a Savonius type. In most cases, a Savonius windmill is placed in the middle of the Darrieus windmill to provide initial momentum or torque to get the unit going with the incidence of low speeds. Once the Darrieus windmill has gained speed and momentum, it requires less wind velocity to keep the unit in continuous motion. The speed to maintain movement in a Darrieus windmill is even lower than its cut-in wind speed. The advantage of the Darrieus system for power generation is that its generator and gearboxes are usually on the ground as opposed to the three-bladed windmill, which must be atop a tower. The other advantage of the vertical axis windmill type is that it does not require any wind vane to guide the unit according to the wind direction. Such types are also called non-directional windmills.

3.6.3.2 Horizontal Axis Windmills

The horizontal axis windmills are the following:

1. Multi-bladed windmills
2. One-, two-, three-, or four-bladed windmills
3. Crosswind Savonius windmills

A horizontal axis windmill's rotor axis is in line with the prevailing wind direction. For an upwind system, the vane is responsible for orienting the unit toward the prevailing

FIGURE 3.5
Photo of various units of windmills for power generation in the northern part of the Philippines.

wind direction. In some publications, these types of windmills are also called directional windmills. Multi-bladed windmills are commonly used for water pumping for either shallow or deep wells, again because water pumping requires high head or torque. A multi-bladed windmill's rotation is quite low, but the torque is quite high. For electrical power production, the one-, two-, three-, or four-bladed types are usually used. Electrical generators would normally require higher speeds—sometimes over a thousand rpm. Windmills are never designed to run this fast, and gearboxes are usually needed to bring the generator rpm to the required higher value in order to generate electrical power. Some gearboxes raise the shafting speed to the generator at a ratio of up to 1:9. One can even visually count the number of turns per minute in a three-bladed 1 MW system running at a very low speed. Figure 3.5 shows a photo of the typical three-bladed windmill for power generation. These units are installed on the shores of Ilocos Norte in the Philippines.

If the shafting of a Savonius windmill is oriented horizontally, it falls under the horizontal axis windmill type and is usually called a crosswind Savonius windmill.

3.7 Wind Speed Measuring Instruments

The most common and relatively cheap wind speed measuring instrument is the mechanical three-cup anemometer shown in Figure 3.5. The little cups catch the wind and spin around at different speeds according to the strength of the wind. A recording device counts the number of rotations during a given period of time and translates these data into actual wind speed. The more accurate wind speed measuring instruments are those that have no moving parts, such as the 3D ultrasonic anemometer, also shown in Figure 3.6 (top unit). This unit has hot wire sensors in three locations that sense the movement of wind and record its speed. These units use ultrasonic sound waves to measure wind speed. There are also acoustic resonance anemometers—the most recent variation of the sonic system. As the name implies, these newer devices use resonating acoustic waves to perform wind speed measurements. The primary advantage of these newer systems is that they are of compact size compared to the 3D sonic anemometers.

The question may be raised on how frequently wind speed should be recorded and stored. Data logging systems have become sophisticated enough that one can get wind speed data every second. The recommendation of the World Meteorological Organization (WMO, Geneva Switzerland), a special agency of the United Nations, is to have wind speed readings at least hourly for accurate wind resource estimates. The standard height of wind measuring instruments is set at 10 meters [32.8 ft]. The WMO suggestion is to have numerous wind speed readings (ideally as many as the measuring instrument can handle, usually every second) 10 minutes before the top of the hour, averaging these values

FIGURE 3.6
Photo of cup anemometers as well as a sonic 3D anemometer.

accordingly (WMO, 2011). It is up to the recorder to set the frequency of measurements and encoding. This will vary according to the capacity of the data logger and the storage capacity of the logging system. For most wind resource calculations, hourly readings are appropriate. For a quick estimate of wind power and wind energy, the average wind speed calculated and Equation 3.6 are used. However, as discussed in earlier sections, the calculation process frequently underestimates the wind power or energy. The next section will provide a method for estimating wind power and wind energy at a given location.

3.8 Wind Power and Energy Calculations from Actual Wind Speed Data

3.8.1 The Rayleigh Distribution

There are numerous statistical distributions that may be used to estimate wind power or energy in a given location. Perhaps the most common is the Rayleigh distribution, given by the probability density function (PDF) shown in Equation 3.8 (Nelson, 2014). In Equation 3.8, c is the value at which the PDF peaks—also called the mode of the distribution. It is important to note that c is not the mean value of the wind speed. The mean value of the velocity is given in Equation 3.9:

$$f(v) = \left(\frac{v}{c^2}\right) exp \left(\frac{-v^2}{2c^2}\right) \tag{3.8}$$

where
v = wind speed (m/s)
c = mode of the distribution (i.e., the value at which the PDF peaks)

$$\bar{v} = c \times \frac{\sqrt{\pi}}{2} \tag{3.9}$$

where
$\bar{v}$ = mean value of the wind speed
c = mode of the distribution

The cumulative density function (CDF) is given in Equation 3.10; the terms have already been defined in Equation 3.8:

$$F(v) = 1 - exp \left(\frac{-v^2}{2c^2}\right) \tag{3.10}$$

The cumulative density function (CDF or $F(v)$) is the probability that the velocity is less than a given value v. If, for example, in a given year, $F(v)$ is 60% at v = 8 m/s [26.24 ft/s], then about 5,256 hours (8,760 × 0.6) is the total number of hours that the velocity is less than 8 m/s [26.24 ft/s]. A wind speed histogram superimposed with a Rayleigh distribution is depicted in Figure 3.7.

The Rayleigh distribution is a simple distribution because it only has one parameter, c. While it seems limited in its shape because of its one-parameter statistical distribution, it is actually quite robust, as shown in Figure 3.8, including other values of the mode of

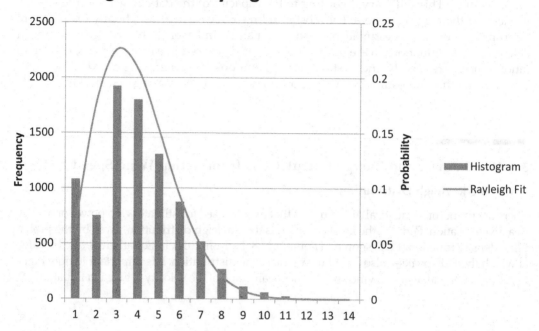

FIGURE 3.7
Wind speed histogram and Rayleigh distribution curve fit.

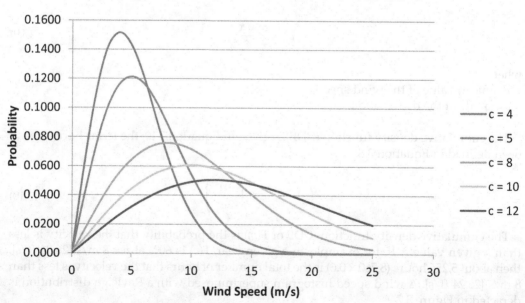

FIGURE 3.8
Variations of Rayleigh distribution mode parameter c.

distribution c. This statistical distribution can approximate various histograms, with the shape being skewed to the right as values of c approach 2. It will be shown later that the Rayleigh distribution is in fact a special case of the Weibull distribution, a statistical distribution that better describes most wind speed data.

One way to estimate the value of the constant c in Equations 3.8, 3.9, and 3.10 is the use of the maximum likelihood estimate shown in Equations 3.11 and 3.12. A correction may also be made for the estimate of c shown in Equation 3.13. A simple problem is given in Examples 3.5 and 3.6:

$$c^2 \approx \frac{1}{2N} \sum_{i=1}^{N} x_i^2 \tag{3.11}$$

$$c' \approx \sqrt{\frac{1}{2N} \sum_{i=1}^{N} x_i^2} \tag{3.12}$$

$$c = c' \frac{4^N N!(N-1)!\sqrt{N}}{(2N)!\sqrt{\pi}} \tag{3.13}$$

Example 3.5: Estimation of Rayleigh Distribution Parameter c

Estimate the value of c and the corrected c, representing the constants in Equations 3.8, 3.9, and 3.10, for the following wind speed data: $N = 57$; the sum of the squares of all data points was 954.52. Data points used for these values are shown in the table below, reading from left to right.

2	7.1	0.6	3.5	3.2	2	3	0.9	1.8
2.5	4.5	2.6	1.0	4.5	5.0	3.2	5.1	1.6
5.5	8.1	2.8	2.1	3.5	2.0	3.9	7.0	0.8
4.0	5.1	4.9	10.0	0.5	0	0	2.0	0.1
8.0	5.6	1.8	2.8	7.0	1.8	4.8	5.5	4.4
2.8	0.7	4.8	4.0	0.6	3.0	3.5	5.1	7.0
1.4	1.6	1.6						

SOLUTION:

a. The c is calculated using Equation 3.12 as follows:

$$c' \approx \sqrt{\frac{1}{2N} \sum_{i=1}^{N} x_i^2} \approx \sqrt{\frac{1}{2 \times 57} \times 954.52} = 2.8936 = 2.9$$

b. The correction is calculated using Equation 3.13:

$$c = c' \frac{4^N N!(N-1)!\sqrt{N}}{(2N)!\sqrt{\pi}} = 2.8936 \times \frac{4^{57} 57!(57-1)!\sqrt{57}}{(2 \times 57)!\sqrt{3.1416}} = 2.8981 = 2.9$$

c. The difference between the estimated value of c and the corrected value is quite insignificant.

Example 3.6: Estimation of Average Velocity Using Rayleigh Distribution

Estimate the value of average velocity for Example Problem 3.5 using the estimated formula given in Equation 3.9.

SOLUTION:

a. The average wind speed is calculated from Equation 3.9:

$$\bar{v} = c \times \frac{\sqrt{\pi}}{2} = 2.8981 \times \frac{\sqrt{3.1416}}{2} = 2.5684 = 2.6$$

b. Note that the average value of wind speed is slightly less than the value of c. If one were to take the average value of all the wind speed data points in the above example, the average, using simple summation, is equal to 3.40. If we were to use this average value by simple summation, we would be overestimating the power and energy from the wind.

3.8.2 The Weibull Distribution

Most wind speed numbers, when plotted, are skewed to the left and not normally distributed. As such, using the mean velocity or mean wind speed to estimate energy and power from the wind is not practical. An example of actual wind speed data for one whole year is shown in Figure 3.9 for College Station, Texas. Note that the data have more values with lower wind speeds (skewed to the left). This behavior is perhaps due to the fact that this region in central Texas is considered to be in the lowest category (Category 1) of wind energy resource. If the data were normally distributed, then the average value of the wind speed would be slightly to the right of this histogram. One may think that the process of using the average wind speed to estimate wind power would provide a higher value of wind power, but this is not actually so. The reason is

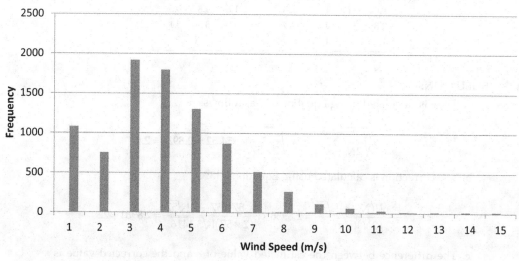

FIGURE 3.9

Histogram of actual wind speed data in College Station, Texas, for the 2005 at 10-meter height.

simply because wind power is proportional to the cube of wind speed and not with the average wind speed. To reiterate the earlier example, the use of average wind speed will actually underestimate wind power.

The PDF for the Weibull distribution is given in Equation 3.14:

$$f(v) = \left(\frac{k}{c}\right)\left(\frac{v}{c}\right)^{(k-1)} exp\left[-\left(\frac{v}{c}\right)^k\right] \tag{3.14}$$

where
v = the wind speed (m/s)
k = the shape parameter or shape factor
c = the scale parameter or scale factor

The Weibull distribution is a two-parameter function, as opposed to a single-parameter distribution like the Rayleigh distribution (Nelson, 2014). The CDF is given in Equation 3.15 with similar constants as in the PDF:

$$F(v) = 1 - exp\left(-\left(\frac{v}{c}\right)^k\right) \tag{3.15}$$

One would think that solving for the shape and scale factor in a Weibull distribution would be difficult. However, this is actually quite simple with the following derivations. If one were to use Equation 3.15, separate the constants from the wind speed parameter v, and take the natural logarithm twice, the resulting relationship is shown in Equation 3.16:

$$ln\left(-ln\left(1 - F(v)\right)\right) = k \times ln(v) - k \times ln(c) \tag{3.16}$$

Note the similarity of this equation with the linear equation shown in Equation 3.17:

$$Y = mX + b \tag{3.17}$$

Equation 3.16 is a simple linear relationship with the slope equal to the shape parameter k and an intercept equal to the parameter $-k \ln(c)$. To analyze wind speed data and fit this in a Weibull distribution, the following stepwise calculation should be made:

a. The ranges (or bins) of wind speed is designated (e.g., 0-1 m/s, 1-2 m/s, 2-3 m/s, and so on). In Microsoft Excel, one would simply use the histogram option within the data analysis option.

b. The histogram of the actual wind seed data is then made based on the earlier designated ranges (or bins).

c. If the counts for each wind speed range are divided by the total number of data points, the PDF of the wind speed data is then calculated.

d. If the PDF of each wind speed range is added up to the maximum wind speed observed, the CDF (or $F(v)$ in Equation 3.15) is then calculated.

e. The CDF is subtracted from 1, and a natural log is taken. A negative sign is placed, and a second natural logarithm is taken to conform to the first parameter in Equation 3.17. This is then designated as the parameter Y.

f. The parameter X in Equation 3.17 is simply the natural log of the wind speed in question ($\ln(v)$). For each wind speed range, this wind speed, v, is the midpoint of the wind speed range (i.e., for the wind speed range between 0 and 1 m/s, this value is 0.5 m/s and so on).

g. The next step is then to perform a linear regression between X ($\ln(v)$) and Y ($\ln(-\ln(1 - F(v)))$).

h. The slope of the regression is automatically the value of the shape factor parameter k, while the intercept is equal to $-k\ln(c)$. Hence, the scale factor c is equal to the exponent of the ratio between the intercept and $-k$ as shown in Equation 3.18:

$$Scale\ Parameter,\ c = exp\left(\frac{intercept}{-k}\right) \tag{3.18}$$

Figure 3.10 shows the actual wind speed histogram and the superimposed Weibull distribution curve fit. The average velocity may also be calculated from the data, and this relationship is shown in Equation 3.19. The gamma function is given in Equation 3.20. The gamma function may be easily solved in Microsoft Excel spreadsheet software, while a quick and simple way to approximate the value of average velocity is with the use of a simpler formula as shown in Equation 3.21. Note that the average value of wind speed is slightly higher than the value of the scale parameter c. The series of examples (Examples 3.7 and 3.8) show how Equations 3.14 to 3.21 are used:

$$\bar{v} = c\Gamma\left(1 + \frac{1}{k}\right) \tag{3.19}$$

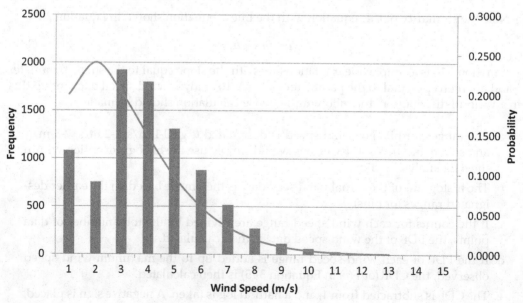

FIGURE 3.10
The wind speed histogram using raw data and the Weibull distribution curve fit.

$$Gamma\ Function,\ \Gamma(x) = \int_0^\infty e^{-1}t^{x-1}\,dt \tag{3.20}$$

$$\bar{v} = c\frac{\pi}{2} \tag{3.21}$$

Example 3.7: Estimation of Various Weibull Distribution Parameters

The hourly wind speed data (in m/s) for one year in a given location were analyzed to establish the following: Weibull shape factor $k = 2$ and scale factor $c = 6$. (a) What is the average velocity for this site? (b) Estimate the number of hours per year that the wind speed is 6 m/s. (c) Estimate the number of hours per year that the wind speed is less than 6 m/s.

SOLUTION:

a. The average velocity for this site is calculated using the simple formula given in Equation 3.21 as follows:

$$\bar{v} = c\frac{\pi}{2} = 6 \times \frac{3.1416}{2} = 9.43\frac{m}{s}$$

If one were to use the Microsoft Excel spreadsheet software to perform the calculations, the answer will be 5.31, which is quite a big difference from using the estimated equation.

b. Equation 3.14 is used to estimate the number of hours per year that the wind speed has a value of 56 m/s. The Microsoft Excel spreadsheet software may also be used with the syntax "Weibull(5,2,6,False)*8760". This will give a value of 1,074 hours:

$$f(6) = \left(\frac{2}{6}\right)\left(\frac{6}{6}\right)^{(2-1)} exp\left(-\left(\frac{6}{6}\right)^2\right) \times 8,760 = 1,074\ hrs$$

c. The CDF Equation 3.15 will be used to estimate the hours that wind speed is less than 6 m/s:

$$F(v) = 1 - exp\left(-\left(\frac{6}{6}\right)^2\right) = (1 - 0.3679) = 0.6321$$

$$Time \times F(v) = Time \times 1 - exp\left(-\left(\frac{6}{6}\right)^2\right) = (1 - 0.3679) \times 8,760 = 5,537\ hrs$$

If one were to use Microsoft Excel spreadsheet software, the answer will be using the syntax "Weibull(6,2,6,True)*8760" equal to 5,537 hours.

Example 3.8: Estimation of Average Velocity from Weibull Distribution

In linear regression calculations, the following results were provided: intercept = −3.5542; X variable = 2.041. Calculate the parameters k and c from these given data. Estimate the average value of wind speed for this location.

SOLUTION:

a. Following Equation 3.17, we know that the parameter k is simply equal to the slope of the linear regression analysis or the x variable. Hence, $k = 2.041$.

b. To calculate for the constant c, Equation 3.18 will be used:

$$c = exp \left(\frac{-3.5542}{-2.041} \right) = exp \ (1.7414) = 5.705$$

c. The average value of wind speed for this location may be found using the approximate relationship given in Equation 3.21 as shown below:

$$\bar{v} = c\frac{\pi}{2} = 5.705 \times \frac{3.1416}{2} = 8.96 \frac{m}{s}$$

3.9 Wind Design Parameters

3.9.1 Cut-In, Cut-Out, and Rated Wind Speed

Windmill designers and installers must realize that wind blades will not move with very low wind speeds in the area. This is because wind blades must overcome friction and weight to initiate movement. The wind speed at which the turbine starts to generate power is called the "cut-in" wind speed. On the other hand, the wind speed at which the wind turbine is shut down to keep loads and generators from being damaged is called the "cut-out" wind speed. The rated wind speed is the wind speed at which the wind turbine reaches its rated turbine power. This is typically the maximum power delivered by the turbine (Manwell, et al., 2002). Some manufacturers designate a specific rated wind speed whereby the rotation of the wind blade is made constant even when the wind speed increases. This will make the power output constant and prevent excessive vibration of the wind rotor at higher wind speeds. This is accomplished by having the blade turn almost parallel to the direction of wind to slow down the rotation of the blades. In this case, the pitch of the wind blades is adjusted. At high wind speeds during storms, the blades are made completely parallel to the wind direction to prevent the wind turbine from moving. This will protect the whole structure from vibrating or collapsing at very high wind speeds. Figure 3.11 shows the simple diagram illustrating the differences between cut-in, cut-out, and rated wind speeds, including the amount of power produced at each stage. As such, there is still room for windmills designed to move at very low wind speeds. The design would be a combination of reducing mechanical and friction losses, developing very lightweight materials for the blades, reducing friction between the air and the blade surfaces, and ultimately increasing the durability of all windmill component parts such that the cut-out wind speed is maximized.

3.9.2 General Components of Horizontal Axis Windmills for Power Generation

The major components of a horizontal axis windmill for power generation are as follows:

a. Blades

b. Tower

c. Low-speed shaft

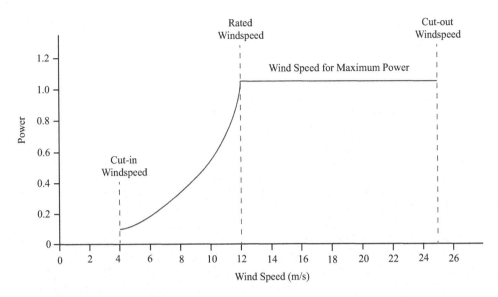

FIGURE 3.11
Diagram showing the relationship between cut-in, cut-out, and rated wind speed.

d. Gearbox

e. High-speed shaft

f. Generator

g. Controller

h. Wind vane and anemometer

i. Yaw drive and motor

j. Pitch drive and motor

k. Outer casing or housing for components (Nacelle)

The most important component parts of a wind machine are those that will protect the assembly for sustainable use and damage. The pitch drive and motor is one of these and is important in ensuring that the wind turbine moves constantly regardless of wind speed. This component is also used during storms to keep the blade parallel to the wind direction to stop the rotation of the wind mill. The controller also monitors wind speed from data received from the anemometer, including the amount of power produced, and performs actions when the wind machine requires control of pitch, yaw, or other electrically related parametric controls. Because the wind turbine moves at a lower speed and the generator requires about a thousand rpm, the gearbox is a ubiquitous component in a wind machine. This is a component that will likely require replacement after several years of use.

Nowadays, many wind turbines are remotely controlled and monitored, which simplifies their operation and maintenance. Vestas, the world's largest manufacturer of wind turbines, based in Denmark, reported that it is able to monitor all its installed wind machines from its head office in Denmark. There are over 550 wind-related manufacturing facilities in the United States (AWEA, 2016) spread across 44 states. GE Energy still tops the list of U.S. manufacturers and has reported that its delivery of a wind turbine order would take at least a year. The popularity of wind machines is at its highest in years, and this trend will likely continue as the price of electricity and the demand for power increases in the United States.

3.9.3 Wind Speed Variations with Height

Wind speed data reported in most countries are measured from a standard height of 10 meters. These data sets are usually taken from regional airports. However, wind machines are installed at much higher elevations to take advantage of higher wind speeds. Unfortunately, no wind speed data are available at the height at which wind machines are installed (except for wind machines already installed with their own anemometer), and one would need extra tools to make an estimate of the wind speed at various heights beyond the standard 10-meter reference point. The most common way to do this is to use the power law. The power law is given in Equation 3.22. In this equation, the velocity at any given height z ($V(z)$) may be estimated by knowing the wind speed at the reference height of 10 meters ($V(z_r)$) and the reference height used (Z_r). The power law coefficient alpha (α) is given in Equation 3.23, and this equation is a function of roughness coefficient z_o (in units of mm). Table 3.1 shows the variations of the roughness coefficient with terrain. Note that the units used in Table 3.1 are in mm, while the parameter in Equation 3.23 is required to be in meters. An example on how to use the power law equation is shown in Example 3.9:

$$\frac{V(z)}{V(z_r)} = \left(\frac{z}{z_r}\right)^{\alpha} \tag{3.22}$$

$$\alpha = 0.096 \times log_{10}z_o + 0.016 \times \left(log_{10}z_o\right)^2 + 0.24 \tag{3.23}$$

Example 3.9: Wind Speed Estimation at Different Heights

The wind speed at a reference height of 10 meters [32.8 ft] is 6 m/s [19.68 ft/s]. Predict the expected wind speed at 30 meters [98.4 ft] assuming the windmill is installed in a forest or in woodlands.

TABLE 3.1

Surface Roughness Coefficients for Various Terrain Descriptions (Nelson, 2014)

Terrain Description	Surface Roughness, z_o (mm)
Very smooth ice or mud	0.01
Calm open sea	0.20
Blown sea	0.50
Snow surface	3
Lawn grass	8
Rough pasture	10
Fallow field	30
Crops	50
Few trees	100
Many trees, hedges, few buildings	250
Forests and woodlands	500
Suburbs	1,500
Centers of cities with tall buildings	3,000

SOLUTION:

a. From Table 3.1, the roughness coefficient was found to be 500 mm. From Equation 3.23, we calculate the value of alpha as follows (after converting units to meters):

$$\alpha = 0.096 \times log_{10}z_o + 0.016 \times \left(log_{10}z_o\right)^2 + 0.24$$

$$\alpha = 0.096 \times log_{10}\left(\frac{500}{100}\right) + 0.016 \times \left(log_{10}\left(\frac{500}{100}\right)\right)^2 + 0.24 = 0.3295$$

b. Thus, using the power law Equation 3.22, the velocity at 3 = 0-meter height was found to be as follows:

$$V(z) = V(z_r) \times \left(\frac{z}{z_r}\right)^{\alpha} = 6\frac{m}{s} \times \left(\frac{30}{10}\right)^{0.3295} = 8.6\frac{m}{s}$$

$$V(z) = 8.6\frac{m}{s} \times \frac{3.28\ ft}{1\ m} = 28.21\frac{ft}{s}$$

3.9.4 Wind Capacity Factor and Availability

Utility scale wind turbines for land-based wind farms come in various sizes in terms of their rotor diameter and tower height. Rotor diameters of 50 to 90 meters [164 to 295.2 ft] are very common, with tower heights of roughly the same size (50 to 90 meters, respectively). A 90-meter diameter wind machine with a 90-meter [295.2-ft] tower would have a total height from the tower base to the tip of the rotor of approximately 135 meters [442 feet]. Offshore turbine designs that are currently under development will have rotors 110 meters [360.8 ft] in diameter. Windmills this large are easier to transport by ship and could be even bigger. When wind blades are transported by land, size is a limitation. Small wind turbines for residential or small business applications will have rotors of 8 meters [26.24 ft] in diameter or less and would be mounted on towers 40 meters [131.2 ft] high or less. In the United States, any structure above 60 meters [200 ft] is subject to regulations by the Federal Aviation Administration (FAA). The FAA must be notified, and permits may be required, especially if the installation is near an airport. There are really no limits to windmill installation height as discussed by FAA spokesperson John Page (Zipp, 2012). What the FAA is concerned with is the exact location of installation, and they make decisions based on a potential installation's hazard to aviation. In the latest report, there was a limit of 152 meters [500 ft].

The towers of wind machines are mostly made of tubular steel. The blades are mostly made of fiberglass-reinforced polyester or wood epoxy. There are numerous new materials being used that are of even lighter weight and can withstand the strength of storms. Many are also designed to prevent harm to birds and other flying animals. They are also designed to prevent damage from lightning.

There are several technical parameters that describe the performance of wind machines. The two most important parameters are the capacity factor and availability or reliability. The capacity factor is defined as the actual amount of power produced over time by a wind machine divided by the power that would be produced if the turbine were operated at its maximum output 100% of the time. This parameter is sometimes referred to as load factor. A conventional utility power plant uses fuel, and it could run much of the time unless idled by maintenance problems. Because of this, their capacity factors are higher and may be between 40% and 80% of their rated capacity. The capacity factors for wind machines are only about 25% to 40%. These wind machines may operate between 65% and 90% of the time, but their output capacity is significantly less at most times (Bhattacharya, 1983):

$$CF = \frac{Electricity\ Production\ During\ the\ Period\ (kWh)}{Installed\ Capacity\ (kW) \times Number\ of\ Hours\ in\ Period\ (hrs)} \qquad (3.24)$$

The availability or availability factor of a wind machine is a measurement of the reliability of the unit to generate power throughout its life. Technically, reliability refers to the percentage of time that the wind machine is ready to generate power (i.e., not out of service for maintenance or repairs). Modern wind machines have an availability of more than 98%, higher than most power plants. Unfortunately, most of the time, they produce below their rated power output. This low production rate is primarily due to the variability of wind speed in any given area.

Wind machines consist of small individual modules and can be easily made smaller or larger without drastic changes in their reliability or capacity factors (Manwell, et al., 2002).

3.10 Comparative Cost of Power of Wind Machines

The cost of electricity of a wind machine is a function of the average wind speed in a given area. As average wind speed increases, the cost to sell electricity lowers. Likewise, the cost of electricity is a function of the size of wind farm installed. A 3 MW wind farm may have an electricity cost of below $0.059/kWh, while a 50 MW wind farm may bring this cost down to $0.036/kWh. It also varies by location and region and is also dependent on the historical cost of power. Some reports have shown that a 50 MW wind farm, at a site with an average wind speed of 13 to 17 miles per hour (or a Class 4 area), would have a capital cost of around $65 million and would generate an annual production of 150 million kWh, assuming a capacity factor of 35%. The financing could be 60% debt and 40% equity. If the power purchase price is 4 cents/kWh, the gross annual revenue could be as high as $6 million. If so, the simple payback period could be within 10 years for the capital cost alone. This is still shorter than payback periods for solar PV systems. There are operational and maintenance expenses, which could be around 8% of the total cost, and distribution charges of around 22%. If the land is rented, there may be an additional 5% cost toward property taxes. There may also be management and insurance fees of about 5%. If 6% of the capital cost is loaned from a bank, a 15-year annuity and 9.5% interest is common. There will also be a 5-year depreciation on wind instruments and equipment. In some instances, there may be credit adjustments for inflation during the first 10 years of operation. This credit adjustment may be about 1.5 cents/kWh (AWEA, 2018).

Windustry.org estimates that small wind turbines cost about $3,000 to $5,000/kW, which is in line with the estimate of the National Association of Homebuilders (NAHB) that a 10 kW system should cost between $40,000 and $50,000 before tax incentives. The AWEA (2016) reports that a wind turbine large enough to power a home costs an average of $30,000 but may range in price from $10,000 to as high as $100,000. Installation costs could be as high as 30% of the capital cost of the wind machine. Example 3.10 shows a simple way of calculating the payback period for a wind machine.

In this example, one can fathom that investing in a small wind machine without government subsidies would take too long to recover the initial investment. Added to this is the fact that the capacity factor of a wind machine is fairly low. A family needing 10,000 kWh/year of energy may need 15 to 20 kW systems over a 10 kW system due to this factor.

Example 3.10: Estimation of Simple Payback Period for Small Wind Turbines

Estimate the simple payback period for a small wind turbine with the following data. According to the U.S. Energy Information Administration (US EIA, 2018), the average annual electricity consumption for a U.S. residence was 11,280 kWh, or an average of 940 kWh/month, in 2011. In 2012, the average price per kWh in the U.S. residential sector was 11.88 cents/kWh. Assuming a 10 kW wind turbine costing $40,000, determine the following:

a. How much does each household pay for electricity in a year?
b. What is the payback period using these assumptions?

SOLUTION:

a. The amount being paid by each household is simply the product of its consumption and how much is paid per kWh. This is simply calculated below:

Yearly payment for electricity = $0.1188/kWh × 11,280 kWh = $1,340.6/yr

b. If the payback period is simply based on the payment for electricity, then the calculation will be the ratio of the capital cost and the amount of electricity the household pays each year as follows:

Payback period = $40,000/$1,340.06/yr = 29.85 years

3.11 Conclusion

There are other reports on the future of wind power, and most are very optimistic. Like any other renewable energy technology, the acceptability is always based upon economics. The NREL of the USDOE has published a formula for calculating the profitability of a wind turbine farm. Refer to the article by Johnson (2009) for a very complete review of this formula and example calculations. The formula has the following components:

a. The fixed charge rate for the power (FCR)
b. The initial capital cost (ICC) and all costs including balance of power cost
c. The levelized replacement cost (LRC) or costs for overhauls and replacements
d. The operation and maintenance cost (O&M)
e. The land lease cost (LLC)
f. The annual energy production (AEP) in units of kWh

The NREL-USDOE formula will provide the total net annual profit. From here, one may be able to calculate the return on investment (ROI) and the break-even point (BEP). The analysis reported by Johnson (2009) shows that, indeed, the break-even point for a 1 MW wind turbine is around 24 years, very close to the much simpler example we have presented in this chapter. Wind power is still highly attractive in the United States because of numerous federal grants an investor can avail to reduce the rate of return.

A goal for wind power technologies is a reduction in overall costs, especially the balance of system costs. In many developing countries where the price of electricity is rather high, the economic indicators are much better, and the return on investment would be much less than 20 years, which is about the life of a commercial wind turbine machine.

3.12 Problems

3.12.1 Kinetic Energy from Wind

P3.1 How much kinetic energy is contained in 10 cubic meters of air moving at a speed of 10 m/s? Assume the air density to be 1.22 kg/m^3.

3.12.2 Power from the Wind

P3.2 Determine the power that could be derived from a 4 m diameter windmill at a location with an average wind speed of 5 m/s. Assume the air density to be 1.22 kg/m^3, a power coefficient of 0.45, and an efficiency of conversion to power of 60%. Express your units in watts or kW.

3.12.3 Power Differential as Wind Speed Is Doubled

P3.3 Compare the power difference in watts for a location with an average wind speed of 5 m/s doubled to 10 m/s. Assume that the other parameters are the same, such as the air density of 1.22 kg/m^3, area of 100 m^2, and overall efficiency of 50% (that includes the power coefficient and conversion efficiency for mechanical and electrical power).

3.12.4 Actual Power from Windmill

P3.4 Determine the actual power generated from a windmill with an effective diameter of 6 meters, air density of 1.225 kg/m^3, average wind speed of 12 m/s, and overall conversion efficiency of 30%.

3.12.5 Rayleigh Distribution Estimate

P3.5 Estimate the value of c and the corrected c, representing the constants in Equations 3.8, 3.9, and 3.10, for the following wind speed data: $N = 52$; the sum of squares of all data points was 860.16.

2	7.1	0.6	3.5	3.2	2.0	3.0	0.9	1.8
4.5	2.6	1.0	4.5	5.0	3.2	1.6	8.1	2.1
3.5	2.0	3.9	7.0	0.8	4.0	5.1	10.0	0.5
0	0	2.0	0.1	8.0	5.6	1.8	2.8	7.0
1.8	4.8	5.5	4.4	2.8	0.7	4.8	4.0	0.6
3.0	3.5	5.1	7.0	1.4	1.6	1.6		

3.12.6 Estimating Average Wind Speed from Rayleigh Distribution

P3.6 Estimate the value of average velocity for Example Problem 3.5 using the estimated formula given in Equation 3.9.

3.12.7 Average Wind Velocity for a Given Site and Hours of Occurrence

P3.7 The hourly wind speed data (in m/s) for one year at a given location was analyzed to establish the Weibull shape factor $k = 2.04$ and scale factor $c = 5.7$. (a) What is the average velocity for this site? (b) Estimate the number of hours per year that the wind speed is 5 m/s. (c) Estimate the number of hours per year that the wind speed is less than 5 m/s.

3.12.8 Estimate Weibull Parameters *k* and *c* from Linear Regression Data

P3.8 In linear regression calculations, the following results were provided: intercept = −3.3172; X variable = 2.1588. Calculate the parameters k and c from these given data. Also estimate the average value of wind speed for this location.

3.12.9 Wind Speed at Different Elevation

P3.9 The wind speed at a reference height of 10 meters was 8 m/s. Predict the expected wind speed at 40 meters, assuming the windmill is installed in an area with a few trees.

3.12.10 Payback Period for Wind Machine

P3.10 Estimate the simple payback period for a small wind turbine with the following data: the average annual electricity consumption was 12,000 kWh, the average price per kWh was 10 cents, and the household invested in a 15 kW system that costs $91,500. Determine the following:

a. How much does each household pay for electricity in a year?

b. What is the payback period using these assumptions? Comment on this value assuming a lack of government subsidy.

References

Aggeliki, K. 2011. Calculating wind power. Bright Hub Online Magazine. Edited by Lamar Stone. January 19. Available online at: https://www.brighthub.com/environment/renewable-energy/articles/103592.aspx. Accessed April 18, 2018.

AWEA. 2016. American Wind Energy Association US Wind Industry First Quarter 2016 Market Report. AWEA Data Services, Washington, DC. Available at: https://emp.lbl.gov/sites/default/files/2016_wind_technologies_market_report_final_optimized.pdf. Accessed April 19, 2018.

AWEA. 2018. The Cost of Wind Energy in the US. American Wind Energy Association. AWEA Data Services, Washington DC. Available at: https://www.awea.org/falling-wind-energy-costs. Accessed April 19, 2018.

Burton, T., D. Sharpe, N. Jenkins and E. Bossanyi. 2001. Wind Energy Handbook. John Wiley and Sons, West Sussex, England.

Busby, R. L. 2012. Wind Power: The Industry Grows Up. PenWell Corporation, Tulsa, OK.

Elliot, D. L., C. G. Holladay, W. R. Barchet, H. P. Foote and W. F. Sandusky. 1986. Wind Energy Resource Atlas of the United States. DOE/CH 10093-4. October 1986 DE86004442 UC Category 60. Pacific Northwest Laboratory, Richland, WA, and prepared for the U.S. Department of Energy Wind/Ocean Technologies Division, Golden, CO. Available at: http://rredc.nrel.gov/wind/pubs/atlas/titlepg.html. Accessed April 18, 2018.

Johnson, T. 2009. How much does a wind turbine cost? Let's calculate some breakeven and profit points. Windpower Engineering & Development Newsletter Online. September 14. Available at: https://www.windpowerengineering.com/projects/windpower-profitability-and-break-even-point-calculations/. Accessed April 19, 2018.

Khun, L. T. 2010. Welcome to Riso National Laboratory for Sustainable Energy, Riso DTU. Seminar presented at the 3D Workshop held on July 6, 2010, at Riso DTU, Technical University of Denmark. Available at: http://www.der-ri.net/index.php?id=93. Accessed April 8, 2018. (Click on one of the fact sheets.)

Manwell, J. F., J. G. McGowan and A. L. Rogers. 2002. Wind Energy Explained: Theory, Design and Application. John Wiley and Sons, West Sussex, England.

Mathew, S. 2006. Wind Energy: Fundamentals, Resource Analysis and Economics. Springer-Verlag, Berlin, Germany.

Moloney, C. 2014. Small wind turbine: what is the payback period? Poplar Network. Available at: http://www.poplarnetwork.com/news/small-wind-turbine-what-payback-period. Copyright Poplar Network, NY. Accessed June 27, 2019.

Nelson, V. 2014. Wind Energy: Renewable Energy and the Environment. 2nd Edition. CRC Press, Taylor and Francis Group, Boca Raton, FL.

NREL, USDOE. 2018. National Renewable Energy Laboratory of the U.S. Department of Energy Wind Speed Maps. Available at: https://www.nrel.gov/gis/wind.html. Accessed April 18, 2018.

USDOE. 2018. Wind Vision: A New Era for Wind Power in the United States. March 12, 2015. Wind and Water Power Technologies Office, U.S. Department of Energy, Golden CO, and Washington, DC. Available at: https://www.osti.gov/. Accessed April 18, 2018.

US EIA. 2018. US Energy Information Administration, Annual Energy Outlook 2018. February 6. Available at: https://www.eia.gov/outlooks/aeo/pdf/AEO2018.pdf. Last accessed April 18, 2018, and https://www.eia.gov/electricity/ for electricity consumption and use.

World Meteorological Organization (WMO). 2011. Review of the tropical cyclone operational plan. WMO and Economic and Social Commission for Asia and Pacific (ESCAP), WMO/ESCAP Panel on Tropical Cyclones Thirty Eighth Session, New Delhi, India, February 21–25.

Zipp, K. 2012. To be precise, the FAA has no 500-feet limit on turbine towers. Windpower Engineering and Development Online Newsletter. April 6. Available at: https://www.windpowerengineering.com/construction/to-be-precise-the-faa-has-no-500-ft-limit-on-turbine-towers/. Accessed April 18, 2018.

4

Biomass Energy

Learning Objectives

Upon completion of this chapter, one should be able to:

1. Enumerate the various biomass resources that may be used for energy conversion.
2. Describe the various conversion pathways for biomass feedstock.
3. Describe the various procedures for characterization of biomass.
4. Describe the various conversion efficiencies that define each biomass conversion pathway.
5. Relate the overall environmental and economic issues concerning biomass energy conversion systems.

4.1 Introduction

In the future, biomass resources may contribute significantly to satisfying the world's energy requirements. However, in using biomass for energy, we must ensure that we do not also use up food for human consumption. There are various biomass residues that may be used for heat, fuel, and electrical power. There are also numerous biomass energy conversion technologies that are now being commercialized. Economics—considering the overall cost of energy conversion, including feedstock costs and transport costs—is the determining factor affecting the widespread use of biomass for fuel and electrical power. Biomass has been used as fuel in many developing countries; this trend will continue because of particular living conditions and the countries' ability to pay for labor. The biomass residues that will likely contribute to a country's energy mix are those that have zero or negative value, such as municipal solid wastes, municipal sewage sludge, and animal manure. Commercial technologies for the conversion of these feedstocks into heat, fuel, and electrical power are slowly being introduced into the market. Deforestation as well as scarcity of land for widespread production affect the use of biomass residues in some developing countries. Following the basic definition of a renewable energy resource, biomass resources must not be used at a rate faster than they are produced. As such, if fuel wood becomes a feedstock for electrical power generation, there must be, in conjunction, a massive reforestation program to replant biomass resources that are continually used each year. For example, for every MW of electrical power energy produced, about 600 hectares [148.2 acres] of forest are needed, assuming 50% conversion efficiency and the productivity of wood as shown in Example 4.1. Thus, a 50 MW facility would require more than

30,000 hectares [7,410 acres] of land, which may be too extensive for some countries. If the conversion efficiency is reduced further because of varying power generation routes (such as gasification), the required area may even double.

Example 4.1: Required Area to Generate Electrical Power Output from Biomass

Determine the area needed to supply the yearly fuel requirement of a 1 MW power plant that uses wood as feedstock. The power plant should operate continuously for at least 350 days in a year. The heating value of wood is found to be 20 MJ/kg [8,617 Btu/lb], and the conversion efficiency via steam generation is assumed to be 50%. The yield of the wood plantation is around 5 dry tonnes/hectare/year [2.23 tons/acre/yr].

SOLUTION:

a. The amount of biomass required may be calculated using the efficiency equation as follows:

$$Efficiency\,(\%) = \frac{Input}{Output} \times 100\%$$

$$50\% = \frac{Input}{1\ MW} \times 100\%$$

$$Input\,(MW) = \frac{1\ MW}{0.50} = 2\ MW$$

b. Hence, about 2 MW of input feedstock is needed to operate this power plant.

c. The next step is to determine the daily and yearly requirement of the biomass, assuming the heating value as well as the length of operation of the power plant in a year. This is shown below:

$$Biomass\left(\frac{tonnes}{day}\right) = 2\ MW \times \frac{MJ}{1\ MW-s} \times \frac{kg}{20\ MJ} \times \frac{3,600\ s}{hr} \times \frac{24\ hrs}{day} \times \frac{1\ tonne}{1,000\ kg} = 8.64\ \frac{tonnes}{day}$$

$$Biomass\left(\frac{tons}{day}\right) = 8.64\frac{tonnes}{day} \times \frac{1,000\ kg}{1\ tonne} \times \frac{2.2\ lbs}{1\ kg} \times \frac{ton}{2,000\ lbs} = 9.504\ \frac{tons}{day}$$

$$Biomass\left(\frac{tonnes}{yr}\right) = 8.64\frac{tonnes}{day} \times \frac{350\ days}{year} = 3,024\frac{tonnes}{year}$$

$$Biomass\left(\frac{tons}{yr}\right) = 9.504\frac{tons}{day} \times \frac{350\ days}{year} = 3,326.4\frac{tons}{year}$$

d. The next step is to estimate the required area given the productivity of the wood fuel that will be used for the power plant as shown below:

$$Area\left(\frac{hectares}{year}\right) = 3,024\frac{tonnes}{yr} \times \frac{hectare-year}{5\ tonnes} = 605\ hectares$$

$$Area\left(\frac{acres}{year}\right) = 605\ hectares \times \frac{2.47\ acres}{1\ hectare} = 1,494.35\ acres$$

e. Hence, more than 600 hectares [1,500 acres] are required to satisfy the yearly requirement of this power plant fueled by wood. If the power output is several MW, say 50 MW, then more than 30,000 hectares [74,100 acres] would be needed.

4.2 Sources of Biomass for Heat, Fuel, and Electrical Power Production

There are numerous biomass resources that may be used for heat, fuel, and electrical power production. Especially important are biomass resources that are cheap or have virtually no harvesting cost. For example, in many communities, each household pays for the collection and disposal of its daily refuse. Each household normally pays a tipping fee for the disposal of municipal solid wastes. These tipping fees go toward processing and segregating the wastes such that they may be used beneficially. The biomass types projected to make major impacts on future conversion systems are those that have negative prices, such as municipal solid wastes (MSW), municipal sewage, and animal manure. At present, these biomass wastes have little to no value. As newer conversion systems are introduced into the commercial market, we will observe an increase in these wastes' value. The biomass resources are discussed below and arranged according to their perceived viability.

4.2.1 Municipal Solid Wastes

MSW are ideal for conversion due to their reported high energy content value, ranging from 15 to 20 MJ/kg [6,463 to 8,617 Btu/lb] (Capareda, 2014). However, as they are collected, MSW contain a wide range of materials that are not combustible or, if combusted, would cause emission problems, like plastics or metals such as aluminum. It is likely that when MSW are to be used for heat, power, or fuel production, they would have to be segregated first. Recycling of MSW is now very common, primarily to conserve valuable components such as aluminum, plastics, and some glass types.

In the United States, the per capita MSW production was reported by the EPS to be around 2 kg/person/day [4.38 lbs/person/day] (USEPA, 2014). However, in making a conversion estimate of a biomass resource, one has to use the amount that is truly recoverable and combustible. Of this amount, around 34.5% is typically recycled, and the remaining garbage goes to a landfill facility. Example 4.2 shows the potential of MSW for generating electric power via gasification.

Example 4.2: Electrical Power from MSW

An analysis of segregated MSW by Balcones Resources of Dallas, Texas, showed that the energy content of the segregated and combustible biomass from the garbage was about 25 MJ/kg [10,771 Btu/lb]. In a city like Dallas, Texas, with a population of 1.258 million people, estimate the amount of garbage produced in tonnes per day if the per capita production is 2 kg/person/day [4.4 lbs/person/day]. If this biomass is converted into electrical power with an overall conversion efficiency of 15%, what would be the output power in MW assuming 24 hr/day operation?

SOLUTION:

a. The amount of garbage generated per day in tonnes is calculated as shown:

$$MSW\left(\frac{tonnes}{day}\right) = 1,258 \times 10^6 \ heads \times \frac{2 \ kg}{hd-day} \times \frac{tonne}{1,000 \ kg} = 2,516 \ tonnes$$

$$MSW\left(\frac{tons}{day}\right) = 2,516 \ tonnes \times \frac{1.1 \ tons}{tonne} = 2,767.6 \ tons$$

b. On a given day, the total energy contained in the biomass is calculated as follows:

$$MSW\left(\frac{MJ}{day}\right) = 2,516 \frac{tonnes}{day} \times \frac{1,000 \ kg}{1 \ tonne} \times \frac{25 \ MJ}{kg} = 62.9 \times 10^6 \frac{MJ}{day}$$

$$MSW\left(\frac{10^6 \ Btu}{day}\right) = 62.9 \times 10^6 \frac{MJ}{day} \times \frac{1 \times 10^6 \ J}{1 \ MJ} \times \frac{Btu}{1,055 \ J} = 596.2 \times 10^6 \frac{Btu}{day}$$

c. The electrical power output is calculated using the conversion efficiency provided and knowing continuous 24-hour operation:

$$Power(MW) = \frac{62.9 \times 10^6 \ MJ}{day} \times \frac{1 \ day}{24 \ hrs} \times \frac{1 \ hr}{3,600 \ s} \times \frac{MW-s}{MJ} \times 0.15 = 109 \ MW$$

d. The output power is 109 MW. For modular systems around the city of Dallas, more than 100 units of modular 1-MW system is possible, the size of one gasification company in Dallas has built in the past.

4.2.2 Municipal Sewage Sludge

Municipal sewage sludge is also a potential biomass resource for fuels in the future. Many communities in the United States have centralized handling of their wastewater. All solids are collected in the wastewater treatment facility in the form of sludge. This sludge is rich in energy and is generated daily. There are currently farmers who spread this sludge on their agricultural land as additional fertilizer. Like MSW, this biomass also has negative cost and is likely one of the first types of biomass that may be turned into useful fuel and products.

4.2.3 Animal Manure

Animal manure is also an abundant biomass resource. In concentrated animal feeding operations (also termed CAFOs), the manure accumulates in large volumes and, unless utilized, would adversely affect the environmental conditions in the area. There is currently no widespread use for this biomass resource. The main detriment to the use of animal manure is the presence of high amounts of ash. This is an issue when manure is used for thermal conversions. For high-temperature combustion processes, animal manure is not feasible to use, primarily because of slagging and fouling problems with high-ash biomass. Therefore, pyrolysis and gasification may be the only logical choices for thermal conversion. Because of the high moisture content of animal manure, the easiest conversion pathway is via anaerobic digestion to produce biogas, a gas mixture of methane and carbon dioxide. However, anaerobic digestion processes also generate voluminous amounts of sludge that would subsequently have to be converted into other useful energy to limit storage issues as well as prevent overloading agricultural land with nitrogen, phosphorus, and potassium. Some agricultural areas in the United States have already been overloaded with these nutrients, which in large concentrations may also be damaging to the crop.

4.2.4 Ligno-Cellulosic Crop Residues

Ligno-cellulosic material has been hailed as the holy grail of biofuels. This biomass resource does not theoretically compete with the food requirements of the human population. Ligno-cellulosic materials are abundant, and in some cases, the amount of ligno-cellulosic materials (e.g., corn stover and corn cobs) generated is greater than that of the main crop (in this example, corn kernels). It has been reported that corn residues exceed corn production by a factor of at least 2—for every kg of corn kernel produced, there is more than 2 kg [4.4 lbs] of biomass residue generated. In recent years, research into newer ligno-cellulosic feedstocks has increased in order to generate higher tonnage per hectare, such as the high-tonnage sorghum biomass being developed at Texas A&M University (Rooney, et al., 2012).

4.3 Biomass Resources That May Have Competing Requirements

Many other biomass resources that can potentially be used for heat, fuel, and power production do compete with food and shelter requirements of the human population. These include oil crops, sugar and starchy crops, fuel wood for shelter, and some aquatic biomass (seaweeds). Thus, there must be careful consideration in planning the widespread utilization of these crops.

4.3.1 Oil Crops

Vegetable oil has been the major source of biodiesel fuel in many developed countries, especially those with a large surplus of this agricultural commodity. However, there are instances where this competes with food needs, and oil feedstock for use in commercial fuel must be carefully identified. It is important to continue further research toward developing new oil crops that will not compete with food. Micro-algae constitute a viable feedstock option. This oil crop could generate maximum oil production per unit area per unit of time, provided certain key factors. Micro-algae need substantial solar radiation and enough water to grow, and enough carbon dioxide must be present. There are regions in the world where these inputs are readily available. Newer designs of photo-bioreactors are now being developed and could generate more than 3 g/L [0.025 lbs/gal] of biomass (Bataller, 2018). The only disadvantage of such systems is the need for cheap power to continuously provide light to grow the micro-algae. More recently, specific algae species are grown for their high-value chemicals. An example is the strain of *Haematococcus pluvialis*, a fresh water species of *Chlorophyta* known to have high content of the strong oxidant astaxanthin. We should witness more developments from this field as research grows in popularity.

4.3.2 Sugar and Starchy Crops

Sugar and starchy crops are primarily grown for human consumption. Again, it can be tricky when these are also used to generate fuels. Research into newer, high-yielding sugar and starchy crops that would not compete with food requirements must be initiated. An excellent species for this purpose is hybrid sorghum. The sugar component may be maximized for one variety of hybrid sorghum and the starch component maximized for another. Grain sorghum, a species high in starch, is an excellent example. Plant breeders have been able to generate a variety that is high yielding and drought resistant while producing starch that is easily digestible (Hernandez, et al., 2009). This particular variety of grain sorghum is definitely superior to corn when converted into bioethanol. Sweet sorghum varieties are also being developed to yield crops with high sugar content. Studies at Texas A&M University have shown that some hybrid varieties would generate as high as 26% sugar content (Rooney, 2015). Such a type might be an ideal energy crop of the future.

4.3.3 Fuel Wood

Numerous developing countries still use fuel wood for cooking, heating, and drying of crops—much to the detriment of forestry cover and the environment. This practice will have to be made more sustainable so that forest covers are maintained. In Guatemala, wood is primarily used for drying cardamom (USDA, 2014) and causes the deterioration of Guatemalan forests. In Colombia, it was reported that for drying each metric tonne of cardamom, some 13.514 cubic meters [476.9 ft^3] of wood must be used, projecting a total of 404,420 cubic meters [14,270,992 ft^3] of wood devoted to this specialty crop per harvest season (USDA, 2014).

In the United States, numerous timber-related facilities still use fuel wood for processing lumber. However, they primarily use wood shavings and the bark of harvested wood. If one limits the type of wood used and regulates its use, fuel wood may even become a sustainable source of feedstock for biofuel and electrical power production.

4.3.4 Aquatic Biomass

A biomass resource that is hardly being looked into is aquatic biomass. Over three-quarters of the earth is composed of water (i.e., seawater), and tapping this vast resource may be possible. There are certainly issues that will have to be addressed, primarily concerning the harvest of aquatic biomass from seas and oceans. There will likely be projects that can culture high-yielding aquatic biomass such as giant kelp and persistent seaweeds. Note that some varieties of seaweed are also consumed by humans as food, especially in countries like Japan, and this competing demand must be considered. Studies conducted in the 1980s by the U.S. Department of Energy (Kresovich, et al., 1982) will also have to be updated and investigated in order to develop new biomass crops for fuel. As much as 50 dry tonnes equivalent (DTE)per hectare per year may be produced from just one species of aquatic plant. This yield is much higher than many high-tonnage crops being developed on land.

4.4 Various Biomass Conversion Processes

There are numerous biomass conversion pathways. The processes may be subdivided into three general areas:

a. Physico-chemical conversion processes
b. Biological conversion processes
c. Thermal conversion processes

Physico-chemical processes require neither microbes nor high operating temperatures. Biological processes make use of microbial populations for the conversion of biomass into useful fuel products, while thermal conversion processes require elevated temperatures for conversion. A detailed description of each conversion pathway follows below. Figure 4.1 shows the various biomass conversion pathways to be discussed in this chapter.

4.4.1 Physico-Chemical Conversion Processes

4.4.1.1 Biodiesel Production

Biodiesel production is perhaps the simplest way to make biofuels compatible with commercial engines. In this process, no microbiological organisms are involved and the temperature required for conversion is low. The feedstocks used are oil crops, including animal fats and oils. In commercial biodiesel production facilities, the feedstock is usually refined, bleached, and deodorized into what are called "RBD" oils. The pre-processing of oils is done elsewhere. In tracking the processes involved when oil seed crops are harvested, the steps are as follows:

a. The oil seeds are cleaned and prepared for storage prior to oil extraction.
b. The oil seeds may need to be processed further, which includes tempering, cracking, dehulling, cleaning, and separating unwanted materials and debris.

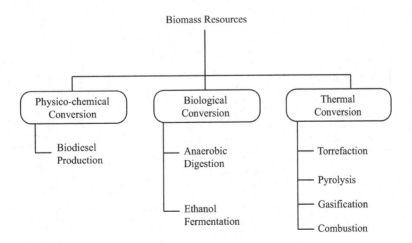

FIGURE 4.1
Various biomass conversion pathways.

c. Oil extraction follows and may be done in several ways, including solvent extraction, with the use of mechanical oil seed presses, and through ultra-sonication of micro-algae samples.

d. Refining of the oil includes neutralization, dewaxing, and degumming, followed by bleaching and deodorization to produce RBD oil.

The transesterification process for vegetable oils and fats has the following governing equation (Equation 4.1):

$$100 \; kg \; RBD \; oil + 10 \; kg \; methanol \rightarrow 100 \; kg \; biodiesel + 10 \; kg \; glycerin \qquad (4.1)$$

Commercial biodiesel producers should be able to achieve mass balance, provided they use high-quality refined, bleached, or deodorized oil (RBD oil). The methanol is usually mixed with another catalyst component, such as sodium hydroxide (NaOH), to generate sodium methoxide, which is the actual catalyst for the transesterification process. The physical nature of the transesterification is shown in Figure 4.2. In this visual, the vegetable oil or triglyceride is composed of a glycerol backbone with three fatty acids attached. They are connected together by what is called the ester linkage, shown in Figure 4.3.

Note that the fatty acids attached to the glycerol may be either of the same type (e.g., all oleic acid) or of different types (oleic + linoleic + oleic). The resulting ester is named after the major type of vegetable oil used or the major type of fatty acid contained in the oil. For example, if soybean oil is used, the resulting product is soybean methyl ester (SME). In Europe, where the major oil used for making biodiesel is rapeseed, the resulting ester is called rapeseed methyl ester (or RME).

These methyl esters derived from catalytic conversion of vegetable oil or triglycerides would have to pass the ASTM standards (ASTM D 6751) before they may be used as fuel blends. Sulfur was previously used as an additive to commercial diesel fuel to improve

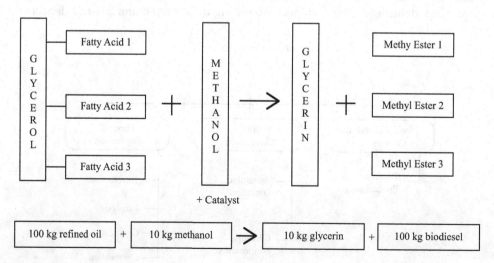

FIGURE 4.2
The physical depiction of the transesterification process and the governing mass balance.

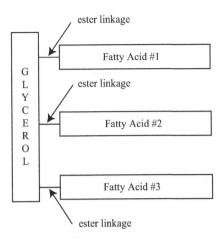

FIGURE 4.3
The action of catalysts on the ester linkage to generate glycerin and the ester of the vegetable oil.

engine lubricity. However, this gave rise to increased emissions of sulfur dioxide, the primary gas compound responsible for the formation of smog. When regulations for emissions were put in place, sulfur additives were strongly discouraged, and biodiesel became the first candidate for their replacement. Five percent biodiesel, called B5, can sufficiently replace the additive abilities of sulfur in diesel composition. However, at higher mixtures of biodiesel, one will observe the increased production of NOx, another environmental pollutant, and this must also be handled appropriately (Santos and Capareda, 2015). Currently, numerous engine manufacturers have warranted the use of as high as 20% biodiesel (B20). Note that biodiesel or the ester of oil is technically a solvent, and there are some engine component parts that may be easily damaged with high concentrations of this biofuel.

Example 4.3 shows how to estimate the feedstock requirement for a given size of biodiesel plant.

Example 4.3: Feedstock Requirement for a 3.785 Million-Liter-Per-Year [1 MGY] Biodiesel Plant

Make an estimate of the vegetable oil requirement (in kg or lbs per year) for a 1-million-gallon-per-year (3.785 million-liter-per-year) biodiesel facility using soybean oil. Use the governing Equation 4.1 for your calculations. Assume that the density of the biodiesel is 7.3 lbs/gal [877 kg/m³].

SOLUTION:

a. Since the governing equation is in units of weight and the biodiesel plant capacity in units of volume, the density units are important. One should begin to convert the volume requirement into weight units as follows:

$$Biodiesel\left(\frac{kg}{yr}\right) = \frac{3.785 \times 10^6 L}{yr} \times \frac{877\ kg}{m^3} \times \frac{1\ m^3}{1,000\ L} = 3.32 \times 10^6 \frac{kg}{yr} \left[7.3 \times 10^6 \frac{lbs}{yr}\right]$$

b. The above calculation means that one would need equal amounts of soybean oil in a year as well.

$$Biodiesel\left(\frac{tonnes}{yr}\right) = 3.32 \times 10^6 \frac{kg}{yr} \times \frac{tonne}{1,000\ kg} = 3,320\ tonnes/yr\left[3,652\frac{tons}{yr}\right]$$

c. Hence, one would need at least 3,320 metric tonnes [3,652 tons] of soybean oil each year for this biodiesel plant.

4.4.2 Biological Conversion Processes

The two major conventional fuels used worldwide are gasoline and diesel. The two major biofuels currently being mass-produced worldwide are bio-ethanol and biodiesel. Biodiesel is now usually mixed with diesel fuel, while bio-ethanol is a regular additive to gasoline fuels. The two major biological conversion processes are fermentation (into alcohol) and anaerobic digestion (to produce biogas). Biogas is a gas mixture that is comprised of primarily methane (CH_4) and carbon dioxide (CO_2), with a ratio of around 65% methane and 35% carbon dioxide (Capareda, 2014).

4.4.2.1 Bio-Ethanol Production

Bio-ethanol is produced in various ways, depending upon the major composition of sugar, starch, or ligno-cellulosic components in a given agricultural crop. Figure 4.4 shows the schematic flow chart when converting biomass into bio-ethanol. If the feedstock is purely in sugar form, one would simply need to use fermentation microbes such as *Saccharomyces cerevisiae* to convert the sugar into ethanol. For starchy materials, the starch must be first converted into sugars prior to the fermentation process. The most common microbial populations used to convert starch into sugars are various types of mold, such as *Aspegillus niger* or *Aspergillus awamori*. These microbes produce enzymes that break down starch and convert them into sugars in a process called saccharification. If the feedstock is rich in ligno-cellulosic materials, the celluloses must be converted into sugars via different sets of microbes that generate enzymes that are able to break down the cellulose bonds to produce sugars. Some species of *Trichoderma reesei* are able to produce these types of enzymes. *Aspergillus niger* also produces enzymes that are able to break down cellulose into sugars. However, the conversion efficiencies may be quite low. Note that cellulose is composed of long chains of sugar glucose. Once the material is converted into sugars, the usual fermentation process should proceed with the presence of the alcohol fermentation microbes enumerated earlier.

After the fermentation process, the ethanol is separated from the fermentation broth by distillation. The other products of fermentation are carbon dioxide (CO_2), the respiration gas of the microbes, and solid residue called dry distiller's grain (DDG). DDG has recently become popular feed material for animals, particularly beef cattle and dairy cows.

When distillation is used to separate alcohol from beer, one cannot generate pure, 100% ethanol because of the azeotropic behavior of water and alcohol—they behave as one compound with a unique boiling point and would be difficult to separate. The maximum ethanol concentration one would achieve is only about 96% (Capareda, 2014). Another step

Ethanol Production Process Flow Chart

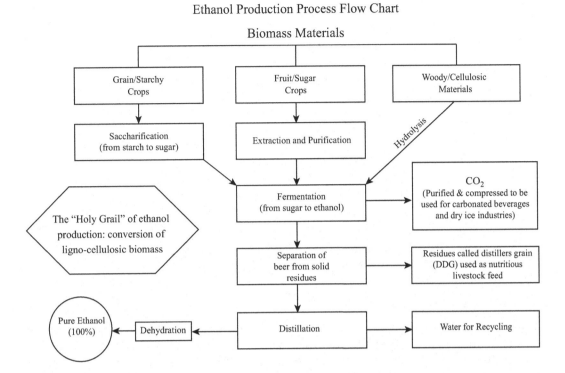

FIGURE 4.4
Ethanol production process flow chart.

must be taken to purify the alcohol for engine use. This would involve using tools such as size-selective molecular sieves. The particularly-sized sieve is designed to separate the bioethanol from the water by molecular size as opposed to boiling point.

The governing equation for ethanol (C_2H_6O) production from sugar or glucose ($C_6H_{12}O_6$) is shown in Equation 4.2:

$$C_6H_{12}O_6 + yeast \rightarrow 2C_2H_6O + 2CO_2 + Heat \tag{4.2}$$

In this equation, 2 moles of ethanol (C_2H_6O) are produced for every mole of glucose used. Two moles of carbon dioxide are also produced, plus some form of heat. The rule-of-thumb ratio for sugar to ethanol production is around 1.5 kg of sugar for every liter of ethanol produced [13 lbs of sugar for every gallon of ethanol produced] as shown in Example 4.4. A good ratio to remember is around 15 lbs of sugar (even though the theoretical calculation is around 13 lbs] for every gallon of ethanol produced [about 2 kg of sugar per liter of ethanol produced].

Example 4.4: Amount of Sugar Needed to Generate a Given Volume of Ethanol

Determine the amount of sugar needed for the volume of ethanol produced using the governing Equation 4.2. Use both English and metric units. Assume the density of ethanol to be 0.789 kg/L.

SOLUTION:

    a. The amount of sugar needed (kg) for every L of ethanol produced is shown below:

$$\frac{kg\ sugar}{L\ ethanol}\left(\frac{kg}{L}\right) = \frac{1\ mol\times\left[(12\times6)+(1\times2)+(16\times6)\right]}{2\ mol\times\left[(12\times2)+(1\times6)+(16\times1)\right]} = \frac{180\ kg}{92\ kg}\times\frac{0.789\ kg}{L}$$

$$= 1.54\frac{kg}{L}$$

    b. In English units, the calculations are as follows:

$$\frac{lbs\ sugar}{gallon\ ethanol}\left(\frac{lbs}{gallon}\right) = 1.54\frac{kg}{L}\times\frac{3.785\ L}{gallon}\times\frac{2.2\ lbs}{kg} = 12.8\frac{lbs}{gallon}$$

    c. The rule-of-thumb ratio for ethanol production from sugar is around 2 kg sugar per liter of ethanol [or 15 lbs sugar per gallon ethanol].

4.4.2.2 Biogas Production

The technical term for biogas production is anaerobic digestion. The microbes responsible for producing biogas ($CH_4 + CO_2$) are anaerobic microbes—microbes that thrive under an oxygen-free environment. The primary feedstock for this process is the manure of ruminants. Ruminants (cows, horse, sheep, goat, deer, and so on) are animals that have naturally occurring microbial flora in their stomachs that are able to hydrolyze ligno-cellulosic materials into useful energy and into methane. The detailed processes involved in anaerobic digestion are the conversions of complex organics into simpler organic compounds. The process of conversion is called hydrolysis, performed by the enzymes excreted by these anaerobic microbes. The organic compounds are then converted into organic acids, such as propionic acids and butyric acids and the like. Some other simpler organic compounds are converted into acetic acid—the most popular organic acid in an anaerobic reactor or an anaerobic digester. These organic acids are then converted into methane by methane-producing microorganisms. The two general types of microbes involved in anaerobic digestion processes are acid-producing and methane-producing microbes. These two types of microbes thrive under various conditions of pH. The acid-producing microbes can withstand low pH (or very acidic conditions), while the methane-producing microbes cannot, so if the reactor is dominated by acid-producing microbes, the methane production may cease, and this can cause sour-smelling phenomena. Therefore, care must be taken to ensure that an ecological balance between acid-producing and methane-producing microbes exists. The simplest way to control this is to neutralize the pH level of the reactor. This is simply done by adding neutralizing agents such as lime (calcium carbonate) or sodium hydroxide (NaOH).

The simplest way to design the anaerobic digester reactor is to set up an experiment that gauges the amount of methane produced by a given biomass feedstock and manure. This setup is shown in Figure 4.5. In this setup, a mixture of equal amounts of fresh manure and water is placed in an enclosed reactor. The reactor is then carefully sealed using a rubber stopper. A tube is inserted into the sealed reactor such that the gas is collected in the second container. The second container is filled with tap water and sealed with a rubber stopper with one port connected to the reactor and another port to a graduated cylinder.

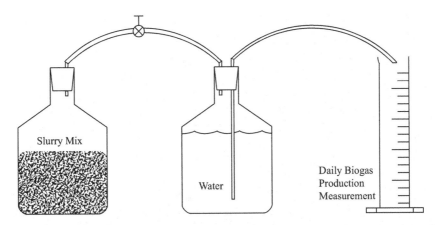

FIGURE 4.5
Batch digestion setup for designing anaerobic digestion reactors.

Note the arrangement of the port should be such that the gas produced from the reactor pushes the water out of the second container into the graduated cylinder. This arrangement will ensure that oxygen from the environment will not come into contact with either the reactor or the second container. Each day, one can simply measure the amount of water collected in the graduated cylinder, which should be equal to the amount of gas produced. When the water in the second container becomes empty, it should be replenished. Note that during replenishment, a valve must be placed between the first and second containers and sealed well to prevent the entry of air into the first chamber. Make sure that this valve is opened once the second reactor, full of water, is in place. Otherwise, gas trapped in the first chamber will blow out the rubber stopper and make the experiment inconclusive. The experiment should be stopped when the amount of biogas produced each day becomes marginal. A graph may be plotted for daily gas production. At some point, a peak in the data may be observed. This peak indicates that there is no longer enough biomass for conversion into methane gas, and fresh biomass and manure samples must be introduced. This peak corresponds to what we call solids retention time (SRT). In real plug-flow reactors, this is the period at which the biomass must be discarded because of marginal gas production. This parameter may also be used to design the size of the biogas reactor, as shown in Equation 4.3:

$$Digester\ Size\left(m^3\right) = Loading\ Rate\left(\frac{m^3}{day}\right) \times SRT\left(days\right) \qquad (4.3)$$

The loading rate (m³/day) is simply the amount of manure and biomass slurry produced each day and placed into the biogas digester. The amount of biogas varies according to the type of manure used, the inoculant used, and the amount of biomass used for what we call co-digestion. Stout (1984) provided several tables for manure generation and biogas production for major farm animals (Chapter 5, Tables 5.21 and 5.22). The reason biomass is added to the raw manure is to improve the important anaerobic digestion parameter called carbon-to-nitrogen ratio (or C:N ratio). Applied research has shown that the ideal C:N ratio is around 30. The C:N ratio of cow manure is around 20 (Maramba, 1978) and must be enhanced with additional biomass that has high carbon content. A scaled-up pilot biogas digester is shown in Figure 4.6.

FIGURE 4.6
Pilot anaerobic digester with floating gas holder.

Example 4.5: Sizing Anaerobic Digesters

Determine the size of the digester for an experiment that has the following results: retention time of 45 days, and type of manure comes from dairy manure without co-digestion. The dairy farm has a population of 350 heads, and approximately 25 kg [55 lbs] of fresh manure is generated each day. Use equal amounts of water to be mixed with fresh manure for input into the digester. Assume the density of slurry to be around 1.1 kg/L.

SOLUTION:

a. The amount of slurry generated and to be fed to the digester is shown below:

$$Slurry\left(\frac{kg}{day}\right) = Manure\left(\frac{25\ kg}{head-day}\right) \times 350\ heads \times 2 = 17,500\frac{kg}{day}$$

$$Slurry\left(\frac{lbs}{day}\right) = 17,500\frac{kg}{day} \times \frac{2.2\ lbs}{1\ kg} = 38,500\frac{lbs}{day}$$

b. The digester size is calculated as follows:

$$Digester\ Size\left(m^3\right) = 17,500\frac{kg}{day} \times 45\ days \times \frac{L}{1.1\ kg} \times \frac{1\ m^3}{1,000\ L} = 716\ m^3$$

$$Digester\ Size\left(ft^3\right) = 716\ m^3 \times \left(\frac{3.28\ ft}{1\ m}\right)^3 = 25,266\ ft^3$$

c. Some free-board must be added to the digester size to ensure that the reactor will not be overfilled during times when the manure production is high.

4.4.3 Thermal Conversion Processes

Thermal conversion processes are biomass conversion processes that utilize high temperatures or heat to convert biomass into useful products. There are basically four types according to the amount of air introduced as well as the temperature used for the reaction. The four thermal conversion processes include (a) torrefaction, (b) pyrolysis, (c) gasification, and (d) combustion. Torrefaction and pyrolysis are reactions that use no oxygen or air, while gasification and combustion both utilize amounts of oxygen or air. Torrefaction is done at lower temperatures, usually below 300°C [572°F], while pyrolysis is done above 300°C [572°F] (Capareda, 2014). Torrefaction is simply a biomass conditioning process— it hardly produces any new product but rather enhances the quality of the biomass in the form of char. Gasification uses incomplete amounts of air (or that below stoichiometry), whereas combustion uses excess amounts of air. As such, the operating temperature for gasification systems is much lower than that of combustion systems. The three thermal conversion systems will be discussed in succeeding sections.

4.4.3.1 Pyrolysis

Pyrolysis is also called destructive distillation. This is defined as a thermal conversion process in the complete absence of oxygen. The biomass is placed in a container and heated externally. Provisions must be taken to remove the gas produced during the reaction; otherwise, it can cause an explosion if not removed from the heated reactor. If this gas is passed or quenched through a condenser, some liquid product will form. At the end of the process, three important products are formed: the solid product called biochar, the liquid product called bio-oil, and the gaseous product called synthesis gas (or syngas). The amount of each of these co-products varies according to the temperature used as well as the rate of reaction. Generally, at lower temperatures, more solid biochar is produced, while at higher temperatures, great volumes of syngas are produced. Liquid bio-oil is maximized at intermediate temperatures. The quality and quantity of the products also vary with rate of conversion or what we call residence time (Equation 4.4). This is defined as the amount of time that the biomass is exposed to a higher reaction temperature. An example using a fluidized bed pyrolyzer is shown in Example 4.6:

$$Residence\ Time(s) = \frac{Reactor\ Volume(m^3)}{Flow\ Rate\left(\dfrac{m^3}{s}\right)} \tag{4.4}$$

Example 4.6: Residence Time of Biomass in a Fluidized Bed Reactor

Determine the residence time for the biomass being thermally converted in a fluidized bed pyrolyzer. The reactor size is 7.62 cm [3 in] by 45.72 cm [18 in] with a volumetric flow rate of 4.17 L/s [8.83 cubic feet per minute, cfm].

SOLUTION:

a. Using Equation 4.4, the units must be the same, and the reactor volume in cubic meters is calculated:

$$Reactor\ Volume(m^3) = \frac{\pi d^2}{4} = \frac{3.1416 \times (7.62\ cm)^2}{4} \times 45.72\ cm = 2,085\ cm^3$$

b. Then, the flow rate is also converted into the same unit:

$$Flow\ Rate\left(\frac{cm^3}{s}\right) = 4.17\frac{L}{s} \times \frac{m^3}{1,000\ L} \times \left(\frac{100\ cm}{1\ m}\right)^3 = 4,170\frac{cm^3}{s}$$

c. The residence time in seconds is calculated as follows:

$$Residence\ Time\ (seconds) = \frac{2,085\ cm^3}{4,170\frac{cm^3}{s}} = 2\ seconds$$

The residence times for many pyrolyzer types are reported as follows: about 1 second for fast pyrolyzers, 10 to 30 seconds for intermediate pyrolyzers, and hours to days for slow pyrolyzers (Bridgewater, 2012).

4.4.3.2 Gasification

Gasification is a thermal conversion process with insufficient amounts of oxidant. Designing the gasification process for biomass begins by estimating the chemical formula of the biomass feedstock to use for the gasification process and setting up the stoichiometric combustion equation (balanced equation for combustion). The ultimate analysis of biomass is determined via ultimate analyzer equipment. The equipment is designed to measure the major elemental components of the biomass—carbon (C), hydrogen (H), oxygen (O), nitrogen (N). and sulfur (S), together with ash. The total weight of the biomass is comprised of these six components. Example 4.7 shows how the chemical formula of a biomass is estimated. Once the chemical formula is established, a complete combustion equation is set up to determine the amount of air that may be used for complete combustion. This process is shown in Example 4.8. Once the amount of air for complete combustion is known, only between 30% and 70% of air is actually used for the gasification process (Maglinao, 2015). This process produces intermediate gases, the majority of which are carbon monoxide (CO) and hydrogen (H_2). These gases are also called synthesis gases, or syngas (as in the pyrolysis process). Of course, there are other combustible gases that are co-produced, including methane (CH_4), ethane (C_2H_4), ethylene (C_2H_4), and other low-molecular-weight or light hydrocarbon gases. In Example 4.8, only 1.45 to 3.4 kg [3.19 to 7.48 lbs] of air may be used per kg of biomass thermally converted in the gasification mode. Otherwise, if excess air is used, only CO_2 and H_2O are produced, along with heat of combustion and no combustible gas, as shown in the complete combustion equation. The energy produced during the complete combustion process is equivalent to the heat of combustion of the biomass or its heating value. The heating value of biomass ranges from a low of 15 MJ/kg [6,463 Btu/lb] to a little over 20 MJ/kg [8,617 Btu/lb]. For rice hull, which has a heating value of 15.4 MJ [6,635 Btu/lb], about 15.4 MJ [6,635 Btu/lb] is released in the form of heat during complete combustion (Capareda, 2014).

Example 4.7: Determining the Approximate Chemical Formula of a Given Biomass

Determine the approximate chemical formula for rice hull. About 100 g of bone-dry rice hull was placed in the ultimate analyzer, and the analysis showed that the carbon, hydrogen, oxygen, nitrogen, sulfur, and ash content was found to be as follows: C = 35 g, H = 5.2 g, O = 39.4 g, N = 0.5 g, S = 0.1 g, and ash = 19.8 g.

SOLUTION:

a. The chemical formula is established by calculating the number of moles of a particular element based on the gravimetric elemental analysis:
   Carbon = 35 g/12 g/mol = 2.92 mol
   Hydrogen = 5.1 g/1 g/mol = 5.2 mol

Oxygen = 39.4 g/16 g/mol = 2.46 mol
Nitrogen = 0.5 g/14 g/mol = 0.36 mol
Sulfur = 0.1 g/32 g/mol = 0.003 mol

b. Hence, the approximate chemical formula of the biomass is as follows:

$$C_{2.92}H_{5.2}O_{2.46}N_{0.36}S_{0.003}$$

c. From the chemical formula, the balanced (or stoichiometric combustion equation) may be calculated. This is shown in the next example.

Example 4.8: Air-to-Fuel Ratio Calculations

Determine the amount of air needed per unit weight of rice hull during complete combustion of this biomass. Use the approximate chemical formula of the biomass as follows:

$$C_{2.92}H_{5.2}O_{2.46}N_{0.36}S_{0.003}$$

SOLUTION:

a. The stoichiometric combustion equation is shown below. Note that air is assumed to be composed primarily of 79% nitrogen and 21% oxygen and that the ratio is 3.76 as shown in the coefficient for nitrogen:

$$C_{2.92}H_{5.2}O_{2.46}N_{0.36}S_{0.003} + 2.99(O_2 + 3.76N_2)$$

$$\rightarrow 2.92CO_2 + 2.6H_2O + 2.99 \times 3.76N_2 + 0.003S$$

b. The air-to-fuel (AFR) ratio is calculated as follows:

$$\frac{A}{F} = \frac{2.99(32 + 105.28)}{\left[(2.92 \times 12) \times (5.2 \times 1) \times (2.46 \times 16) \times (0.36 \times 14) \times (0.003 \times 32)\right]} = \frac{410.5\ kg}{84.74\ kg}$$

$$= 4.844 \frac{kg}{kg}$$

c. Thus, it would take around 4.8 kg of air per kg of rice hull burned in the combustion mode or [4.848 lb of air per lb of rice hull].

Why should we have to gasify or go through the gasification process to generate heat instead of complete combustion? There are two simple answers: first, not all biomass may be combusted for longer periods and in great amounts due to slagging and fouling problems; second, heat energy is not the only form of energy required as a result. Liquid fuel or electrical power is required in many instances. However, there are few biomass items that may be thermally converted in the combustion mode, such as wood species that have low ash content. Firewood is still combusted in many regions of the country without any issues. When agricultural residues such as corn stover and wheat straw are combusted, the ash inherent in the biomass may generate slag due to the high combustion temperature. The inorganic ash content would melt and upon cooling would generate slag on the conveying surfaces of combustion. When slag accumulates in pipes and conveying surfaces, it may cause fouling or complete blockage of passageways. Slagging and fouling are definitely two unwanted processes that may occur during the combustion of many agricultural residues. There are also a few agricultural residues that may be thermally converted in the combustion mode, such as cotton hulls and sugarcane bagasse. While these biomasses have appreciable amounts of ash, the particular type of ash result does not melt at elevated temperatures of combustion. During any thermal conversion processes, one would need to evaluate biomass based on its eutectic point, or the melting point of its ash component. The coal industry has been using the slagging and fouling factors or R_s and R_f respectively. Many agricultural residues do not behave like coal, however, and while their R_s and R_f values indicate no slagging and fouling, they inevitably occur after thermal conversion at elevated temperatures. There are other parameters one may use to determine the propensity of a biomass to generate slag or encounter fouling problems during thermal conversion.

4.4.3.3 Eutectic Point of Biomass

In coal combustion, the slagging factor R_s and fouling factor R_f are calculated as follows: (a) the ash from a particular biomass is generated, (b) the complete inorganic ash compositions are determined, and (c) the slagging and fouling indicators are found. Example 4.9 shows how R_s and R_f are calculated from the complete inorganic ash analysis. The slagging factor R_s is given by Equation 4.5, and the slagging factor R_f is given by Equation 4.6:

$$R_s = \left(\frac{Base}{Acid}\right) \times \% \, Sulfur \ on \ dry \ coal \qquad (4.5)$$

$$R_f = \left(\frac{Base}{Acid}\right) \times \% \, Na_2O \qquad (4.6)$$

Agricultural residues, however, do not behave like coal, and the more appropriate melting point or eutectic point indicators are the following:

a. Base-to-acid ratio, $R_{b/a}$

b. Biomass agglomeration index (BAI)

c. Alkali index (AI)

The base-to-acid ratio is calculated based on Equation 4.7. The numerator is the sum of all basic compounds found in the ash, while the denominator is the sum of all acid compounds found in the ash:

$$R_{b/a} = \frac{\% \left(Fe_2O_3 + CaO + MgO + K_2O + Na_sO\right)}{\% \left(SiO_2 + TiO_2 + Al_2O_3\right)} \qquad (4.7)$$

If the base-to-acid ratio is more than the unity, slagging and fouling are certain to occur.

The BAI is shown in Equation 4.8. This is the ratio of the iron-based compounds to the potassium- and sodium-based compounds:

$$BAI = \frac{\% \left(Fe_2O_3\right)}{\% \left(K_2O + Na_sO\right)} \qquad (4.8)$$

Bed agglomeration is sure to occur when the BAI values are lower than 0.15.

The AI is the ratio of the amount of K_2O and Na_2O compounds divided by the heating value of the material in GJ as shown in Equation 4.9:

$$AI = \frac{kg \left(K_2O + Na_sO\right)}{GJ} \qquad (4.9)$$

When the AI is below 0.17 kg/GJ [0.4 lb/MMBtu], the biomass is projected to have fairly low slagging risk. For values above 0.34 kg/GJ [0.8 lb/MMBtu], slagging risk is high (Maglinao and Capareda, 2010). The use of the equations listed above are shown in Example 4.9.

Example 4.9: Eutectic Point of Biomass

Determine the base-to-acid ratio, the BAI, and the AI for the biomass; the ash profile is given in the table. The heating value of this biomass is around 7,000 Btu/lb [16.25 MJ/kg]. The percentage of ash in the biomass is 25%. Comment on the values obtained.

Basic Compounds	Values	Acidic Compounds	Values
Fe_2O_3	0.7	SiO_2	41.8
CaO	10.8	TiO_2	0
MgO	3.3	Al_2O_3	3.1
K_2O	10.5		
Na_2O	0.6		
Totals	25.9	Totals	44.9

SOLUTION:

a. The base-to-acid ratio is calculated using Equation 4.7:

$$R_{b/a} = \frac{\%(Fe_2O_3 + CaO + MgO + K_2O + Na_sO)}{\%(SiO_2 + TiO_2 + Al_2O_3)}$$

$$R_{b/a} = \frac{\%(0.7 + 10.8 + 3.3 + 10.5 + 0.6)}{\%(41.8 + 0 + 3.1)} = \frac{25.9}{44.9} = 0.58$$

The values are less than the unity; hence, slagging is expected, and fouling events are minimal.

b. The BAI is shown below using Equation 4.8.

$$BAI = \frac{\%(Fe_2O_3)}{\%(K_2O + Na_sO)}$$

$$BAI = \frac{\%(0.7)}{\%(10.5 + 0.6)} = 0.063$$

Bed agglomeration may occur due to a low BAI.

c. The AI is given by Equation 4.9, and calculations are shown below. Note that the % K_2O and Na_2O are 10.5% and 0.6%, respectively, and for every kg of ash material, around 11.1% is the alkali. For every kg of biomass, 25% is ash; the amount of this alkali per kg is the product of 25% and 11.1%, which is 0.02775 kg (0.25 × 0.111). Thus, the heating value of GJ is simply the heating value in MJ divided by 1,000 as shown:

$$AI = \frac{kg(K_2O + Na_sO)}{GJ}$$

$$AI = \frac{kg(K_2O + Na_sO)}{GJ} = \frac{0.02775}{16.25/1,000} = 1.71\frac{kg}{GJ}$$

The value of AI is higher than 0.34, and therefore this biomass has high slagging and fouling risk. In the calculations above, two out of the three parameters show that the biomass has potential for slagging and fouling, and, conclusively, one should be careful when converting this biomass through thermal means. The occurrence of agglomeration, slagging, and fouling is very likely at high temperatures of operation.

4.4.3.4 Combustion Processes

Combustion of biomass continues to be a predominant activity in both developed and developing countries. Fuel wood is perhaps the most convenient biomass to use for heating and cooking in many developing countries. Developed countries use fuel wood for heating and processing wood products as well as for power generation. Wood chips, shavings, and tree bark are continually used in many sawmills to generate steam during the processing of fiberboard and many other wood products. Fuel wood is convenient to use because of its low ash content. Its propensity for slagging and fouling events is minimal, unless one uses bark with somewhat different ash composition.

In many developing countries, the use of wood-based or biomass-based cook stoves are prevalent. The U.S. Department of Energy continues to organize the gathering of cook stove experts, as there are now numerous high-efficiency cook stoves marketed worldwide.

Other biomass materials, such as agricultural crops and especially animal manure, are difficult to combust fully due to their high probability of slagging and fouling. A way around this is to mix agricultural biomass and animal manure with solid fuels that have low slagging and fouling propensity, like coal. If fuel wood biomass is used for power generation, large tracts of land may be needed. Example 4.10 shows the simple calculations for estimating the area required to build a large biomass power plant. The factors affecting this acreage include productivity of land for biomass, conversion efficiency, and moisture of biomass used. Southern Power Company, a subsidiary of Southern Power in Alabama, operates a 100 MW biomass power plant in Sacul, Texas (in northwestern Nacogdoches County). The plant has been in operation since 2012.

Example 4.10: Area Needed for Wood Power

Determine the amount of area needed to produce 25 MW of electrical power from the combustion of wood chips. Assume the heating value of the wood chips to be 20 MJ/kg [8,616 Btu/lb] and the productivity of land to be about 5 dry tonnes per hectare per year (1,000 kg is 1 tonne) [2.23 tons/acre/yr]. Further, assume the overall conversion efficiency of about 50% using the steam cycle. The plant will be operated continuously throughout the year, or 365 days in a year.

SOLUTION:

a. The input power is first calculated using the given conversion efficiency:

$$Input\ Biomass = \frac{Output\ MW}{Overall\ Efficiency\ (decimal)} = \frac{25\ MW}{0.50} = 50\ MW$$

b. Then, input biomass in tonnes per hr is calculated:

$$Biomass\ Input\left(\frac{tonnes}{hr}\right) = 50\ MW \times \frac{MJ}{MW-s} \times \frac{kg}{20\ MJ} \times \frac{3600s}{hr} = 9,000\frac{kg}{hr} = 9\frac{tonnes}{hr}$$

$$Biomass\ Input\left(\frac{tons}{hr}\right) = 9\frac{tonnes}{hr} \times \frac{1.1\ ton}{1\ tonne} = 9.9\frac{tons}{hr}$$

c. Finally, the acreage required is calculated:

$$Area(ha) = 9\frac{tonnes}{hr} \times \frac{24\ hrs}{day} \times \frac{365\ days}{yr} \times \frac{ha-yr}{5\ tonnes} = 15,768\ ha$$

$$Area(acres) = 15,768\ hectares \times \frac{2.47\ acre}{1\ ha} = 38,947\ acres$$

d. Over 15,000 hectares [39,000 acres] may be needed for this project, or a little over 600 hectares [1,482 acres] per MW of power output.

4.5 Economics of Heat, Fuel, and Electrical Power Production from Biomass

The economics of biomass conversion depends much on the price of the feedstock used for the conversion and production of a specific product. For example, it is common knowledge in the biodiesel community that 85% of the biodiesel cost comes from the cost of the refined oil used for making it (Capareda, 2014). Hence, if the refined oil cost is greater than the diesel fuel cost, the process is not economically sound. Likewise, if the ethanol production cost is higher than the cost of the gasoline fuel that will be used for blending the biomass-based fuel, it would not be economical to produce this fuel from biomass. There are, however, biomasses ligno-cellulosic in nature, not used for food, that can be used to make the fuel economically. Biodiesel and bio-ethanol are the two most important fuels produced from biomass, primarily because each of these fuels is blended with diesel and gasoline fuel, respectively. One other type of fuel, biogas, developed through anaerobic digestion of wastes and manure, is also gaining popularity. This fuel can replace natural gas, which is ubiquitous in households for cooling and heating purposes.

4.5.1 Biodiesel Economics

In biodiesel manufacturing, the cost of feedstock comprises 85% of the overall cost of the final product being sold, and the production of biodiesel is also dependent upon other chemicals used, such as the catalyst. In a report by Willits (2007), the conversion and processing cost of biodiesel in the United States is approximately $0.06/L [$0.22/gal]. The other costs go toward depreciation and manpower. The capital cost of a commercial biodiesel production facility may be divided into major components such as the process equipment, piping, electrical, building and structures, automation, and others.

The overall cost of a biodiesel plant varies according to the type of facility and equipment used to separate glycerin and biodiesel. Expensive facilities make extensive use of centrifuges, while cheaper facilities will use gravity for most separation processes. Small-scale biodiesel facilities can simply separate biodiesel from glycerin by decantation, or physically separating the glycerin and biodiesel, since their makeup prevents them from mixing with each other. A design by Crown Iron Works (Waranica, 2007) is popular for gravity separation, while a design by Desmet Ballestra (Willits, 2007) makes use of centri-fuges for separation—a much quicker and effective (but slightly costly) way of separating biodiesel and glycerin.

Glycerin is an important and expensive material. However, it has to be purified to close to 99.9% in order to be sold to the pharmaceutical industry. This is an additional cost to the biodiesel producer and is typically not performed at a commercial biodiesel facility. In some biodiesel plants, there is a surplus of these glycerin co-products if there is no guar-anteed market. Some researchers are looking to turn this glycerin by-product into other valuable products with limited initial investment. Glycerin can be an excellent substrate for making biogas via anaerobic digestion processes (Capareda, 2014).

4.5.2 Ethanol Economics

Ethanol produced in the United States comes primarily from corn. In Brazil, sugarcane is the main substrate. There was initially controversy over the production of etha-nol from corn due to the perceived negative net energy balance. This controversy has since been disproven by research and numerous publications by the U.S. Department of Agriculture (USDA) (Shapouri, et al., 2006a). In their report, ethanol production from Brazil is still lower, at $0.21/L [$0.81/gal], compared to ethanol production in the United States using wet milling at $0.27/L [$1.03/gal] and dry milling at $0.28/L [$1.03/gal]. The feedstock cost for the Brazilian ethanol is 37% and the remaining 63% is the production cost. In the United States, wet milling costs are comprised of about 39% feedstock and 61% production cost. Half of the cost of dry milling goes toward feedstock and the other half toward production. If the ethanol in the United States is produced from sugarcane, the cost is $0.63/L [$2.40/gal]. At the current price of gaso-line of below $2.00/gallon in 2015, it will be quite difficult to make ethanol from this feedstock competitive with gasoline. The crude oil prices in some portions of 2015 were at all-time lows of close to $45/barrel. Accordingly, biomass-based fuels struggle when crude oil prices are lowest. Cellulosic ethanol could potentially break this price hurdle if the price of enzymes lowers.

The state of bioethanol in the United States is very much dependent on federal and gov-ernment interventions, including indirect forms of subsidy. Corn production in the United States is still at an all-time high, with a guaranteed surplus each year. Corn, however, is used as input for many food products, and its price cannot go lower. It is believed that cel-lulosic ethanol should take up the slack by developing newer production pathways with lower input costs. There are newer thermal conversion systems that are being investigated in the hopes of bringing down the cost of biomass-based fuel so that it can compete with gasoline, diesel, or aviation fuel.

The United States will continue to increase the production of biomass-based biofuels. The U.S. Environmental Protection Agency (U.S. EPA) just released the proposed vol-ume requirements under the Renewable Fuels Standard Program for cellulosic biofuels, advanced biofuels, and total renewable fuel for calendar year 2019. They also proposed the biomass-based diesel volume standards for calendar year 2020 amounting to 2.43 billion gallons [9.2 billion liters]. (Lane, 2018). Table 4.1 shows the proposed and final renewable fuel volume requirements for 2018-2010.

TABLE 4.1

Proposed and Final Renewable Fuel Volume Requirements for 2018-2020 in the United States

Type of Biofuel	2018	2019	2020
Cellulosic biofuel (million gallons) (ML)	288 (1090)	381(1442)	N/A
Biomass-based diesel (billion gallons) (billion liters)	2.1 (7.95)	N/A*	2.43 (9)
Advanced biofuel (billion gallons) (billion liters)	4.29 (16)	4.88 (18.5)	N/A
Renewable fuel (billion gallons) (billion liters)	19.29 (73)	19.88 (75)	N/A
Implied conventional biofuel (billion gallons) (BL)	15 (57)	15 (57)	N/A

*The biomass-based diesel standard for 2019 was set at 2.1 billion gallons [7.95 billion liters] in 2018 and cannot be changed.

4.6 Sustainability Issues with Biomass Energy Use

The use of biomass for fuel and power has created an issue with the use of biomass for feeding the growing world population. This is coupled with the scarcity of water to grow biomass. A number of research reports have been focused on the food, energy, and water nexus. Most reports have shown that the use of biomass as a fuel source does affect the price of other commodities that use the same feedstock for other purposes (e.g., corn for fuel and corn as food). We must then think through the choice of feedstock and its long-term viability for fuel and power. The use of ligno-cellulosic biomass seems a logical choice. However, the recalcitrant nature of this particular feedstock causes various issues in pre-treatment and pre-processing, such as the need for expensive enzymes, biological organisms, and heat to break it down. The bottom line is the issue of sustainability. The USDOE and USDA's parameters for sustainability demand that the product is economically competitive, conserves the natural resource base, and ensures social well-being. (USDOE and USDA, 2009). Various methods have been developed to describe the sustainability of the use of certain biomasses for fuel and power.

The first set of calculations deals with the energy used to make the fuel. The two common parameters are the net energy ratio (NER) and the net energy balance (NEB) shown in Equations 4.10 and 4.11. The NER is defined as the energy content of the biofuel divided by the energy required to produce the biofuel. If the value is greater than 1, then the production of the biofuel is deemed sustainable. The NEB, on the other hand, is the difference between the heating values of the biofuel produced with the energy required to produce the fuel. If the value is likewise positive, then the process is said to be sustainable. For example, the wet milling of corn was shown to have an NER of 1.02, while the value was 1.10 for dry milling. The NEB for wet milled corn for ethanol was reported to be 0.513 and that for dry milling was 1.871 (Shapouri, et al., 2006b). The values are all positive with wet milling having the value closest to zero:

$$NER = \frac{Energy\ content\ of\ biofuel\,(MJ)}{Energy\ required\ to\ produce\ the\ biofuel\,(MJ)} \qquad (4.10)$$

$$NEB = Biofuel\ HV\,(MJ) - energy\ required\ to\ produce\ biofuel\,(MJ) \qquad (4.11)$$

Finally, many other software programs have been developed to evaluate the sustainability or life cycle analysis (LCA) of making fuel or power from biomass. One of the most popular pieces of software is GREET, developed at the Argonne National Laboratory (ANL) of the U.S. DOE. GREET stands for greenhouse gases, regulated emissions, and energy use in transportation vehicles. This is freely accessible software that evaluates the total energy used to make a fuel compared with other conventional processes and projects greenhouse gas emissions produced when the fuel is made. The production pathways of basic fuels from biomass, such as ethanol and biodiesel, have already been established.

4.7 Conclusion

In many developing countries, biomass satisfies a large portion of the energy mix over fossil fuels. Because biomass comes from fuel wood, forest cover repletion and reforestation efforts are affected. The use of fuel wood in the long term is not sustainable. Other biomass resources dedicated to making new fuel must be developed. Biomass also competes with certain food resources. The two most common biofuels (bio-ethanol and biodiesel) come from sugar and starchy crops and make use of vegetable oils and animal fats. Accordingly, the price of the feedstock affects the price of food.

The fuels produced from biomass will always be compared to fuels from fossil sources. As the price of crude oil goes down, so does the price of conventional fuels, such as gasoline and diesel. The bio-ethanol and biodiesel produced from biomass will likewise have to compete with fluctuations in world crude oil prices. It is clear from the various biomass conversion pathways that the price of the final product is much dependent on the price of feedstock used to make the final product. It can be inferred that the types of biomass that will ultimately be used to make fuel or power are those with low initial feedstock costs. None of the agricultural residues are considered cheap on a per tonne basis and may need for the price of crude oil to go above 100/barrel to be competitive. The biomass with a negative initial cost will have the edge as a potential resource for making biofuels or producing power. Municipal solid wastes, municipal sludge, and animal manure are such resources that may have very low or negative cost. Numerous technologies using these biomass-based feedstocks should be developed. The goal will always be to make the production cost as low as possible. Hence, there is always room for improved conversion efficiencies and cost-effectiveness for future research undertakings.

4.8 Problems

4.8.1 Area Required to Build a Power Plant

P4.1 Determine the area needed to supply the yearly fuel requirement of a 3 MW power plant using wood as feedstock. The power plant should operate continuously for at least 365 days in a year. The heating value of wood was found to be 22 MJ/kg, and the conversion efficiency via steam generation was assumed to be 50%. The yield of the wood plantation is around 4 tonnes/hectare/year.

4.8.2 Electrical Power from MSW

P4.2 An analysis of segregated municipal solid waste (MSW) by Balcones Resources of Dallas, Texas, showed that the energy content of the segregated and combustible biomass from the garbage was about 20 MJ/kg. In a city like Dallas with a population of 1.3 million people, estimate the amount of garbage produced in tonnes per day if the per capita production is 2.5 kg/person/day. If this biomass were converted into electrical power with an overall conversion efficiency of 15%, what would be the output power in MW, assuming you operate 24 hrs/day?

4.8.3 Feedstock Requirement for a 3 MGY Biodiesel Plant

P4.3 Make an estimate of the vegetable oil requirement (in kg or lbs per year) for a 3-million-gallon-per-year (11.355-million liter-per-year) biodiesel facility using soybean oil. Use the governing Equation 4.1 for your calculations. Assume that the density of biodiesel is 7.3 lbs/gal [877 kg/m³].

4.8.4 Sugar Needed to Produce Ethanol

P4.4 Determine the amount of ethanol produced per kg of sugar needed using the governing Equation 4.2. Use both English and metric units. Assume the density of ethanol to be 0.8 kg/L.

4.8.5 Biogas Digester Sizing

P4.5 Determine the size of the digester for an experiment that has the following results: retention time is 45 days, and the type of manure is dairy manure without co-digestion. The dairy farm population is 350 heads, and approximately 25 kg [55 lbs] of fresh manure is generated each day. Use equal amounts of water mixed with fresh manure in the digester. Assume the density of slurry to be around 1.1 kg/L.

4.8.6 Residence Time for Biomass Conversion in Fluidized Bed Reactors

P4.6 Determine the residence time for the biomass being thermally converted in a fluidized bed pyrolyzer. The reactor size is 7.62 cm [3 in] and 45.72 cm [18 in] with a volumetric flow rate of 4.17 L/s [8.83 cubic feet per minute, cfm].

4.8.7 Chemical Formula for Biomass

P4.7 Determine the approximate chemical formula for corn stover. About 100 g of bone-dry corn stover was placed in the ultimate analyzer, and the analysis showed that the carbon, hydrogen, oxygen, nitrogen, sulfur, and ash content was found to be as follows: C = 40 g, H = 5.9 g, O = 46.2 g, N = 1.3 g, S = 0.2 g, and ash = 6.3 g.

4.8.8 Air-to-Fuel Ratio (AFR) Calculations

P4.8 Determine the amount of air needed per unit weight of corn stover during complete combustion of this biomass. Use the approximate chemical formula of the biomass as follows:

$$C_{3.3}H_{5.9}O_{2.9}N_{0.1}S_{0.01}$$

4.8.9 Eutectic Point of Biomass

P4.9 Determine the base-to-acid ratio, the BAI, and the AI index for the biomass with the ash profile given in the table below. The heating value of this biomass is around 15.4 MJ/kg [6,635 Btu/lb]. The percentage of ash in the biomass is 12.5%. Comment on the values obtained.

Basic Compounds	Values	Acidic Compounds	Values
Fe_2O_3	1.0	SiO_2	73.2
CaO	5.0	TiO_2	0
MgO	1.5	Al_2O_3	5.1
K_2O	8.4		
Na_2O	0.4		
Totals	16.3	Totals	78.3

4.8.10 Area Needed for Wood Power

P4.10 Determine the amount of area needed to produce 25 MW of electrical power from the combustion of high-tonnage sorghum via gasification. Assume the heating value of high-tonnage sorghum to be 16 MJ/kg and the productivity of land to be about 30 tonnes per hectare per year (1,000 kg is 1 tonne). Further, assume the overall conversion efficiency of about 15% using the steam cycle. The plant will be operated continuously throughout the year (365 days in a year).

References

Bataller, B. 2018. Design, fabrication and assessment of an internally-illuminated concentric-tube airlift photobioreactor for the cultivation of Spirulina platensis. PhD Dissertation, Biological and Agricultural Engineering Department, College of Agriculture and Life Sciences, Texas A&M University, College Station, May.

Bridgewater, A. V. 2012. Review of fast pyrolysis of biomass and product upgrading. Biomass and Bioenergy 38: 68–94.

Capareda, S. C. 2014. Introduction to Biomass Energy Conversion. CRC Press, Taylor and Francis Group, Boca Raton, FL.

CRS-Guatemala. 2014. Rapid Economic Feasibility Study in Guatemala: Cardamom, Cinnamon, Macadamia Nut and Nutmeg. Final Report. May 2014. Catholic Relief Services (CRS) Guatemala, Baltimore, MD.

Hernandez, J. 2009. Simultaneous saccharification and fermentation of dry-grind highly digestible grain sorghum lines for ethanol production. Master of Science Thesis, Office of Graduate Studies, Texas A&M University, College Station, May.

Kresovich, S., C. K. Wagner, D. A. Scanland, S. S. Groet and W. T. Lawton. 1982. The utilization of emergent aquatic plants for biomass energy systems development. USDOE Report Under Contract No. EG-77-C-01-4042. USDOE, Golden, CO.

Lane, J. 2018. EPA lifts renewable fuels in 2018–2010 proposal: industry welcomes volumes, warns on refinery waiver impact. Biofuels Digest. June 27. Available at: http://www.biofuelsdigest.com/bdigest/2018/06/26/epa-lifts-renewable-fuels-in-2018-2020-proposal-industry-welcomes-volumes-warms-on-refinery-waiver-imapct/. Accessed June 27, 2018.

Maglinao, A. L., Jr. 2009. Instrumentation and evaluation of a pilot scale fluidized bed biomass gasification system. Master of Science Thesis, Biological and Agricultural Engineering, Office of Graduate Studies, Texas A&M University, College Station, May.

Maglinao, A. L., Jr. and S. C. Capareda. 2010. Predicting fouling and slagging behavior of dairy manure (DM) and cotton gin trash (CGT) during thermal conversion. Transactions of the ASABE 53 (3): 903–909.

Maramba, F. D. 1978. Biogas and Waste Recycling: The Philippine Experience. Regal Printing Company, Manila, Philippines.

Rooney, W. R., J. Blumenthal, B. Bean and J. E. Mullet. 2007. Designing sorghum as a dedicated bioenergy feedstock. Biofuels, Bioproducts and Biorefining 1: 147–157.

Rooney, W. R. 2015. Personal Communication. Texas A&M University, College Station.

Shapouri H., M. Salassi and J. N. Fairbanks. 2006a. The economic feasibility of ethanol production from sugar in the United States. USDA Report for the Office of Energy Policy and New Uses (OEPNU), Office of the Chief Economics (OCE), Louisiana State University and the U.S. Department of Agriculture, Washington, DC. July.

Shapouri, H., M. Wang and J. A. Duffield. 2006b. Net energy balancing and fuel-cycle analysis. In: (Dewulf, J. and H. Van Langenhove (Editors). Renewable-Based Technology Sustainability Assessment. John Wiley & Sons, New York.

Santos, B. S. and S. C. Capareda. 2015. A comparative study on the engine performance and exhaust emissions of biodiesel from various vegetable oils and animal fat. Journal of Sustainable Bio-Energy Systems 5: 89–103.

Stout, B. A. 1984. Energy Use and Management in Agriculture. Breton Publishers, North Scituate, MA.

Tramper, J. and Y. Zhu. 2011. Modern Biotechnology: Panacea or New Pandora's Box? Wageningen Academics Publishers, Wageningen, Netherlands.

U.S. EPA. 2014. Municipal solid wastes generation, recycling and disposal in the United States. Facts and figures for 2012. Available at: https://www.epa.gov/sites/production/files/2015-09/documents/2012_msw_fs.pdf. Accessed May 20, 2019.

USDA. 2014. Cardamom—The 3Gs—Green Gold of Guatemala. USDA Foreign Agricultural Service. Global Agricultural Information Network. June 29. GAIN Report No. GT-1404. Washington, DC.

USDOE and USDA. 2009. Sustainability of biofuels: future research opportunities. Report from the October 2008 Workshop, DOE/SC-0114, U.S. Department of Energy Office of Science and U.S. Department of Agriculture. Available at: https://genomicscience.energy.gov/biofuels/sustainability/. Accesses May 20, 2019.

Waranica, G. 2007. Equipment and technology for biodiesel production. Lecture presented at the 4th Practical Short Course on Biodiesel and Industrial Applications of Vegetable Oils, August 19–22, Texas A&M University, Organized by the Food Protein Research and Development Center (FPRDC), Texas Engineering Experiment Station, College Station.

Willits, J. 2007. Biodiesel plant design: planning, site selection and critical parameters. Lecture presented at the 4th Practical Short Course on Biodiesel and Industrial Applications of Vegetable Oils, August 19–22, Texas A&M University, Organized by the Food Protein Research and Development Center (FPRDC), Texas Engineering Experiment Station, College Station.

5

Hydro Power

Learning Objectives

Upon completion of this chapter, one should be able to:

1. Describe the hydrologic cycle.
2. Enumerate the various ways of generating useful energy and power from water resources.
3. Estimate available power from water bodies.
4. Describe the various impellers used for many water conversion devices.
5. Describe the various conversion efficiencies.
6. Describe the hydraulic ram and its design.
7. Relate the overall environmental and economic issues concerning hydro power systems.

5.1 Introduction

Humankind has harnessed power and energy from water bodies over centuries. This renewable energy resource has perhaps been exploited by man on the largest scale. The hydrologic cycle is one of the earth's fundamental processes—water from the ocean evaporates, turns into clouds, and is brought back to land in the form of rain. This cycle is depicted in Figure 5.1. As the earth's temperature rises, snow melts from higher latitudes and creates runoff, feeding lakes and streams with fresh water. Some water percolates through the ground and is temporarily stored in those areas. As water drops from higher elevations, energy-related events occur owing to the kinetic energy of water and the drop in potential energy.

Hydro power supplies about 16% of the world's electricity requirement from all sources (WEC, 2017). Hydro power is the leading renewable energy source for electrical power worldwide and was found to make up 71% of all renewable electricity. The installed capacity (WEC, 2017) by region is as follows and represents the values at the end of 2015. The English equivalent of GW is translated into Quad/yr. One Quad is equivalent to 1×10^{15} Btu or 1.055 EJ (10^{18} Joules):

a. Asia = 511 GW [15.27 Quad/yr]
b. East Asia = 381 GW [11.39 Quad/yr]

The Hydrologic Cycle

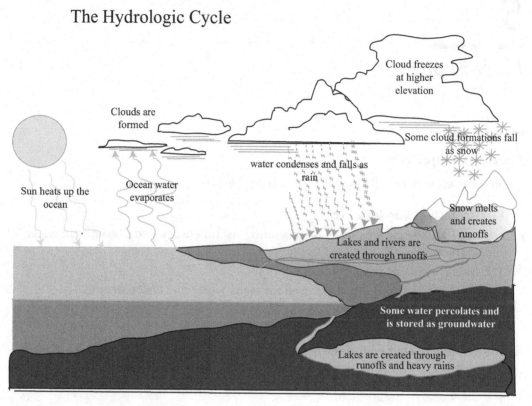

FIGURE 5.1

The hydrologic cycle and the transport of water in a continuous cycle.

   c. Europe = 293 GW [8.76 Quad/yr]

   d. North America = 193 GW [5.77 Quad/yr]

   e. Latin America and the Caribbean = 159 GW [4.75 Quad/yr]

   f. South and Central Asia = 72.3 GW [2.16 Quad/yr]

   g. Southeast Asia and the Pacific = 57.8 GW 1.73 Quad/yr]

   h. Africa = 25.3 GW [0.76 Quad/yr] and

   i. Middle East and North Africa = 20.6 GW [0.62 Quad/yr]

China is the leading nation in hydro power generation, followed by the United States, Brazil, Canada, India, and then Russia. Statistical data vary from different sources, but the general trend remains. One does have to be very careful in reporting data such as installed capacity and power output based on capacity factor. As mentioned in the first chapter, China had the largest installed capacity in its Three Gorges Dam unit but garnered a lower power output than a smaller facility in Brazil.

In the United States, there are over 2,000 hydro power plants with a total capacity of over 79,985 MW. The top five states are Washington (27%), California (13%), Oregon (11%), New York (6%), and Alabama (4%). The largest hydro power plant in the United States is the 7,600 MW Grand Coulee Power Station along the Columbia River in Washington State (EIA, 2018).

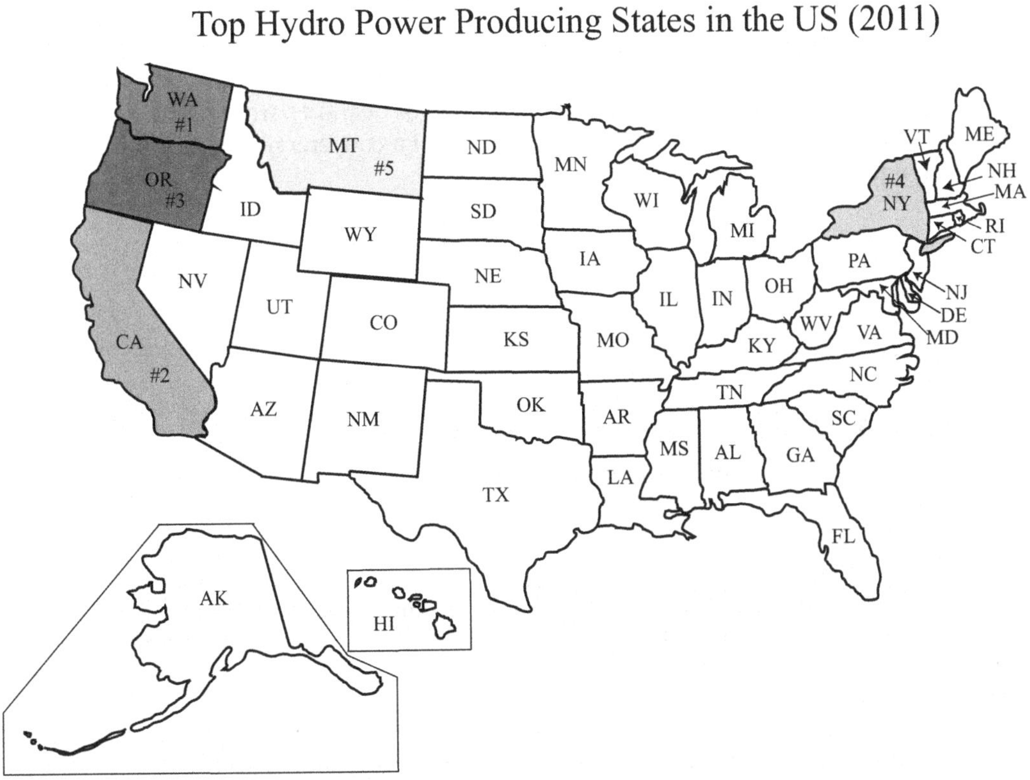

Top Hydro Power Producing States in the US (2011)

FIGURE 5.2
Top hydro power–producing states in the United States in 2011.

The top hydro power–producing state in the United States is Washington, followed by California, Oregon, and New York. This is depicted in Figure 5.2.

Hydro power is the renewable energy source that produces the most electricity in the United States. Power from water accounted for 6% of the total U.S. electricity generation and 60% of power generated from renewables in 2010. This supplies 28 million households with an electricity equivalent to nearly 500 million barrels of oil. One barrel of oil is equivalent to 42 U.S. gallons [159.0 liters]. The United States consumes around 18.8 million barrels of crude oil per day (EIA, 2018).

5.2 Power from Water

The theoretical power derived from water is a consequence of the potential energy of water as it is dropped from a given height. The basic equation simply finds the product of volumetric flow rate (kg/s in metric system) and the dynamic head (in meters) as shown in Equation 5.1. Example 5.1 shows the units used in Equation 5.1 to come up with the unit of power in watts. Note that the acceleration due to gravity of 9.8 m/s² [32.2 ft/s²] must be used as well as Newton's Law relationship for force (in Newton), mass (in kg or lb mass), and acceleration (in m/s² or ft/s²). Note that in the example, the unit of watts is quite a small number, and it is more convenient to use kW, which is 1,000 watts [1.34 hp]. Subsequent formulas are presented to calculate theoretical hydro power (in kW or hp) using other units and other combinations such that a constant is introduced to make the calculation much easier. In these sets of formulas, various units of Q and H are presented with their corresponding constants. Thus, one will simply select a convenient unit of choice and the particular equation that is applicable:

$$Theoretical\ Power(Watts) = Q \times H \qquad (5.1)$$

where
  Q = volumetric flow rate (kg/s)
  H = dynamic head (meters)

Example 5.1: Theoretical Power from Water Streams

Determine the theoretical power in watts for water flowing at 200 kg/s and dropped from a dynamic head of 20 meters.

SOLUTION:

a. The use of Equation 5.1 is straightforward, making use of the acceleration due to gravity constant of 9.8 m/s² as well as Newton's Law for the units of force as a function of mass and acceleration:

$$P_t(Watts) = \frac{200\ kg}{s} \times \frac{20\ m}{1} \times \frac{N-s^2}{kg-m} \times \frac{9.8\ m}{s^2} \times \frac{W-s}{N-m} = 39,200\ Watts$$

$$P_t(hp) = 39,200\ W \times \frac{hp}{746\ W} = 52.55\ hp$$

b. Hence, this amount of water has a potential energy of around 39,200 watts [52.55 hp].

Other Useful Equations for Calculating Hydro Power

a. The theoretical power in kW using the metric system of units:

$$P_t(kW) = \frac{Q \times H}{102} \qquad (5.2)$$

In this equation, Q is in units of kg/s and H in meters.

b. The actual power in kW:

$$P_a(kW) = \frac{Q \times H \times E_o}{102} \qquad (5.3)$$

In this equation, the overall conversion efficiency (E_o) is introduced. This conversion efficiency accounts for mechanical and electrical losses as well as the type of conversion device used. The constant 102 is derived from using the acceleration due to gravity of 9.8 m/s² as well as the conversion from watts to kW as shown below:

$$\frac{1}{102} = \frac{kg}{s} \times \frac{m}{1} \times \frac{N-s^2}{kg-m} \times \frac{9.8\ m}{s^2} \times \frac{W-s}{N-m} \times \frac{kW}{1,000\ W}$$

c. The theoretical power using the English system, with horsepower (hp) as the unit:

$$P_t(hp) = \frac{Q \times H}{33,000} \tag{5.4}$$

where
  Q = volumetric flow rate (lbs/min)
  H = dynamic head (ft)

In this equation, the unit of conversion of 33,000 is the equivalent of a horsepower; that is, there are 33,000 ft-lb/min in a horsepower. Recall horsepower as first defined by James Watts—he observed a horse walking at a speed of 2.5 miles per hour, carrying a load of 150 lbs, and called this 1 horsepower. The calculation is shown below.

$$hp = \frac{150\ lbs}{} \times \frac{2.5\ miles}{hr} \times \frac{5,280\ ft}{mile} \times \frac{1\ hr}{60\ min} = 33,000 \frac{ft-lb}{min}$$

d. The actual power in the English System with hp as the unit. In this equation, the overall conversion efficiency E_o is introduced:

$$P_a(hp) = \frac{Q \times H \times E_o}{33,000} \tag{5.5}$$

e. Alternate equation with volumetric flow rate Q in gallons per minute (gpm) and H in feet:

$$P_t(W) = \frac{Q \times H}{5.3} \tag{5.6}$$

f. Alternate actual power equation with Q in gpm and H in ft but the units of power in watts:

$$P_a(W) = \frac{Q \times H \times E_o}{5.3} \tag{5.7}$$

Note that conversion 5.3 is calculated from the illustration below:

$$\frac{1}{5.3} = \frac{gallons}{min} \times \frac{8.34\ lbs}{gallon} \times \frac{feet}{} \times \frac{hp-min}{33,000\ ft-lb} \times \frac{746\ W}{hp}$$

5.3 Inefficiencies in Hydro Power Plants

The energy losses in a hydro power plant are mainly due to the following:

a. Hydraulic loses in conduits and turbines
b. Mechanical losses in bearings and the power transmission system
c. Electrical losses in the generator, station use, and transmission (for hydroelectric power)

Losses due to friction account for the main energy reduction from the theoretical power (through conduit pipes and turbine blades). There are also mechanical losses in bearings and power transmissions as energy is converted from one form to another and as it is conveyed to its final applications. Finally, there are electrical losses in generators and transmission lines. The overall effect reduces the theoretical power by a factor of 0.60 to 0.80. Large-scale hydro power plants have overall conversion efficiencies of 60% to 80%. These values are much higher than those of conventional internal combustion engines and coal power plants. Example 5.2 shows the efficiency of a small commercial micro–hydro power unit, which demonstrates efficiencies much lower than those in large-scale hydro power plants.

Example 5.2: Efficiency of Micro–Hydro Power Systems and Yearly Output Energy

A small commercial micro–hydro power unit had the following data: the volumetric flow rate was 35 liters/s [555 gpm] and the dynamic head was 1.5 meters [4.92 ft]. This unit was reported to produce 200 watts [0.268 hp] of continuous power output at those conditions. Determine the actual efficiency of this unit. Also calculate the yearly output of this small device in units of kWh/year, assuming 24 hours per day and 365 days in a year.

SOLUTION:

a. The volumetric flow rate is converted into units of kg/s:

$$Q\left(\frac{kg}{s}\right) = \frac{35\ L}{sec} \times \frac{1\ kg}{L} = 35\frac{kg}{s}$$

$$Q\left(\frac{lbs}{min}\right) = 35\frac{kg}{s} \times \frac{2.2\ lbs}{1\ kg} \times \frac{60\ s}{min} = 4,620\frac{lbs}{min}$$

b. Then, the theoretical power is calculated:

$$Power(W)_t = Q \times H = \frac{35\ kg}{sec} \times \frac{1.5\ m}{} \times \frac{N-s^2}{kg-m} \times \frac{9.8\ m}{s^2} \times \frac{W-s}{N-m} = 514.5\ Watts$$

$$Power(hp)_t = 514.5\ W \times \frac{hp}{746\ W} = 0.69\ hp$$

c. The efficiency is calculated using the ratio of actual and theoretical power as shown:

$$\eta(\%) = \frac{Actual\ Power}{Theoretical\ Power} \times 100\% = \frac{200\ W}{514.5\ W} \times 100\% = 38.9\%$$

d. The overall conversion efficiency of this unit is well below 60%.

e. The yearly output may then be estimated as follows:

$$Yearly\ Output\left(\frac{kWh}{yr}\right) = 200\ W \times \frac{24\ hrs}{day} \times \frac{365\ days}{year} \times \frac{kW}{1,000\ W} = 1,752\frac{kWh}{yr}$$

$$Yearly\ Output\left(\frac{Btu}{yr}\right) = 1,752\frac{kWh}{yr} \times \frac{3412\ Btu}{1\ kWh} = 5,977,824\frac{Btu}{yr} \cong 6\frac{Million\ Btu}{yr}$$

5.4 Basic Components of a Hydro Power Plant

The basic components of a hydro power plant are shown in Figure 5.3. The components are enumerated below:

a. Water reservoir
b. Dam
c. Intake and control gate
d. Penstock
e. Turbine
f. Generator
g. Power house
h. Transformer
i. Outflow
j. Power lines

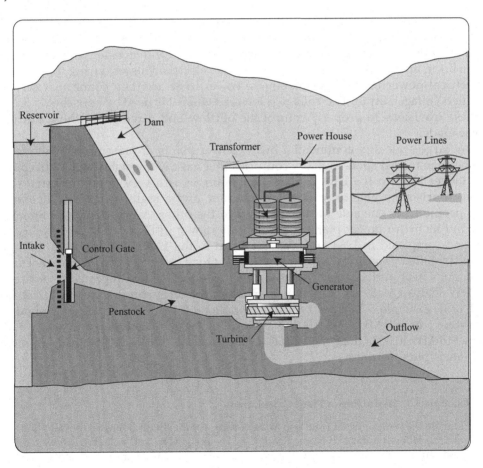

FIGURE 5.3
Basic components of a hydro power plant.

The most important component of a hydro power plant is the reservoir for the water as power is generated. The first step in planning for the installation of a hydro power plant is to evaluate the stream flow in a given water body, such as a lake or stream. The particular water body must provide continuous flow and charge of water over several years and over the life of the facility. Once this is established, a barrier or a dam may be built. The dam must be sturdy enough to withstand the force of water, including additional loads of silt and possible flood flow or water surges. Simulation studies and models are developed prior to construction. For example, during the design process of the Hoover Dam in the United States, a miniature scale model was developed using mercury as the fluid medium. Mercury, which is 13 times heavier than water, was used to simulate loads on the dam to be built, and the miniature model became a means to refine the dam's infrastructure. A hydro power plant must also have an intake port through which incoming water is diverted to the turbine. This intake port must have a control gate to ensure proper control of incoming water. To prevent damage to the turbine blades, provisions must be built to keep the incoming water to a standard of quality. Some debris, such as rocks, stones, and other plant materials, must not be allowed to enter the penstock or the conveying duct to the turbine blades. The turbine impeller, once moved by incoming water, will rotate. Coupled to a generator, this should immediately produce electrical power. The generator should be housed properly and safely protected (in a power house) while still accessible and easily maintained. In large-scale power plants, the start-up typically takes less than 30 minutes. Generator power output is usually of lower voltage. A transformer is usually necessary to bring up the voltage in order to minimize electrical losses during the transport of electrical power over longer distances. Power lines are then connected, carrying very high voltage output. The voltage is lowered when it is used by households or local utilities. Provisions to properly control the outflow and direction of the water must also be made.

Many large-scale and commercial hydro power plants are fascinating to visit and have become tourist spots in many countries that have well-established hydro power. The Hoover Dam in the United States, for example, is a very popular tourist destination. Tourists are able to walk along the dam, and a highway crosses the dam to bring people from one side of the reservoir to the other. A photograph of Bhumibol Dam and its hydro power plant in Thailand is shown in Figure 5.4. This facility was built and completed in 1964 and was upgraded in the 1990s. The reported total generating installed capacity was 749 MW [0.022 Quad/yr] of electrical power. During the author's visit in the early 1980s, the power production was reported to be around 535 MW [0.016 Quad/yr], generating an annual energy production of 2.2 B kWh [7.5 T Btu]. Example 5.3 shows typical calculations for this facility. The original turbines were 6 to 76.5 MW [6 to 0.002 Quad/yr] Francis-type turbines, and newly added were a 115 MW [0.003 Quad/yr] Pelton turbine and a 175 MW [0.005 Quad/yr] reversible Francis pump-turbine.

Example 5.3: Hydro Power Plant Calculations

Estimate the power output from one of the turbines of the Bhumibol hydro power plant with the following data: dynamic head = 150 meters [492 ft] and volumetric flow rate in one turbine measured as 65,025 kg/s [143,055 lbs/s] Assume an overall conversion efficiency of 80%.

FIGURE 5.4
The Bhumibol Dam and hydro power plant in Thailand.

SOLUTION:

a. Since the units for Q and H are those defined in Equation 5.2, the calculation of theoretical power is straightforward as shown:

$$P_t(kW) = \frac{QH}{102} = \frac{65,025\ kg}{sec} \times \frac{150\ m}{102} = 95,625\ kW\ [95.625\ MW]$$

$$P_t(hp) = 95,625\ kW \times \frac{1,000\ W}{1\ kW} \times \frac{hp}{746\ W} = 128,183\ hp$$

b. Then, using the overall conversion efficiency, the actual power is calculated:

$$P_a(MW) = 95.625\ MW \times 0.80 = 76.5\ MW$$

$$P_a(hp) = 76.5\ MW \times \frac{1,000,000\ W}{1\ MW} \times \frac{hp}{746\ W} = 102,684\ hp$$

c. This power is the output of only one of the six turbines in the Bhumibol Dam hydro power plant. The six turbines would generate a total power of 459 MW [0.014 Quad/yr], close to the original reported values.

5.5 Water Power–Generating Devices

There are three general types of water power–generating devices. The first set of devices include hydraulic rotating prime movers such as water wheels, tub wheels, and turbines. The other two types are mechanical devices such as hydraulic rams and hydraulic air compressors.

5.5.1 Water Wheels and Tub Wheels

There are three types of water wheels: the overshot, the undershot, and the breast wheel. These are horizontal-axis water wheels used in earlier periods when simple rotating mechanical devices needed actuators. In overshot water wheels, the water is directed on top of the water-turning device. An undershot wheel has water directed to the bottom of the water-powering device. Breast wheels have water directed to the middle of the rotating device. Tub wheels are vertical-axis water wheels, with the impellers usually placed along the path of the river or stream. The rotating shaft protrudes vertically so that mechanical devices may be attached for whatever purpose. In the past, water wheels were used to power households for mechanical and rotating actions needed, such as for grain milling and grinding. Older sawmills in the United States also utilized water wheels for processing some agricultural products.

5.5.2 Turbines

There are two types of turbines:

   a. Impulse turbines
   b. Reaction turbines

In impulse turbines, the momentum (or impulse) of water jets is used to move the turbine blades, resulting in a rotary motion of the shafting where the blades are attached. Since the turbine is spinning, the force acts over a distance (work), and the diverted water flow is left with diminished energy. Newton's Second Law (force = mass × acceleration) describes the transfer of energy in impulse turbines. Impulse turbines are most often used in very high head applications. Prior to hitting the turbine blades, the water's pressure (its potential energy) is converted into kinetic energy by a nozzle focused on the turbine. No pressure change occurs in turbine blades, and the turbine actually does not require housing for this operation.

Reaction turbines, on the other hand, are acted on by water, which changes pressure and gives up energy as it moves through the turbine. The turbines must be either encased to contain the water pressure (or suction) or fully submerged in the water flow. Newton's Third Law (law of reciprocal reactions or conservation of momentum) describes the transfer of energy for reaction turbines. Most reaction turbines are used in low and medium head applications.

The most common types of water turbines for power generation are enumerated below, including their description, range of heads, as well as specific speeds:

   a. Francis Turbine. The majority of large hydroelectric power plants use Francis turbines. Typical head ranges for this turbine type are 300 to 500 meters [984 to 1,640 ft]. Specific speeds for these turbines range from 60 to 400 rpm.
   b. Kaplan Runner. The Kaplan runner is a propeller-type turbine, similar to those found in axial flow fans or rotors for pump boats. The dynamic head for Kaplan runners ranges from 2 to 70 meters [6.56 to 229.6 ft]. Its specific speeds range from 300 to 1,100 rpm.
   c. Pelton Runner. This is an example of an impulse turbine with blades that look like cups. The water jet is impinged, moving the blades in the opposite direction and turning the shafting where the blades are attached. Pelton runners are very common

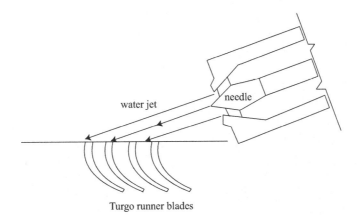

water jet

needle

Turgo runner blades

FIGURE 5.5
A schematic of Turgo wheel action.

standby emergency power platforms for hydro power plants. They are usually fed with a pipe of around a fraction of a meter (about 1 ft) and may generate about a MW or a fraction of a MW. The dynamic head for Pelton runners ranges from 50 to 1,300 meters [164 to 4,264 ft]. The specific speeds range from 4 to 70 meters.

d. Cross Flow Turbines. Examples of cross flow turbines are the Banki turbine and Mitchell or Ossberger turbine. These turbine types use nozzles and blades instead of buckets or cups as in the Pelton runner. These turbines are very similar to those of overshot water wheels.

e. Turgo Wheels or Runners. Figure 5.5 is a schematic of a Turgo wheel's mechanism. These runners are usually at the bottom of the hydro power structure, and the shafting that holds the blades is vertically oriented. The generator is usually on top and the nozzle inlet situated in between the generator and the blades at an angle, as shown.

5.5.3 Specific Speeds for Turbines

Specific speeds (N_s) are used to compare different impellers. This is a dimensionless parameter that specifies a type of impeller as well as suggests what type of turbine to use. The relationship is given in Equation 5.8 using the metric system:

$$N_s = \frac{\omega q^{0.5}}{\left(gh\right)^{0.75}} \tag{5.8}$$

where
  ω = pump shaft rotational speed (radians per second)
  q = flow rate at the point of best efficiency (m³/s)
  h = total head (m)
  g = acceleration due to gravity (9.8 m/s²)

The specific speed describes the speed of the turbine at its maximum efficiency with respect to the power and flow rate. The specific speed is derived to be independent of turbine size. If one is given the fluid flow conditions and the desired shaft output speed, the specific speed can be calculated and an appropriate turbine design selected.

The specific speed, using the fundamental formulas, can be employed to reliably scale an existing design of known performance to a new size with its corresponding performance. Example 5.4 shows how the specific speed is calculated.

Example 5.4: Specific Speeds of Pumps and Turbines

Given a pump rotational speed of 1,760 rps, flow rate of 1,500 m³/s, and head of 100 m, calculate the specific speed.

SOLUTION:

a. Equation 5.8 is used:

$$N_s = \frac{\omega q^{0.5}}{(gh)^{0.75}} = \frac{1,760 \times 1,500^{0.5}}{(9.8 \times 10)^{0.75}} = 2,188$$

$$N_s = \frac{1760 * 1500^{(1/2)}}{(9.8 \times 10)^{(3/4)}} = 2188$$

b. This is the specific speed of a typical centrifugal pump.

The specific speeds for various types of pumps and impellers are as follows: radial flow turbines have specific speeds ranging from 500 to 4,000, mixed flow turbine speeds range from 2,000 to 8,000, and axial flow turbines have specific speeds ranging from 7,000 to 20,000. Radial flow turbines include centrifugal impeller pumps with radial vanes. The turbine blades could be single- or double-suction. The Francis vane also allows for a higher specific speed range. Axial flow turbines are used along with propeller-type axial fans or blades. Figure 5.6 shows various blade shapes against various head specifications. Low head applications usually employ ultra-fast turbines, while high head applications are paired with devices designed for slow specific speeds, such as Francis turbines.

5.5.4 Turbine Selection

A monograph is usually used to plan a project and select a turbine. An example is shown in Figure 5.7. This figure shows the combination of head (in meters) and flow (in m³/s). Each enclosed gray colored region displays the type of turbine recommended. As shown in the figure, Francis turbines are used for up to 1,000 MW [0.03 Quad/yr] of power, with heads ranging from 10 to 1,000 meters [32.8 to 3,280 ft]. Kaplan turbines can have as much as close to 100 MW [0.003 Quad/yr] of power, with flows ranging from 1 to 1,000 m³/s [15,837 to 15,837,053 gpm]. Pelton turbines have heads of 1,000 meters [3,280 ft] and flows up to 60 m³/s [950,223 gpm].

Turgo turbines have from 1 to 10 m³/s [15,837 to 158,371 gpm] of flow and around 50 to 250 meters [164 to 820 ft] of head. Turgo turbines have much lower head but a similar range of flow. The power output of Turgo wheels is higher than that of cross flow turbines.

Of the described types, Francis and Pelton turbines have the highest potential power output. They are most suited to large-scale hydroelectric power plants.

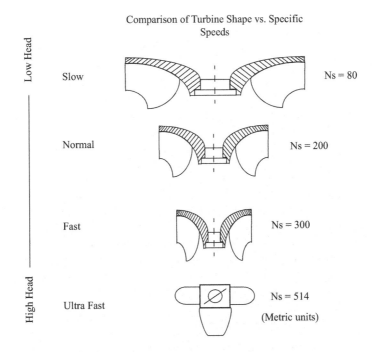

FIGURE 5.6
Blade configuration comparison for different specific speeds.

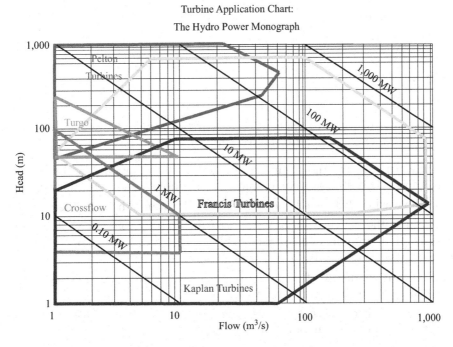

FIGURE 5.7
Turbine application chart: a monograph for turbine type selection.

5.6 Hydraulic Ram

A hydraulic ram is a water power–generating machine that mainly serves to bring a certain volume of water from its source to a much higher elevation. Hydraulic rams are efficient, simple, and durable to use in locations with favorable topographical conditions. The energy stored in descending water is itself utilized to raise a portion of the water to a much higher level—2 to 20 times the height of the original fall.

The machine was originally invented in 1772 by John Whitehurst of Cheshire in the United Kingdom. The device was manually controlled and was originally considered as some sort of pulsation engine. With his prototypes, Whitehurst was able to raise water to a height of about 5 meters [16 feet]. In 1873, he was also able to install another unit in Ireland but failed to patent his work. It was Joseph Michael Montgolfier (who was, in fact, a co-inventor of the hot air balloon) who invented an automatically operated system in 1796. A friend of Montgolfier's availed of a British patent on his behalf in 1797. The automated ram uses a water hammer effect to develop pressure, which allows a portion of input water to be lifted to a point higher than the primary source. The unit ingeniously requires no outside power source besides the kinetic energy of flowing water.

5.6.1 Construction and Principles of Operation

A hydraulic ram has two major parts: the waste valve and the delivery valve. The waste valve, also known as the "clack" valve because of the sound it makes, is either spring-loaded or weight-loaded. Through the delivery valve, water is conveyed from its source to a much higher elevation. Hydraulic rams are cheap to build, easy to maintain, and quite reliable. The basic components are the following:

a. Inlet drive pipe

b. Outlet delivery pipe

c. Waste valve

d. Delivery check valve

e. Pressure vessel

Figure 5.8 shows the basic components of a hydraulic ram and how the components are arranged. The sequence of operations is as follows:

1. As the water flows through the inlet pipe, the counterweight is raised until the waste valve closes. This does spill some water—an amount considered as an initial loss.

2. With the waste valve closed, the water is forced through the outlet delivery pipe.

3. This opens the delivery check valve, and the water goes through the pressure vessel and out into the delivery pipe, which raises the water to a higher elevation. Excess pressure is absorbed by the pressure vessel.

4. This water is continually forced, and with increasing pressure against gravity, the flow slows down and ultimately reverses.

5. The delivery check valve then closes, and excess force from the pressure vessel aids the flow of water out of the delivery pipe.

6. Gravity brings the counterweight down temporarily, but it is ultimately raised because of the incoming water pressure from the source. This cycle is repeated.

Basic Components of a Hydraulic Ram

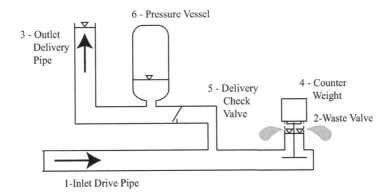

FIGURE 5.8
Basic components of a hydraulic ram.

Note that the pressure vessel may be removed from the system without affecting the cycle of flow. However, this would reduce the device's overall efficiency. The pressure vessel allows for uniform flow through the delivery pipe.

Figure 5.9 demonstrates the typical installation and complete configuration of a hydraulic ram in a field. The numbered components are the following:

1. River or source of water
2. Feed pipe
3. Feed tank
4. Drive pipe
5. Ram shelter

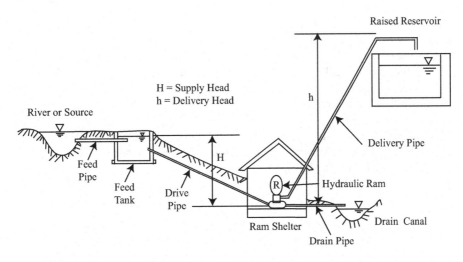

FIGURE 5.9
Typical hydraulic ram installation.

6. Hydraulic ram
7. Drain pipe
8. Drain canal
9. Delivery pipe
10. Raised reservoir

As detailed in the figure, note that there is a downward slope from the river or source (1) to the hydraulic ram unit. This is labeled as H. While it is true that a horizontal pipe would work in a hydraulic ram installation, the slope with the dynamic head H can help sustain the flow of water, giving the water the kinetic energy it needs to rise to a higher elevation. A feed pipe (2) is important such that a continuous flow of water enters the feed tank (3). The feed tank acts as an equalizer tank that ensures that the dynamic head H is always constant. One can also note that the delivery pipe is quite long. This is typical of hydraulic rams, and it would be difficult to install a workable unit with the lack of ample space. The hydraulic ram (6) must then be placed in a shelter (5) to be protected from storms and water surges. Since the ram will have a continuous overflow of wastewater, some means to collect it must be in place. A drain pipe (7) must direct the wastewater to a drain canal (8), which ultimately returns the wastewater to a lower elevation downriver. The delivery pipe (9) has a smaller diameter than the drive pipe, but it is sometimes greater in length. Finally, a water reservoir must be installed at a raised elevation.

5.6.2 Hydraulic Ram Calculations

In calculating energy, volumetric efficiency, and the amount of water that can be raised, it is important to remember the nomenclature shown in Figure 5.10. The efficiency, also called D'Aubuisson's efficiency, for a hydraulic ram is given by Equation 5.9.

$$E_{DA} = \frac{\rho q h}{\rho (Q+q) H} \times 100\% = \frac{q h}{(Q+q) H} \times 100\% \tag{5.9}$$

where
E_{DA} = the D'Aubuisson's efficiency
ρ = the density of water (1,000 kg/m^3)
q = the volume of water delivered by the hydraulic ram (cubic meters, m^3)
Q = the volume of water taken from the source (cubic meters, m^3)
h = the effective delivery head (meters, m)
H = the supply head (meters, m)

The volumetric efficiency and the water use efficiencies are given in Equations 5.10 and 5.11, respectively. Examples 5.5 and 5.6 show the difference between the volumetric efficiency and energy efficiency of hydraulic rams:

$$E_v = \frac{q}{(Q+q)} \times 100\% \tag{5.10}$$

$$E_w = \frac{q h}{Q H} \times 100\% = \frac{energy\ of\ water\ delivered}{energy\ of\ water\ source} \times 100\% \tag{5.11}$$

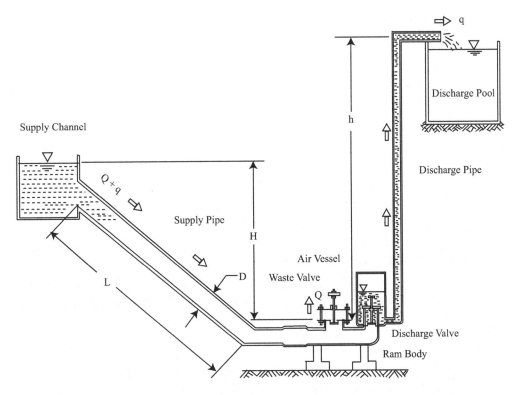

FIGURE 5.10
Nomenclatures for the design and calculation of hydraulic ram efficiencies.

The typical overall efficiency of hydraulic ram ranges from 60% to 80%. These values are not to be confused with volumetric efficiency. Volumetric efficiency relates the amount of water delivered at a given height to the water taken from source; this could be around only 20%. If, for example, the source is 10 meters [33 ft] above the ram and the water is lifted 20 meters [66 feet] above the ram, only about 20% of the supplied water is available. The other 80% is expended through the waste valve. The actual amount of water delivered is further reduced by friction losses in the valves.

Example 5.5: Volumetric Efficiency of Hydraulic Rams

Calculate the volumetric efficiency of a hydraulic ram with water delivered to a higher elevation at the rate of 3,785 liters per minute [1,000 gpm]. A volume of 19 cubic meters per minute [5,015 gpm] is taken and diverted from the water source.

SOLUTION:

a. Equation 5.10 is used as shown below:

$$E_v = \frac{q}{(Q+q)} \times 100\% = \frac{3,785}{(19,000+3,785)} \times 100\% = 16.6\%$$

b. The volumetric efficiency is less than 20%.

Example 5.6: Energy Efficiency of Hydraulic Rams

Calculate the energy efficiency of a hydraulic ram with a volume delivered to a higher elevation at a rate of 3,785 liters per minute [1,000 gpm], with a vertical lift of 20 meters [65.6 ft] and a horizontal fall of 6 meters [19.68]. A volume of 19 cubic meters per minute [5,015 gpm] is taken and diverted from the water source [19,000 liters per minute].

SOLUTION:

a. Equation 5.11 is used as shown below:

$$E_w = \frac{qh}{QH} \times 100\% = \frac{3,785 \times 20}{19,000 \times 6} \times 100\% = 66.4\%$$

b. The energy efficiency is more than 60%.

5.6.3 Design Procedures for Commercial Rife Rams

In designing commercial hydraulic rams, such as Rife rams, in the United States (Nanticoke, Pennsylvania), the following information is necessary. Refer to Figure 5.11 for the description of nomenclatures:

1. The flow of water at the source of supply (in gpm or Q in L/s)
2. The vertical fall (in feet) from the source to ram (or supply head)
3. The vertical elevation above the ram to which water is to be pumped
4. The distance between the point of supply and ram location and the length of the pipe line over which the water is to be delivered
5. The number of gallons required per day

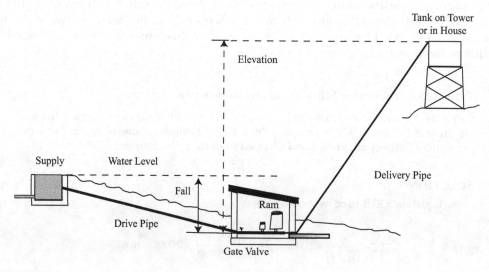

FIGURE 5.11
Nomenclatures for the design of commercial Rife rams.

TABLE 5.1

Performance Chart for Rife Hydraulic Ram

Fall (ft)	Vertical Lift (in ft) Including Delivery Pipe Friction								
	8	16	25	50	75	100	150	200	500
4	22.5%	12.5%	8.0%	3.6%	1.6%				
8		22.5%	13.2%	9.6%	6.4%	4.8%	2.7%	2.0%	
12			21.5%	13.2%	9.6%	7.2%	4.8%	3.3%	
16				16.0%	11.7%	9.6%	6.4%	4.8%	
20				18.0%	14.7%	12.0%	8.0%	6.0%	2.0%
25				22.5%	16.7%	13.8%	10.0%	7.5%	2.5%
30					18.0%	15.0%	12.0%	9.0%	3.3%
35					21.0%	17.5%	14.0%	10.5%	4.2%
40						18.0%	14.7%	12.0%	4.8%
50						22.5%	16.7%	13.8%	6.0%

Source: Rife Hydraulic Engineering Manufacturing Co., Nanticoke, PA.

5.6.4 Specifying Pipe Sizes and Discharge Pipe Lengths

The empirical relationship for specifying drive pipe size and length is given in Equation 5.12. This is suggested by Krol (1977). Example 5.7 shows how this equation is used, including ballpark figures for pulsations per minute:

$$L = \frac{900\,H}{N^2 D} \tag{5.12}$$

where
   L = length of the drive pipe in meters
   H = supply head in meters
   N = the number of pulsations per minute
   D = the diameter of the drive pipe in meters

Table 5.1 shows a performance chart for a commercial Rife hydraulic ram. The chart shows the relationship between the horizontal fall against the vertical lift and the corresponding percentage of water delivered. Table 5.2 shows the recommended values for the volumetric flow rate (in liters per second) for various combinations of pipe diameter (in mm) and supply head in meters. Example 5.8 shows how these tables are to be used for the design and specification of hydraulic rams made by Rife.

TABLE 5.2

Recommended Values of Drive Pipe Diameter and Supply Head for a Given Range of Volumetric Flow Rates

Diameter, D (Drive Pipe)	Supply Head (H) in Meters					
	1–2	2.01–5	5.01–10	10.01–20	20.01–30	30.01–40
75 mm (3 in)	2	2	3	5	6	7
100 mm (4 in)	3	4	6	8	10	12
150 mm (6 in)	8	10	15	20	25	30
200 mm (8 in)	15	20	30	40	50	60
250 mm (10 in)	25	35	50	60	70	80

Example 5.7: Specifying Drive Pipe Size and Lengths

Use the empirical Equation 5.12 to calculate the drive pipe length L if the supply head is 5 meters [16.4 ft] and the drive pipe diameter is 0.1 meter [~4 in]. The number of pulsations per minute, N, is about 60 pulsations.

SOLUTION:

a. Equation 5.12 is used and the length of drive pipe is calculated as follows:

$$L = \frac{900H}{N^2 D} = \frac{900 \times 5}{60^2 \times 0.10} = 12.5 \ meters \left[41 \ feet \right]$$

b. Note how long the drive pipe is for this application—about 12.5 meters, or over 40 feet—which is a typical length for drive pipes.

Example 5.8: Calculating Drive Pipe Sizes

Calculate the drive pipe size (diameter in inches) for a hydraulic ram that would take water from a source with a maximum flow of around 240 L/min (63.4 gpm). This water will be brought 50 ft above this source [15.24 m]. A vertical fall of 12 ft [3.66 m] is available. Use the Rife ram data for this problem as shown (in the performance chart). Estimate the amount of water that may be delivered (in gpm) as well as the pipe length, assuming N of 45 beats per second.

SOLUTION:

a. Referring to the performance chart for Rife hydraulic rams, about 13.2% of the water may be delivered, or about 8.4 gpm of water (63.4 × 0.132 = 8.4 gpm).

b. Calculate Q in L/s [(240 L/min)/(60 s/min)] = 4 L/s.

c. The diameter of drive pipe will be approximately 4″ from the table.

d. The length of drive pipe can be calculated as follows:

$$L = \frac{900H}{N^2 D} = \frac{900 \times 3.66}{45^2 (0.102)} = 16 \ m$$

5.6.5 Starting Operation Procedure for Hydraulic Rams

The starting procedure for a newly installed hydraulic ram is explained below:

1. The waste valve is typically found in the raised (or "closed") position. To initiate water flow, the valve must be pushed down manually into the "open" position and released.

2. The waste valve must be pushed down repeatedly until it cycles on its own, usually after three or four manual cycles.

3. Air bubbles must be removed from the system.

4. The delivery check valve must start fully closed, then gradually opened to fill the delivery pipe.

5. The process is similar to priming a hand pump—ensuring that the conveying pipes are filled with water with no trapped air bubbles.

5.6.6 Troubleshooting Hydraulic Rams

When a hydraulic ram is not operating properly, some troubleshooting may have to be performed. Possible common issues are the following:

1. The waste valve is improperly adjusted (a common culprit).
2. Too little air is in the pressure valve.
3. Valve blockage due to debris is common if water is not filtered properly.
4. There is low, insufficient water flow from the inlet delivery pipe if the storage tank is downstream of the ram.
5. Plastic (PVC) pipes are used instead of steel pipes—PVC pipes are not actually ideal despite the perception that plastic pipes would provide lower resistance to flow.
6. Occasional restart is common and one must refer to the starting operation list.

5.7 Types of Hydro Power Plant

Hydro power plants may be classified according to their basis of operation (base and peak load) plant capacity (micro-, medium-, and high-capacity and super power plants), head (low, medium, and high head facilities), their hydraulic features (conventional, pumped storage, and tidal type), and construction features (run-of-the river, valley dam type, diversion canal, and high head diversion). These categories will be discussed briefly in the following sections.

5.7.1 On the Basis of Operation

Based upon its type of operation, a hydro power plant is categorized as either a base load plant or a peak load plant. A base load plant operates 24 hours a day, 365 days in a year, and provides continuous power year-round. They are designed to operate at this base load indeterminately. They are neither allowed to produce power higher than the rated design nor permitted to operate way below their rated capacity. Engineers designing these facilities must have several years of running data of their chosen water source, including forecasts for droughts and heavy downpours.

A hydro power plant may also be designed to operate on a peak load basis. It would operate for staggered periods, timed with peak loads during a given season or for a given time of day. Some mega-cities use too much power at the beginning of a day and require a dedicated peak load power plant to address their intermittent needs. Some peak loads are twice the average requirement and would take quite the toll on a power plant not designed to increase production at such a rate. In such cases, peak load power plants are paired with base load power plants. Peak loads occur during particular seasons. For example, the cotton harvesting season in the United States typically necessitates large power requirements for cotton gins. This occurs immediately following the harvest season in the late summer and early fall. Each cotton gin in the United States would require several MW of electrical power during this peak season, putting a strain on local utilities. The balance of power supply is an issue the world over. In the Philippines, for example, peak power is needed during rice harvest—each rice mill requires at least a

MW of electrical power. In areas where local utilities have no additional power plants to accommodate these peak loads, mills are forced to seek independent means. Many rice mills in the Philippines each have their own diesel-fueled power plant, increasing their operational costs and consequently inflating the price of rice. A rice mill in the Philippines would typically interchange three power plants. One facility may have three units of 350 kW [469 hp] generator sets that operate two at a time, allowing one generator set to be on standby if another unit breaks down.

5.7.2 Based on Plant Capacity

Based on plant capacity, hydro power plants may be classified as micro-capacity (<5 MW) [0.00015 Quad/yr], medium-capacity (5 to 100 MW) [0.00015 to 0.003 Quad/yr], high-capacity (100 to 1,000) [0.003 to 0.03 Quad/yr], and super power plant (>1,000 MW) [>0.03 Quad/yr]. In many power plant operations in developing countries, there is a minimum standard as to the amount of standby power allotted toward major key cities and regions—around 5,000 MW [0.149 Quad/yr]. High-capacity hydro power plants typically cannot account for such an amount of power and turn to base load plants for standby power. Note that this setup is arbitrary and that each country may adopt its own system depending upon the country's power requirement and potential. Some developing countries are also able to deploy mini-hydro power plants (capacity of less than a MW each; around 500 kW or lower) [500 kW = 670 hp], and perhaps in the future, nano–hydro power plants (capacity of less than a watt each) will also exist.

5.7.3 Based on Head

Hydroelectric power plants may be classified according to the dynamic head available for power generation. A low head power plant is said to have a head less than 15 meters [49.2 ft], medium head hydroelectric power plants have dynamic heads between 15 and 50 meters [49.2 to 164 ft], and high head plants are those with dynamic heads greater than 50 meters. These classifications are highly arbitrary and are not sacrosanct.

5.7.4 Based on Hydraulic Features

5.7.4.1 Conventional

Conventional hydroelectric power plants generate electrical power using conventional means, taking advantage of the potential and kinetic energy of falling water bodies. Water is neither recycled nor reused for subsequent power production and instead is allowed to take its course down the level. Dams are usually built through rivers and streams. The stored water is used to generate power continuously throughout the year.

5.7.4.2 Pumped Storage Systems

Pumped storage hydro power plants are used in electric power systems to address peak load problems. In mega-cities, power demands skyrocket at the beginning of the work day. During this time, the demand for power is at its maximum, and if there is no excess power available, brownouts and outages can occur. A pumped storage system is built to ensure that when peak power conditions are experienced, it can deliver the needed power. Once the power requirement goes down, the water released to generate power is brought back to storage in preparation for the next peak power demand.

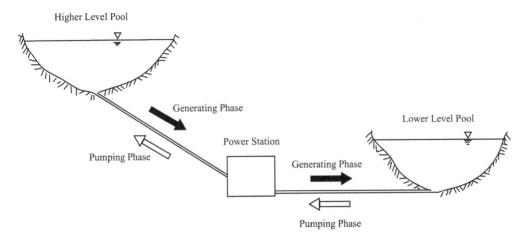

General Arrangement of a Pumped Storage Power Plant

FIGURE 5.12
Schematic for the general arrangement of a pumped storage system showing the higher level pool and the lower level pool.

The basic arrangement for a pumped storage system is shown in Figure 5.12 below. We find a working example of this facility in the Philippines, detailed in Figure 5.13 with a photo in Figure 5.14. The plant has two upper reservoirs—Lumot Lake and Lake Caliraya. Lumot Lake is a natural lake and has a maximum normal elevation of 290 meters [951.2 ft]. It contains water with an average volume of 21×10^6 cubic meters [5,543 million gall]. Lake Caliraya is a man-made lake developed by the National Power Corporation (NPC) of the Philippine government to address the substantial power requirement of metro Manila. This facility is about 65 km [40.4 mi] from metro Manila. The maximum normal elevation of this lake is around 288 meters [944.64 ft], and it has a storage capacity of roughly 22×10^6 cubic meters [5,807 million gall]. A 6-meter [19.68-ft] diameter pipe runs from the lake to the power plant some 291 meters below [954.48 ft]. This water drop is responsible for generating the necessary electrical power during peak hours. Unfortunately, if this is done every day, water would run out at the facility. Examples 5.9 and 5.10 show how long the power generated from this facility lasts as well as how long in hours or days it will last if it is run continuously.

Example 5.9: Pumped Storage Power Production

Determine the power generated from the Kalayaan pumped storage facility if the dynamic head is 290 meters [951.2 ft] and the volumetric flow rate is 280 cubic meters per second [4,434,375 gpm]. Assume overall efficiency of 80%.

SOLUTION:

a. The actual power generated is calculated using Equation 5.3.

b. $P_a(kW) = \dfrac{QH}{102} = \dfrac{280\ m^3 \times 290\ m}{102} \times \dfrac{1,000\ kg}{m^3} \times 0.80 = 636,863\ kW\left[637\ MW\right]$

$P_a(hp) = 636,863\ kW \times \dfrac{1,000\ W}{1\ kW} \times \dfrac{hp}{746\ W} = 853,704\ hp$

c. The Kalayaan pumped storage power Plant generated 620 MW [0.01853 Quad/yr] through two Francis turbines.

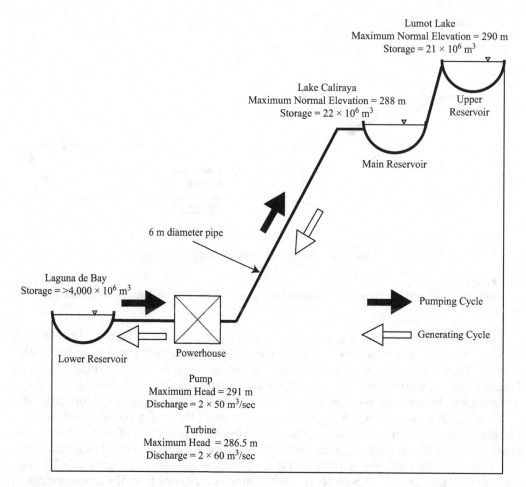

FIGURE 5.13
A pumped storage facility in the Philippines: the Kalayaan pumped storage hydro power plant located in Kalayaan, Laguna.

Example 5.10: Pumped Storage Power Production Water Use

Determine the total maximum time (in hrs and days) for the water from the Caliraya Lake to be emptied when power is generated continuously at the facility. The volumetric flow rate is 280 cubic meters per second [4,434,375 gpm], and Lake Caliraya has a storage volume of 22×10^6 m³ [5,807 million gallons].

SOLUTION:

a. The time (in hours) for Lake Caliraya to be emptied is simply the ratio of the storage volume and the amount of water used for the power plant as shown below:

$$Time\,(hrs) = \frac{Storage\ Volume\left(m^3\right)}{Volumetric\ Flow\ Rate\left(\dfrac{m^3}{s}\right)} \times \frac{1\ hr}{3,600\ s} \times \frac{22 \times 10^6\ m^2}{280\dfrac{m^3}{s}} \times \frac{1\ hr}{3,600\ s} = 21\ hrs$$

FIGURE 5.14
Photos of the Kalayaan pumped storage hydro power plant in the Philippines.

      b. Note that at this storage volume, after 21 hours, water will run out. Water is usually pumped back in the evening in preparation for the next day's event.

Pumped storage hydro power systems are arranged according to type as enumerated below. A schematic of the system types is shown in Figure 5.15:

    a. Recirculating type

    b. Multi-use type

    c. Water transfer type

    d. Tidal type

In the recirculating type, the power is produced during peak load and returned to the upper storage basin at low load—similar to the previous example of the Kalayaan pumped storage hydro power plant in the Philippines. Water is conserved since it is returned to the upper reservoir if there is no recharge for a given day.

In the multi-use type, the water is transferred from a higher elevation dam to a lower elevation dam, as shown in the Figure 5.14. The water may not be returned to the top storage reservoir. The Philippines has numerous types of this plant on its southern island of Mindanao, in Lanao Del Norte and Lanao Del Sur. The power plants are called Agus 1 to 7.

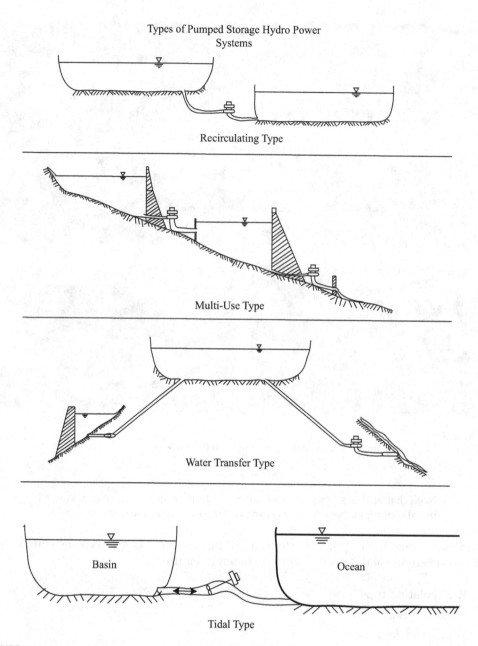

FIGURE 5.15
Schematic of the three basic types of pumped storage power plants.

Each dam power plant feeds the dam beneath the slopes, with Agus 1 being the highest dam and Agus 7 the lowest. The mountain from where the water comes is quite high in elevation, and water cannot be returned. For these systems, power is produced based on the amount of water recharge experienced during a given season. It is brought down in a controlled manner to make sure that power is produced continuously throughout the year. The operation of these types of facilities depends much on rain recharge. On rainy days, very high amounts of power may be generated. In tropical countries like the Philippines, monsoon seasons are conducive to high power production.

In the water transfer type, the reservoir is usually at a higher elevation (as shown in Figure 5.15) and is fed by a much higher elevation water shed. Power is produced by channeling water through either side of the storage facility.

A tidal type is also shown in Figure 5.15. Tidal types are erected near seashores to take advantage of the rise and fall of tides. Tidal types will be discussed more fully in another chapter of this book.

5.7.5 Based on Construction Features

Based on construction features, hydro power plants classify under the following types:

    a. Run-of-river
    b. Valley dam
    c. Diversion canal
    d. High head diversion

A schematic of these types of power plants is shown in Figure 5.16. In the run-of-river type, the power plant is based along the path of the river. Dam construction is minimal. A weir is usually built to measure and control the flow of water. For this type of system, power output is not very high since the dynamic head is high. The power is more proportional to the volume of water flowing in the river.

In diversion canal types, a new canal is made to divert water so that the main river is not obstructed. The power plant is placed in a path parallel to the river, and water is released to the lower part of the river, as shown. This design takes advantage of the peculiar topography of the area; a big drop in head is where the power house is constructed.

The valley dam type is a typical feature of most large-scale hydro power plants. In this system, a dam that serves as a reservoir for the power-generating water is built along the river. The component parts for these systems have already been discussed and enumerated in this chapter.

The high head diversion types are almost always associated with huge waterfalls. Instead of taking away from the beauty of a waterfall, a power canal is built from an intake point on the upper end of the waterfall. This power canal is then directed to the forebay of the power plant, as shown in the figure. The forebay distributes to a penstock, which in turn feeds the turbine impeller and the generator. The power house is located at the bottom of the waterfall such that the tail race is returned to the river below. This strategic design preserves the integrity of a waterfall, which doubles as a tourist spot in many instances. The water directed to the power plant must be controlled so as to not empty the waterfall and render it useless.

There are numerous high head diversion projects around the world, the most famous of which are the power plants developed along the awesome Niagara Falls, straddling Canada and the United States. In the Philippines, the beautiful Maria Cristina Falls also feeds a high head diversion power plant. Through well-designed water diversion, power is produced at the bottom of these magnificent tourist spots.

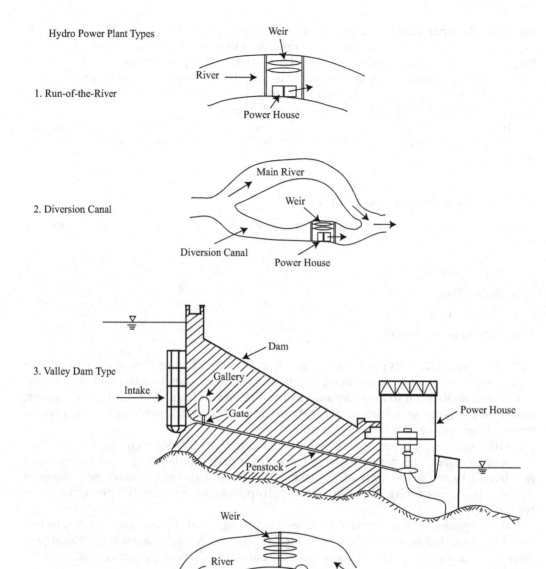

FIGURE 5.16
Schematic of the different types of hydro power plants based on their construction features.

5.8 Environmental and Economic Issues

Hydro power plants compete with geothermal power plants as the cheapest and most viable sources of renewable energy. While hydro power plants can generate continuous and low-cost electrical power, there are numerous environmental issues surrounding their use. One of the most prevalent issues is water quality. When water passes through the penstock, turbine blades, and other conveying components, losses of oxygen along with gains of nitrogen and phosphorus compounds are very common. Of course, if the water is repurposed for agriculture and irrigation purposes, the additional nitrogen and phosphorus compounds may act as fertilizers and may be considered good for the land. Other concerns include damaging the surroundings during the construction of a dam and its associated diversion canals. Easily accessible roads should be built to ease operation and maintenance procedures and prevent further damage to surrounding areas.

Aquatic life within the direct vicinity of a hydro power plant may also be affected, particularly during the construction period and near the penstock entrance when the plant is in operation. While a good power plant would have provisions that screen to protect the beneficial water populations, many of the aquatic species could still make it through the penstock and become macerated by the turbine blades. These aquatic plants present an additional maintenance challenge against the longevity of the turbine blades. Water hyacinth is a very common nuisance to most power plants and must be dealt with accordingly.

One of the more serious and costly environmental issues is siltation. Silt accumulates at the bed bottom of the river, reducing water flow and, consequently, the dynamic head. Several years following the construction of a hydro power plant, the river bed bottom should be dredged by removing the accumulated silt. This process poses a substantial cost to the facility but must be done. Otherwise, the power output of the plant will diminish with time—the silt may clog the penstock, and power may not be produced at all.

Water in rivers are almost always used for irrigation purposes in many countries. Dam construction could affect irrigation scheduling downstream. Agricultural land beneath these dams and power plants could be substantially affected and would have to rely on a controlled distribution of water. Because the power plant takes advantage of the potential energy of the water, the total water quantity is not so affected.

The overflowing of a dam could be an issue for agricultural farms dependent on a water source prior to the existence of a power plant. During typhoons, heavy rains, or tornadoes, a power plant could very likely release a great volume of water it does not need, putting a burden on communities downstream of the dam. Steps must also be taken to prevent damage to surrounding residential communities. Meticulous planning and preparation are necessary when erecting a hydro power plant. Builders of these huge structures must complete an environmental impact assessment (EIA) and prepare a complete analysis of the consequences to all that could be affected—flora and fauna, local residential communities, aquatic plants, and living organisms—for various emergency cases, including typhoons, tornadoes, and other natural phenomena causing inundation or severe lack of water.

While the economics of hydro power plants are quite appealing (as long as a continuous supply of water is available year-round), their environmental impact could be discouraging. Nevertheless, commercial manufacturers worldwide—Billabong Rams in Australia, Cecoco Rams in Japan, Blake Rams in the United Kingdom, Las Gaviotas in

Colombia, Premier Rams in India, Rockfer in Brazil, Schlumf in Switzerland, Sano in Germany, and Rife in the United States—have capitalized on the economic viability of hydro power plants.

5.9 Conclusions

There are numerous locations in the United States where water is released with appreciable head but not enough power is generated even just to recover parasitic loads of facilities. This is true in most water treatment facilities. Hence, small hydro power–generating facilities should be developed, even to simply provide lighting in the evening for security reasons. There are now small-scale hydro power systems for as low as 1 meter [3.28 ft] of dynamic head.

Hydro power is considered one of the oldest sources of energy for producing electrical and mechanical power, beginning thousands of years ago. Hydro power was the source of energy for many grain and timber processing facilities in the United States even before the invention of engines and generators for electricity. We could perhaps revive and update these vintage water-moving devices. There could be thousands of small-scale applications in rural households and villages.

The hydraulic ram is another piece of hydro power technology that could be refined and improved. With an update of this technology, water can easily be brought to higher elevations without the use of any fossil fuel–operated devices. The design is simple, and just about anybody should be able to assemble a unit upon acquiring the components and parts from a simple hardware store or home improvement store (such as Home Depot or Lowe's in the United States).

In many large-scale hydro power plants and some medium-sized facilities, the single most important maintenance issue is siltation. Over years of use, the dynamic head of hydro power facilities will decrease due to siltation and accumulation of debris at the bottom of the waterway. Removing silt is one of the most expensive rehabilitation processes for many hydro power facilities and must be performed on a regular basis. In addition, care must be taken to make sure that incoming water is free of debris or aquatic plants, such as water hyacinth, to ensure that the turbine blades of hydro power facilities are not damaged. The component parts of many hydro power plants are easily repaired and would not need complicated and advanced engineering manufacturing processes.

As a final statement, pumped storage facilities in many developed countries, such as the United States, are in place to be used as backup power during emergencies or calamities. The United States has one of the highest pumped storage generating capacities in the world and is prepared for such an eventuality but is second only to the collective European Union.

5.10 Problems

5.10.1 Theoretical Power from Water

P5.1 Determine the theoretical power in watts for water flowing at 50 kg/s and dropped from a dynamic head of 10 meters.

5.10.2 Actual Efficiencies of Micro Hydro Units

P5.2 A small commercial micro-hydro power unit had the following data: the volumetric flow rate was 70 liters/s, and the dynamic head was 1.5 meters. This unit was reported to produce 500 watts of continuous power output at those conditions. Determine the actual efficiency of this unit. Also calculate the yearly output of this small device in units of kWh/year, assuming 24 hours per day and 365 days in a year.

5.10.3 Hydro Power Plant Calculations

P5.3 Estimate the power output from one of the turbines of the Bhumibol hydro power plant with the following data: dynamic head = 150 meters and volumetric flow rate in one turbine = 97,750 kg/s. Assume an overall conversion efficiency of 80%.

5.10.4 Pump Specific Speed

P5.4 Given a pump rotational speed of 1,800 rps, flow rate of 1,200 m^3/s, and head of 20 m, calculate the specific speed.

5.10.5 Volumetric Efficiency of Hydraulic Rams

P5.5 Calculate the volumetric efficiency of a hydraulic ram with a volume delivered to a higher elevation at 500 liters per minute. The total volume taken and diverted from a water source is 2 cubic meters per minute [2,000 liters per minute].

5.10.6 Energy Efficiency of Hydraulic Rams

P5.6 Calculate the energy efficiency for a hydraulic ram with a volume delivered to a higher elevation at 500 liters per minute, with a vertical lift of 25 meters and a horizontal fall of 8 meters. The total volume taken and diverted from a water source is 2 cubic meters per minute [2,000 liters per minute].

5.10.7 Specifying Drive Pipe Size and Lengths

P5.7 Use the empirical Equation 5.12 to calculate the drive pipe length L if the supply head is 4 meters and the drive pipe diameter is 0.07522 meter [~3 in]. The number of pulsations per minute, N, is about 60.

5.10.8 Specifying Drive Pipe Size Using Rife Ram

P5.8 Calculate the drive pipe size (diameter in inches) for a hydraulic ram that would take water from a source with a maximum flow of around 360 L/min (95.1 gpm). This water will be brought 100 feet above the source—this is the vertical lift = 30.48 m. A vertical fall or supply head of 20 ft [6.10 m] is available.

Use the Rife ram data for this problem as shown in the performance chart. Estimate the amount of water that may be delivered (in gpm) as well as the pipe length assuming N of 60 beats per second.

5.10.9 Pumped Storage Power Production

P5.9 Determine the power generated from the Kalayaan pumped storage facility if the dynamic head is 290 meters and the volumetric flow rate is 120 cubic meters per second. Assume an overall efficiency of 80%.

5.10.10 Pumped Storage Power Production Water Use

P5.10 Determine the total maximum time (in hours and days) in which the water from Caliraya Lake would empty if power is generated by the Kalayaan pumped storage facility continuously. The volumetric flow rate is 120 cubic meters per second, and Lake Caliraya has a storage volume of 80×10^6 m^3 during the rainy season.

References

Bhattacharya, S. C. 1983. Lecture Notes in "Renewable Energy Conversion" Class. Asian Institute of Technology, Bangkok, Thailand.

EIA. 2017. Electric Power Annual 2016. December 2017. US Energy Information Administration, Independent Statistics and Analysis, US Department of Energy, Washington, DC. Available at: https://www.eia.gov/electricity/annual/pdf/epa.pdf. Accessed May 29, 2019.

EIA. 2018. Most U.S. Hydro Power Capacity Is in the West. US Energy Information Administration, Independent Statistics and Analysis, US Department of Energy, Washington, DC. Available at: https://www.eia.gov/energyexplained/index.cfm?page=hydro power_where. Accessed April 19, 2018.

Krol, J. 1976. The automatic hydraulic ram: its theory and design. Paper presented at the Design Engineering Conference and Show, Chicago, IL, April 5–8, 1976. Published by the ASME, United Engineering Center, New York.

Orozco, J. C. 1999. Hydraulic water rams: construction and design. Agricultural Mechanization Development Program (AMDP), College of Engineering and Agro-Industrial Technology (CEAT), University of the Philippines at Los Baños, College, Laguna.

Rife Hydraulic Engineering Manufacturing Co. (2019). A brief history of hydraulic ram. Available at: https://www.riferam.com/. Accessed April 2, 2019.

Tiwari, G. N. and R. K. Mishra. 2012. Advanced Renewable Energy Sources. RSC Publishing, New Delhi, India.

WEC. 2017. World Energy Council—Hydro Power Energy Resources 2016. Available at: https://www.worldenergy.org/data/resources/resource/hydro power/. Accessed April 19, 2018.

6

Geothermal Energy

Learning Objectives

Upon completion of this chapter, one should be able to:

1. Describe the various geothermal energy resources.
2. Enumerate the various ways of generating useful energy and power through geothermal means.
3. Estimate available power from geothermal resources.
4. Describe the various applications of geothermal energy.
5. Describe the various conversion efficiencies in geothermal power cycles.
6. Relate the overall environmental and economic issues concerning geothermal power systems.

6.1 Introduction

The world's geothermal energy output in 2015 was reported to be around 75 TWh, or about 0.27 EJ [0.256 Quad] (WEC, 2017). The Electric Power Research Institute (EPRI) in the United States estimated the total energy stored approximately 3 kilometers beneath the earth's surface to be roughly 43 million EJ [40.76 Quad], which is considerably greater than the world's reported energy consumption of 560 EJ [530.8 Quad] in 2012 (EPRI, 1978). The top five countries with the highest total geothermal generating capacity for electrical power are the United States (3.6 GW) [0.108 Quad/yr], the Philippines (1.93 GW) [0.058 Quad/yr], Indonesia (1.46 GW) [0.044 Quad/yr], Mexico (1.1 GW) [0.033 Quad/yr], and New Zealand (0.979 GW) [0.024 Quad/yr] (WEC, 2017).

In terms of total electricity generation (in GWh) in 2015, the U.S. geothermal energy output was reported at 16,800 GWh [0.057 Quad], followed by the Philippines at 10,308 GWh [0.035 Quad], Indonesia (10,038 GWh) [0.035 Quad], New Zealand (7,258 GWh) [0.025 Quad], and then Mexico (6,000 GWh) [0.0205 Quad]. The United States still produces the highest power output from geothermal electric power systems. The Philippines comes next, followed by Indonesia, in both installed capacity and generating output. In some countries like the Philippines, the electric power coming from geothermal energy comprised a larger percentage of the total power mix because of a lack of fossil fuel–based sources, such as coal and natural gas.

Of the world's geothermal resource bases, North America has the highest geothermal resource potential of approximately 21% of the world's available resources. This was followed by Asia at a similar 21% resource potential. Eastern Europe (17%), South America (14%), and Africa (13%) follow. Other countries, mostly in the Pacific Islands, represent less than 10% of the world's resource potential (WEC, 2017). The countries that have the most geothermal potential lie along the aptly called Pacific Ring of Fire, as shown in Figure 6.1. Japan, the Philippines, and Indonesia lie along this area, and the ring also borders the United States. These countries have already tapped into geothermal power potential and are currently generating a considerable amount of energy from the resource. Technically speaking, it is tricky to classify geothermal energy as renewable because it concerns the cooling of the earth, which in itself not a reversible process. There is no way to replenish geothermal energy once it is used. However, energy is continually produced from the decay of naturally radioactive materials, such as uranium and potassium. Solar energy is a temporal resource and cannot be used for this purpose. The amount of energy currently being removed is very small compared to the magnitude of geothermal resources available. At least for practical purposes, geothermal energy may be considered a renewable energy resource.

6.2 Temperature Profile in Earth's Core

Geothermal energy can be harnessed because the temperature profile of the earth increases as one goes deeper into its core. The temperature profile is illustrated in Figure 6.2. In this figure, the earth is divided into three zones: the mantle, the outer core, and the inner core. The mantle is some 2,900 km [1,802.4 mi] below the earth's crust, while the outer core is around 5,100 km [3,169.7 mi] from the crust. The inner core is some 6,378 km [3,964 mi] from the earth's crust, which is close to 100 km [62.15 mi] in thickness. The lithosphere is the area of the earth that is on the uppermost solid mantle, or the crust. At the bottom of this region, the temperature is approximately 1,750°C [3,182°F].

Going deeper into the earth's core, the bottom of the mantle has an approximate temperature of about 3,090°C [5,594°F], and the top layer of the inner core has an estimated temperature of 4,250°C [7,682°F]. Currently these areas are not being tapped by geothermal energy systems since technology has not yet been developed for those purposes.

The temperature profile of the earth surface is also variable depending upon the geologic nature of a site. Figure 6.3 shows some temperature profiles for three types of gradients. These are enumerated below:

a. Normal conductive gradient
b. High conductive gradient
c. Enhanced convective flow

A normal conductive gradient has temperature that varies linearly and reaches the boiling point of water (100°C) [212°F] at around 5 km [3.10 mi]. A high conductive gradient will have reached 100°C [212°F] in a little over 1 km [0.622 mi]. The enhanced convective

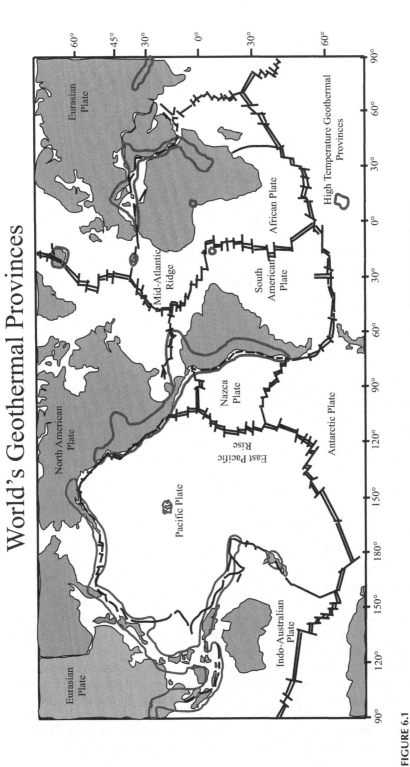

FIGURE 6.1
The world's geothermal provinces along the Ring of Fire in the Pacific Ocean (Dickson and Fanelli, 2004; US GEO, 2018).

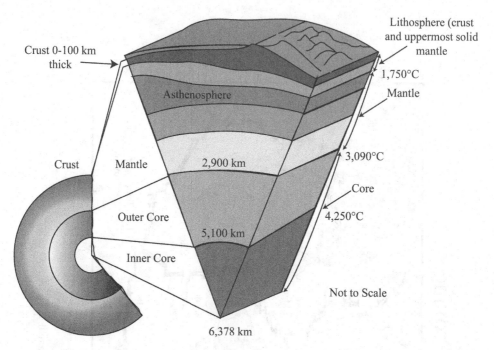

FIGURE 6.2
Temperature profile in the earth's core (Aclife, 2014; Williams, 2016).

flow is most ideal for geothermal systems, and such sites are usually found within the Ring of Fire. Planners of geothermal system projects have devised some means to evaluate whether a site has potential for geothermal applications. Planners typically conduct exploration surveys to evaluate a certain site. The specific activities used to identify geothermal potential in an area may be as follows:

- Satellite imagery and aerial photography
- Volcanological studies
- Geologic and structural mapping
- Geochemical surveys
- Geophysical surveys
- Temperature gradient hole drilling

The presence of volcanoes, whether dormant or active, is the most obvious indication of underground heat. Oftentimes, steam coming out of the ground is visible. Many such areas have already been investigated or perhaps have become tourist spots, like the geysers in California. Satellite imagery is relatively new and requires certain computer techniques and calculations to observe any peculiar thermal discrepancies. Aerial photography and heat-sensitive photographic devices have also been used in the past. The difficult task of finding actual sites and determining temperature profiles may be quite expensive. Activities used to determine viability include drilling and measurement of temperature as a function of depth and checking to see if there are clear changes that can lead to a higher-than-normal temperature gradient.

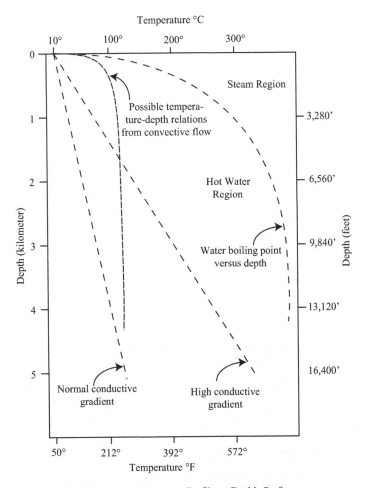

Various Types of Temperature Profile on Earth's Surface

FIGURE 6.3
Temperature profile of earth's surface (Bhattacharya, 1983).

There are also newer techniques to analyze chemical compounds when drilling is performed. Example 6.1 illustrates a simple process to select a potential well site for geothermal applications.

Example 6.1: Well Selection for Geothermal Energy Project

Three wells were dug, and the temperature profile is shown. Select which well is most likely to be a potential geothermal site.

Depth (km)	Well A (°C)	Well B (°C)	Well C (°C)
1	50	85	25
2	100	125	55
3	135	205	85
4	190	265	100
5	220	310	120

SOLUTION:

a. The easiest way to select a certain well is by plotting its temperature profile. This is shown below. Well B is a better candidate since the temperature is higher at lower depths.

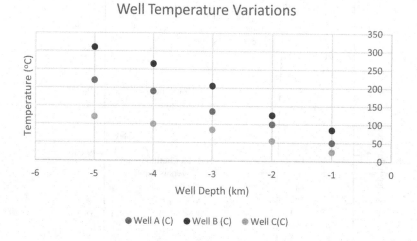

b. If the data are predictable as the graph has shown, upon quick inspection, well B is clearly superior, and data plotting is not necessary.

6.3 Geothermal Resource Systems

There are four distinct types of geothermal resource systems: liquid-dominated systems, vapor-dominated systems, hot dry rock systems, and geo-pressure systems. Each type has its own advantages.

6.3.1 Liquid-Dominated Systems

The liquid-dominated system is the most common type. As the name implies, the working medium is still in liquid form. Thermal energy is transported by liquids (water or brine) from rocks from deep regions to near the surface region with temperatures up to ~360°C [680°F]. These geothermal resource systems are located near visible features like volcanoes, geysers, and other similar geologic features. In these types, pumps are generally not necessary since the liquid should rise to the surface on its own due to its inherent thermal energy. A liquid-dominated system can easily generate output power up to 10 MW (also denoted as MWe) [0.0003 Quad/yr]. Steam is also a by-product and is separated from the liquid via cyclone separators; the liquid is returned to the reservoir for possible reuse. The largest liquid-dominated geothermal power plant is reported to be in Cerro Prieto, Mexico. This facility generates 750 MWe [0.022 Quad/yr] from temperatures reaching 350°C [662°F]. The Salton Sea field in Southern California offers the potential of generating 200 MWe [0.006 Quad/yr] of power.

There are, however, liquid-dominated systems of much lower temperatures, in the range of 120°C to 200°C [248°F to 392°F]. The natural convective flow is not enhanced in these systems and therefore would require some pumping. These systems will require heat exchangers, and the vaporization of organic working fluid (with low boiling points) drives the turbine. The ideal thermodynamic cycle for these systems is the Rankine power cycle using binary liquids. The thermodynamic cycle will be discussed further in later sections. Typical uses of these systems are small applications that require hot water and simple space heating. Most hot water types of geothermal systems in the United States are located in Alaska, California, and Hawaii.

6.3.2 Vapor-Dominated Systems

The vapor-dominated systems are quite rare. These systems are dominated by dry steam fields with temperatures ranging from about 150°C to 220°C [302°F to 428°F]. While steam vapor dominates the system, water may be present underneath. As the name implies, the working fluid is in a vapor state and usually in the superheated region once used in turbines. These systems are ideal for the true Rankine cycle, and generating power is easy because the condition of the working fluid is already ideal for running turbine blades. The hot steam in the superheated condition will simply turn the turbine blades and generate electrical power. The steam is then condensed, and the liquid portion is either re-injected into the ground or treated in a water treatment facility for reuse or disposal.

6.3.3 Hot Dry Rock Systems

Hot dry rock systems have vast volumes available but have not yet been truly explored. These systems require the introduction of cold water from the surface to extract energy from heated water down below. Hot dry rock resources are found at depths from 3 to 5 miles [5 to 8 km] beneath the earth's surface and are found at shallower depths in some geologically perfect areas (NREL, US DOE, 2018). The technology for these systems is similar to that of shale oil well drilling or water fracking, except that geothermal systems are located at even deeper depths. Therefore, it would be quite expensive to have this technology in place for the simple extraction of heat from the ground.

6.3.4 Geo-Pressure Systems

In geo-pressure systems, the fluids are under pressure in enclosed aquifers. The hydrostatic head is quite high, and its weight approaches that of overlying rocks. Most of the time, the sediments lie on top of a non-porous surface, and only an impermeable barrier will have to be broken down. Breaking this barrier is perhaps the most difficult step in harnessing energy from such systems. Drilling equipment must be able to withstand the hardness of the impermeable zone and requires the strongest and hardest drill bits with diamond tips. Geo-pressure is often present in many areas, but depths may vary, affecting how economical each system is.

The design of the power generation system would depend on what type of geothermal resource is available. The simplest is the vapor-dominated system, followed by the liquid-dominated system, and more complex are the hot dry rock and geo-pressure systems.

6.4 Geothermal Resource Potential in Texas

While most geothermal resource bases are west of the North American continent, Texas has its share of geothermal energy resources. Figure 6.4 shows the areas in Texas with known direct geothermal applications. The hydro-thermal, hot dry rock, and geo-pressure systems are all available and found in Texas with various degrees of potential. As illustrated in Figure 6.4, most of central Texas has potential hydrothermal systems. Geo-pressure systems are located along the Texas Gulf Coast, and there are a few hot dry rock systems on the eastern portion of the state. Figure 6.5 shows the proven geothermal basins in Texas.

Areas with hydrothermal potential have temperatures ranging from 90°F to 160°F [33°C to 70°C] at depths from 500 to 5,000 feet [152 to 1,524 m]. In some cases, water is

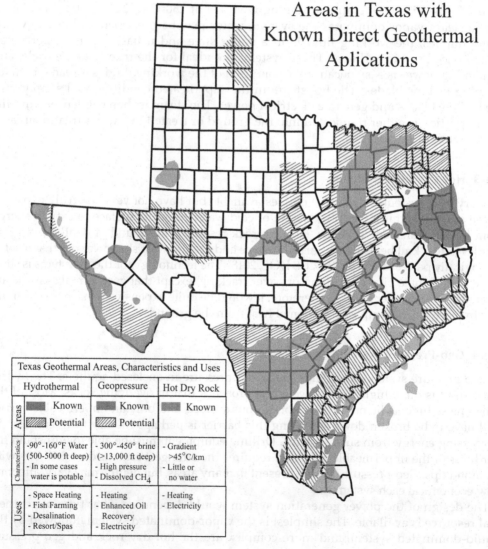

Areas in Texas with Known Direct Geothermal Aplications

Texas Geothermal Areas, Characteristics and Uses		
Hydrothermal	Geopressure	Hot Dry Rock
Areas: Known / Potential	Known / Potential	Known
Characteristics: -90°-160°F Water (500-5000 ft deep) - In some cases water is potable	- 300°-450° brine (>13,000 ft deep) - High pressure - Dissolved CH$_4$	- Gradient >45°C/km - Little or no water
Uses: - Space Heating - Fish Farming - Desalination - Resort/Spas	- Heating - Enhanced Oil Recovery - Electricity	- Heating - Electricity

FIGURE 6.4
Geothermal resource areas in Texas (adapted from SECO, 2008).

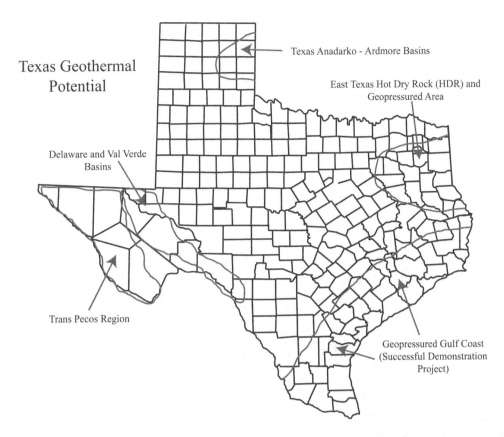

FIGURE 6.5
Regions in Texas with proven geothermal resource bases (SECO, 2008).

found to be potable. These types of resources are applicable for use in space heating, fish farming, and desalination and for resorts or spas.

6.5 Geothermal Power Cycles

The ideal thermodynamic cycle for geothermal power systems is the Rankine Cycle. In this system, the working fluid is alternately vaporized and condensed. This is depicted in Figure 6.6. The dry saturated steam enters the prime mover (either an engine or a steam turbine) and expands to higher pressure. Steam is then condensed at constant pressure and temperature to a saturated liquid. The liquid is then pumped back to the reservoir.

The steam cycle processes are also depicted in the figure. The thermodynamic properties at each point are available from steam tables and charts, except for properties 2 and 4 as shown. The illustrative processes are enumerated below:

1–2: reversible pumping of water to boiler pressure, with volume almost constant

2–2′: reversible isobaric heating of water

2′–3′: reversible isobaric isothermal vaporization of water to saturated steam

3′–3: reversible isobaric superheating of steam

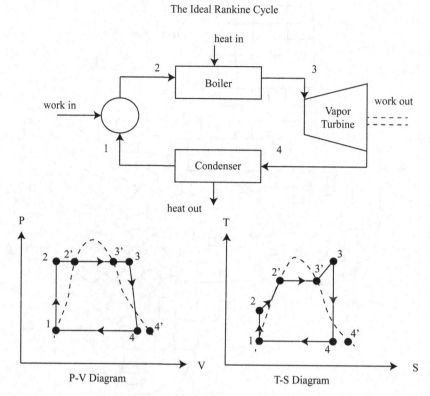

The Ideal Rankine Cycle

FIGURE 6.6
Basic geothermal power cycle.

3–4: reversible adiabatic expansion of steam to wet steam

4–1: reversible isobaric isothermal condensation of steam to saturated water

The component parts in a Rankine Cycle are as follows:

a. Pump

b. Boiler

c. Condenser

d. Vapor turbine

In a geothermal power cycle, the boiler is usually not needed since steam is already produced from the ground. Steam is simply injected into the turbine as vapor, condensed into liquid, and pumped back into the ground or reservoir. The actual cycle points are depicted in Figure 6.7.

6.5.1 Analysis of the Thermodynamic Cycle (Exell, 1983)

For process 1–2, the change in entropy is zero ($\Delta s = 0$), and the change in volume is also negligible ($\Delta v = 0$); thus, the internal energy is also conserved ($\Delta u = 0$). In this process, $v_2 = v_1$ and $u_2 = u_1$ and $s_2 = s_1$, giving Equation 6.1:

$$h_2 = h_1 + v_1\left(p_2 - p_1\right) \tag{6.1}$$

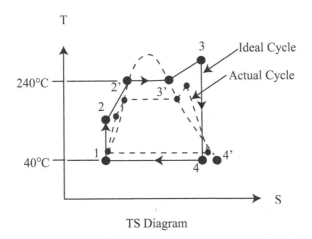

FIGURE 6.7
The TS diagram for geothermal systems.

Process 1–4′ occurs at constant temperature (T) and pressure (p), and we have Equation 6.2:

$$h_1 - T_1 s_1 = h_4 - T_1 s_4 \tag{6.2}$$

Process 3–4 is an isentropic process, and thus $s_4 = s_3$, and this results in Equation 6.3:

$$h_4 = h_1 + T_1 (s_3 - s_1) \tag{6.3}$$

The quality of wet steam at point 4 (or state 4) is given by the vapor-to-liquid mass ratio given in Equation 6.4:

$$\frac{x}{(1-x)} \tag{6.4}$$

Since entropy is an extensive property, x can be found from Equation 6.5

$$x = \frac{(s_3 - s_1)}{(s_4' - s_1)} \tag{6.5}$$

6.5.2 Energy Flows or First Law Analysis

The work input from the pump is given by Equation 6.6:

$$w_{12} = h_2 - h_1 \tag{6.6}$$

The heat input from the boiler is given by Equation 6.7:

$$q_{23} = h_3 - h_2 \tag{6.7}$$

The work output to the turbine is given by Equation 6.8:

$$w_{34} = h_3 - h_4 \tag{6.8}$$

The heat output to the condenser is given by Equation 6.9:

$$q_{41} = h_4 - h_1 \tag{6.9}$$

The energy balance equation is given by Equation 6.10:

$$w_{12} + q_{23} = w_{41} - q_{41} \tag{6.10}$$

Finally, the ideal first law efficiency is given by Equation 6.11:

$$\eta_1 = \frac{net\ work\ output}{heat\ input} = \frac{(w_{34} - w_{12})}{q_{23}} \tag{6.11}$$

Example 6.2 shows how to use the above equations.

Example 6.2: First Law Efficiency of Geothermal Systems

Given the following data for the ideal Rankine Cycle:

a. Boiler pressure, $p_2 = 2$ Mpa $= 2 \times 10^6$ Pa [290 psi]
b. Boiler temperature, $T_3 = 240°C$ [464°F], and
c. Condenser temperature, $T_1 = 40°C$ [104°F]

Determine the first law efficiency (η_1).

SOLUTION:

a. Determine the properties first from tables and charts. One is shown in Figure 6.7.
b. The saturated water property table (see Holman, 1980, Table A-7M) gives the following:
   $p_1 = 7,384$ Pa [1.071 psi]
   $v_1 = 1.0078 \times 10^{-3}$ m^3/kg [0.0162 ft^3/lb]
   $h_1 = 167.57$ kJ/kg [72.2 Btu/lb]
c. From Equation 6.1:

$$h_2 = h_1 + v_1(p_2 - p_1)$$

$$h_2 = 167.57\frac{kJ}{kg} + 1.0078 \times 10^{-3}\frac{m^3}{kg} \times (2,000 - 7.384)\frac{kN}{m^2} = 169.58\frac{kJ}{kg}$$

$$h_2 = 169.58\frac{kJ}{kg} \times \frac{1\ kg}{2.2\ lbs} \times \frac{1,000\ J}{kJ} \times \frac{Btu}{1,055\ J} = 73.06\frac{Btu}{lb}$$

d. From the table of compressed water properties (see Holman, 1980, Table A-10M), we have $h = 169.77$ kJ/kg [73.1 Btu/lb] at 40°C [104°F] and 2.5 Mpa [362.6 psi] to give $T_2 = 40°C$ [104°F] with minimal error.

e. The superheated vapor table is used next (Holman, 1980, Table A-9M), and from this table we can take the following values:

$h_3 = 2{,}876.5$ kJ/kg [1239.3 Btu/lb]

$s_3 = 6.4952$ kJ/kg [2.79845 Btu/lb]

From the previous table (Holman, 1980, Table A-7M), we also have the data for entropy:

$s_1 = 0.5725$ kJ/kg [0.24661 Btu/lb]

$s_{4'} = 8.2570$ kJ/kg [3.55752 Btu/lb]

The data for superheated vapor is also given in Holman (1980).

f. Using Equation 6.2 with $T_1 = 313.5$ K (i.e., 273.15 K + 40°C), we obtain $h_4 = h_1 + T_1(s_3 - s_1)$:

$$h_4 = 167.57\frac{kJ}{kg} + 313.15\ K \times (6.4952 - 0.5725)\frac{kJ}{kg\ K} = 2{,}022.3\frac{kJ}{kg}$$

$$h_4 = 2022.3\frac{kJ}{kg} \times \frac{1{,}000\ J}{1\ kJ} \times \frac{Btu}{1055\ J} \times \frac{1\ kg}{2.2\ lbs} = 871.3\frac{Btu}{lb}$$

g. The quality of steam is then estimated using Equation 6.5:

$$x = \frac{(s_3 - s_1)}{(s_4' - s_1)} = \frac{(6.4952 - 0.5725)}{(8.2570 - 0.5725)} = 0.771$$

h. The energy flows are then calculated as follows:

Work input from pump

$$w_{12} = (h_2 - h_1) = (169.58 - 167.57)\frac{kJ}{kg} = 2.01\frac{kJ}{kg}$$

$$w_2 = 2.01\frac{kJ}{kg} \times \frac{1\ kg}{2.2\ lbs} \times \frac{Btu}{1{,}055\ J} \times \frac{1{,}000\ J}{kJ} = 0.866\frac{Btu}{lb}$$

Heat input from boiler

$$q_{23} = (h_3 - h_2) = (2{,}876.5 - 2{,}022.3)\frac{kJ}{kg} = 2{,}706.92\frac{kJ}{kg}$$

$$q_{23} = 2{,}706.92\frac{kJ}{kg} \times \frac{1\ kg}{2.2\ lbs} \times \frac{Btu}{1055\ J} \times \frac{1{,}000\ J}{kJ} = 1{,}166.3\frac{Btu}{lb}$$

Work output to turbine

$$w_{34} = (h_3 - h_4) = (2{,}876.5 - 2{,}022.3)\frac{kJ}{kg} = 854.2\frac{kJ}{kg}$$

$$w_{34} = 854.2\frac{kJ}{kg} \times \frac{1\ kg}{2.2\ lbs} \times \frac{Btu}{1055\ J} \times \frac{1{,}000\ J}{kJ} = 368\frac{Btu}{lb}$$

Heat output to condenser

$$q_{41} = \left(h_4 - h_1\right) = \left(2,022.3 - 167.57\right)\frac{kJ}{kg} = 1,854.73\frac{kJ}{kg}$$

$$q_{41} = 1,854.73\frac{kJ}{kg} \times \frac{1\ kg}{2.2\ lbs} \times \frac{Btu}{1,055\ J} \times \frac{1,000\ J}{kJ} = 799.1\frac{Btu}{lb}$$

i. Thus, the first law of efficiency will be:

$$\eta_1 = \frac{\left(w_{34} - w_{12}\right)}{q_{23}} = \frac{\left(854.2 - 2.0\right)}{2,706.9} \times 100\% = 31.5\%$$

j. The schematic for the points and their values are shown in Figure 6.8.

In an actual geothermal facility, the boiler is absent, and the energy supplied (q_{23}) is equal to the product of the mass of steam and the enthalpy of that steam at a given temperature and pressure as it comes out of the well. Some pump work (w_{12}) is still needed to pump this steam to the turbine as well as to the ground after the working fluid has condensed.

The ideal cycle and actual cycle have already been shown in Figure 6.6. The main reasons for inefficiencies are as follows:

1. Mechanical inefficiencies in the pump and turbine
2. Heat losses to the environment from the boiler
3. Friction and heat losses in the pipe work

A summary of state points in the ideal Rankine Cycle is shown in Table 6.1.

Examples 6.3 and 6.4 show changes in efficiencies as temperature changes. These calculations show that conversion efficiencies for the ideal systems are already low and that actual efficiencies are even lower. But these calculations also demonstrate that power plants with lower efficiencies can sometimes generate more output than plants with higher efficiencies.

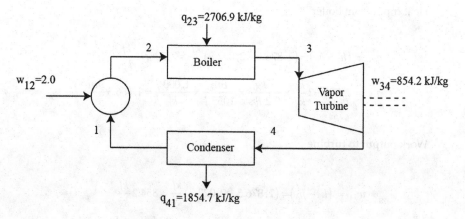

FIGURE 6.8
Values of thermodynamic properties at each major point in the thermodynamic cycle.

TABLE 6.1

Summary of State Points in the Ideal Rankine Cycle Example (Reworked from Exell, 1983)

State Point	Temperature °C	Temperature K	Pressure kPa	v 10^{-3} m³/kg	h kJ/kg	s kJ/kg-K
1	40	313.15	7.384	1.0078	167.57	0.5725
2	40	313.15	2,000	1.0067	169.58	0.5715
2′	240	513.15	3,344	1.2291	1,037.3	2.7015
3′	240	513.15	3,344	59.76	2,803.8	6.1437
3	240	513.15	2,000	108.5	2,876.5	6.4952
4	40	313.15	?	?	2,022.3	6.4952
4′	40	313.15	7.384	19,523	2,574.3	8.2570

Example 6.3: Calculating System Efficiency for Geothermal Systems

Assume that an ideal power plant with 100% isentropic efficiency has a source with a temperature of 120°C [248°F] and a steam rate of 150 kg/s [330 lbs/s] of flow. The cooling water is assumed to enter the power plant at 10°C [50°F] and leave the plant at 20°C [68°F]. Assume further that the power plant is able to cool the geothermal fluid down to 80°C [176°F]. The resulting configuration is schematically shown in the figure. Determine the efficiency of this system.

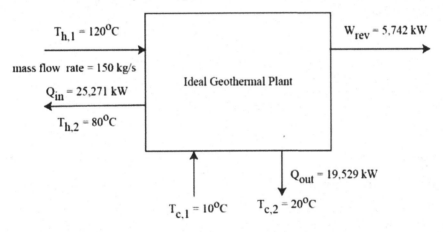

SOLUTION:

a. The efficiency of the ideal geothermal power plant is simply the ratio of the work output to the energy input as follows:

$$\eta_t = \frac{W_{net}}{Q_{in}} \times 100\% = \frac{5,742 \; kW}{25,271 \; kW} \times 100\% = 22.72\%$$

b. The ideal efficiency is rather low at 22.7%.

Example 6.4: Determining Geothermal Systems Efficiencies

Assume the same configuration as Example 6.3—a 100% isentropic efficiency plant with a source having a temperature of 120°C [248°F] and a steam rate of 150 kg/s [330 lbs/s] of flow. The cooling water is assumed to enter the power plant at 10°C [50°F] and leave the plant at 20°C [68°F]. Assume further that the power plant is able to cool the geothermal fluid down to 40°C [104°F] instead of 80°C [176°F]. The resulting configuration and new values for the energy output are schematically shown in the figure below. Determine the new efficiency of this system.

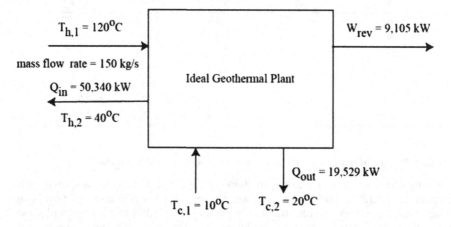

SOLUTION:

a. The new efficiency of the ideal geothermal power plant is, again, the ratio of the work output to the energy input as follows:

$$\eta_t = \frac{W_{net}}{Q_{in}} \times 100\% = \frac{9,105\ kW}{50,340\ kW} \times 100\% = 18.09\%$$

b. The ideal efficiency went down further to 18.09% despite the fact that the power output went up to 9.1 MW [0.00027 Quad/yr].

6.6 Geothermal Heat Pumps

Geothermal heat pumps (GHPs) are commonly used to heat and cool buildings and to provide hot water. GHPs use conventional vapor compression (refrigerant-based) heat pumps to extract low-grade heat for space heating. In the summer, the process reverses, and the underground becomes the heat sink while providing space cooling (i.e., refrigeration). GHPs are used in all 50 states in the United States.

The GHPs may be represented by the vapor refrigeration cycle. The TS diagram is schematically shown in Figure 6.9, while the block diagram of major components is shown in Figure 6.10.

The governing equation for these systems, also called the coefficient of performance (COP), is given in Equation 6.12:

$$COP = \frac{desired\ output}{required\ input} = \frac{refrigeration\ effect}{work\ supplied} = \frac{Q_L}{W} = \frac{(h_1 - h_4)}{(h_2 - h_1)} \tag{6.12}$$

The vapor refrigeration system attempts to duplicate the Carnot refrigeration cycle. Examples 6.5 and 6.6 provide illustrations for calculating this COP.

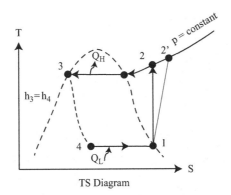

FIGURE 6.9
The TS diagram for the vapor refrigeration cycle.

Example 6.5: COP and Rate of Discharge of Energy to Environment Calculations

A refrigerator is to be maintained at 5°C by removing heat at a rate of 350 kJ/min. If the required power to the refrigerator is 1.9 kW, determine (a) the COP and (b) the rate of discharge of energy to the environment (i.e., the that which houses the refrigerator).

SOLUTION:

a. The COP is given by Equation 6.9:

$$COP = \frac{refrigeration\ effect}{work\ supplied} = \frac{Q_L}{W} = \frac{350,000\ J}{1,900\ W-min} \times \frac{W-s}{J} \times \frac{1\ min}{60\ s} = 3$$

b. The rate of discharge to the environment will be the sum of the work done and the refrigeration effect as follows:

$$Q_L + W = Q_H = 350\frac{kJ}{min} + 1.9\ kW \times \frac{60\ kJ}{kW-min} = 464\frac{kJ}{min}$$

$$Q_H = 464\frac{kJ}{min} \times \frac{1,000\ J}{1\ kJ} \times \frac{Btu}{1,055\ J} = 439.8\frac{Btu}{min}$$

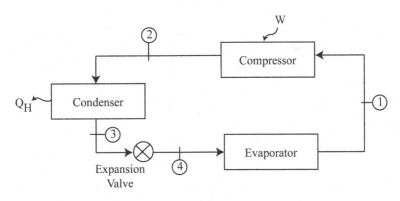

FIGURE 6.10
Vapor refrigeration cycles.

Example 6.6: COP and HP Input Calculations

An ideal vapor refrigeration cycle uses Freon 12 as its working fluid. The saturation temperature in the condenser is 120°F [48.9°C], and the evaporator temperature is 30°F [34.4°C]. Calculate the COP and the horsepower (hp) input necessary to produce the cooling effect of 5 tons. Draw the diagram for this process.

SOLUTION:

a. The ton refrigeration effect is defined as the heat rate to melt a ton of ice in a 24-hour period, equal to 12,000 Btu/hr, or 3.516 kW. The diagram is shown in Figure 6.11.

b. The next step is to consult the properties table for the working fluid. In this example, the working fluid is Freon 12.

c. The properties table (Holman, 1980) gives the following data:
$h_1 = 80.419$ Btu/lb$_m$ [186.65 kJ/kg] (saturated vapor at 30°F) [–1.11°C]
$s_1 = 0.16648$ Btu/lb$_m$°R [0.697 kJ/kg-K]
$h_3 = h_4 = 36.013$ Btu/lb$_m$ [83.59 kJ/kg]
$p_3 = p_2 = 172.35$ psia [1.19 MPa]

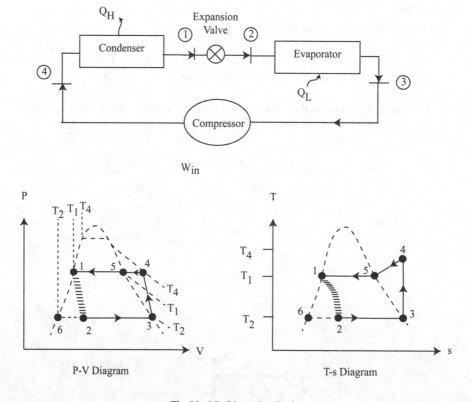

The Ideal Refrigeration Cycle

FIGURE 6.11
Ideal refrigeration cycle.

d. Consult superheated tables/charts (Holman, 1980) to obtain the value of $h_2 = 91.5$ Btu/lb$_m$ by interpolation of given pressure (p_2=172.35 psia). Thus,

$$COP = \frac{Q_L}{W} = \frac{(h_1 - h_4)}{(h_2 - h_1)} = \frac{(80.4 - 36.0)}{(91.5 - 80.4)} = 4$$

e. Properties in the superheated region require inter-/extrapolation if taken from these tables because temperature or pressure values are not given exactly.

In this problem, the pressure was found to be 172.35 psia [1.19 Mpa]; therefore, the value of h_2 must be interpolated between 150 psia [1.03 MPa] and 175 psia [1.21 MPa]. Further, the value of $s_1 = s_2 = 0.16648$ Btu/lb$_m$°R [0.697kJ/kg-K] and must also be interpolated to get $h_2 = 91.5$ Btu/lb [212.4 kJ/kg].
Work input

$$W\left(\frac{Btu}{lb}\right) = h_2 - h_1 = (91.5 - 80.419)\frac{Btu}{lb} = 11.1\frac{Btu}{lb}$$

$$W\left(\frac{kJ}{kg}\right) = 11.1\frac{Btu}{lb} \times \frac{1,055\ J}{Btu} \times \frac{2.2\ lb}{kg} \times \frac{1\ kJ}{1,000\ J} = 25.76\frac{kJ}{kg}$$

Total rate of flow

$$Q_L = \dot{m}(h_1 - h_4)$$

$$\dot{m} = \frac{Q_L}{(h_1 - h_4)} = \frac{(5 \times 12,000)}{(80.4 - 36.0)} = 1,351\frac{lb_m}{hr}$$

$$\dot{m} = 1,351\frac{lb_m}{hr} \times \frac{kg}{2.2\ lb} \times \frac{1\ hr}{60\ s} = 10.23\frac{kg}{s}$$

Total work required

$$W_{total} = \dot{m}(h_2 - h_1)$$

$$W_{total} = 1,351\frac{lb_m}{hr} \times (91.5 - 80.419)\frac{Btu}{lb_m} = 14,970\frac{Btu}{hr}$$

$$W_{total} = 14,970\frac{Btu}{hr} \times \frac{1,055\ J}{Btu} \times \frac{1\ hr}{3,600\ s} \times \frac{W}{J/s} = 4,387\ W = 4.4\ kW$$

$$W_{total} = 14,970\frac{Btu}{hr} \times \frac{hp - hr}{2,544\ Btu} 5.9\ hp$$

$$W_{total} = 5.9\ hp \times \frac{746\ W}{1\ hp} \times \frac{kW}{1,000\ W} = 4.4\ kW$$

6.6.1 Geothermal Heat Pump (Opposite of Refrigeration)

Imagine an air conditioner turned the opposite way. It would heat up the room by extracting energy from the cold environment on the other side. The unit heating cost for a heat pump is cheaper than that of a fossil fuel heater, even though the unit cost of energy purchased is much higher for the heat pump. In most heat pump applications, the initial capital cost of the unit is the determining difference. Example 6.7 shows the calculations for heat pumps. Example 6.8 shows a simple economic analysis of heat pumps.

Example 6.7: Power Consumption and Heat Extraction Rate of Heat Pump

A heat pump is used to meet the heating requirements of a house and maintain its temperature at 25°C [77°F]. On a day when the outdoor temperature is 10°C [50°F], the house is estimated to lose heat at a rate of 50,000 kJ/hr [47,393.4 Btu/hr]. If the heat pump has a COP of 3, determine (a) the power consumed by the heat pump and (b) the rate at which heat is extracted from the cold outdoor air.

SOLUTION:

a. The power consumed by the heat pump is simply the work input:

$$W_{net,in} = \frac{Q_H}{COP} = \frac{50,000\frac{kJ}{hr}}{3} = 16,667\frac{kJ}{hr}[4.6\ kW]$$

b. The rate at which energy is extracted from the house is the same energy rate to keep the temperature at 25°C [77°F] and is equal to 50,000 kJ/hr [47,393.4 Btu/hr]. The difference will be the heat extracted from the cold outdoor air as follows:

$$Q_L = Q_H - W_{net,in} = 50,000\frac{kJ}{hr} - 16,667\frac{kJ}{hr} = 33,333\frac{kJ}{hr}[9.26\ kW]$$

c. Note that if all the heating will come from an electrical resistance heater, a total of 50,000 kJ/hr [47,393.4 Btu/hr], or 13.9 kW, will be needed. As such, a heat pump is a popular option to provide this heating effect, though the initial capital cost of a heat pump can be quite high.

Example 6.8: Cost Comparison and COP

A heat pump using Freon 12 as its refrigerant was used to meet the heating requirements of a facility and maintain its temperature at 120°F [48.9°C]. The outdoor temperature was 40°F [4.44°C]. The compressor efficiency was 85%. The heating requirement is 100,000 Btu/hr [29.3 kW]. The electricity cost was 0.10/kWh. Compare the cost of operating this heat pump with that of a fuel oil heater that would cost $25 per 10^6 Btu. Assume a 70% efficient oil heater. Find COP as well.

SOLUTION:

a. The following parameters are taken from the table:
   h_1 = 81.436 Btu/lb$_m$ [189 kJ/kg] (saturated vapor at 40°F [4.44°C]; see from earlier table)
   s_1 = 0.16586 Btu/lb$_m$°R = s$_2$ [0.6944 kJ/kg-K]
   $h_3 = h_4$ = 36.013 Btu/lb$_m$ [83.59 kJ/kg] (saturated liquid at 120°F) [48.8°C]
   $p_3 = p_2$ = 172.35 psia [400 kJ/kg] (saturation pressure at 120°F) [48.8°C]

b. Using $s_1 = s_2$ = 0.16586 [0.6944 kJ/kg-K] and p = 172.5 psia [1.2 MPa] and by interpolation
   h_2 = 90.64 Btu/lb$_m$ [210.4 kJ/kg]

c. To estimate $h_{2'}$, the compressor efficiency equation is written as follows:

$$0.85 = \frac{(h_2 - h_1)}{(h_{2'} - h_1)} = \frac{(90.64 - 81.436)}{(h_{2'} - 81.436)}$$

$$h_{2'} = 92.264 \frac{Btu}{lb_m}$$

$$h_{2'} = 92.264 \frac{Btu}{lb} \times \frac{1,055\ J}{Btu} \times \frac{2.2\ lbs}{kg} \times \frac{kJ}{1,000\ J} = 214 \frac{kJ}{kg}$$

d. The flow rate is then calculated:

$$Q_H = \dot{m} \times (h_{2'} - h_3) = 100,000 \frac{Btu}{hr}$$

$$Q_H = 100,000 \frac{Btu}{hr} \times \frac{1,055\ J}{Btu} \times \frac{1\ hr}{3,600\ s} \times \frac{W}{J/s} \times \frac{kW}{1,000\ W} = 29.3\ kW$$

$$\dot{m} = \frac{Q_H}{(h_{2'} - h_3)} = \frac{100,000 \dfrac{Btu}{hr}}{(92.264 - 36.013)} = 1,778 \frac{lb_m}{hr}$$

$$\dot{m} = 1,778 \frac{lb_m}{hr} \times \frac{kg}{2.2\ lb_m} \times \frac{1\ hr}{3600\ s} = 0.2245 \frac{kg}{s}$$

e. The work input will be as follows:

$$W = \dot{m} \times (h_{2'} - h_1) = 1,778 \frac{lb_m}{hr} \times (92.264 - 81.436) \frac{Btu}{lb_m} = 19,252 \frac{Btu}{hr} [5.64\ kW]$$

f. The cost for the heat pump will be as follows:

$$C_{HP} = \left(\frac{\$0.10}{kWh} \right) \times 5.64\ kW = \frac{\$0.564}{hr}$$

g. The unit heating cost (per million Btu) will be as follows:

$$\frac{C_{HP}}{10^6\ Btu} = \frac{\$0.564/hr}{0.1 \times 10^6\ Btu/hr} = \frac{\$5.64}{10^6\ Btu}$$

$$\frac{C_{HP}}{10^6\ kJ} = \frac{\$5.64}{1 \times 10^6\ Btu} \times \frac{1\ Btu}{1,055\ J} \times \frac{1,000\ J}{kJ} = \frac{\$5.35}{10^6\ kJ}$$

h. For the oil heater, the following calculations will be made:

$$Efficiency\,(\%) = \frac{Output}{Input} \times 100\%$$

$$Input = \frac{Output}{Efficiency\,(\%)} \times 100\% = \frac{100,000\ Btu/hr}{70\%} \times 100\% = 142,857 \frac{Btu}{hr}$$

$$Input = 142,857 \frac{Btu}{hr} \times \frac{1,055\ J}{1\ Btu} \times \frac{1\ hr}{3,600\ s} \times \frac{W}{J/s} \times \frac{kW}{1,000\ W} = 41.87\ kW$$

i. At $25/10^6$ Btu, the cost will be:

$$C_{FO} = \left(\frac{\$25}{10^6 \ Btu}\right) \times 0.142857 \times 10^6 \frac{Btu}{hr} = \frac{\$3.57}{hr}$$

$$\frac{C_{FO}}{10^6 \ Btu} = \frac{\$3.57/hr}{0.10 \times 10^6 \ Btu/hr} = \frac{\$35.70}{10^6 \ Btu}$$

$$\frac{C_{HP}}{10^6 \ kJ} = \frac{\$35.7}{1 \times 10^6 \ Btu} \times \frac{1 \ Btu}{1,055 \ J} \times \frac{1,000 \ J}{kJ} = \frac{\$33.84}{10^6 \ kJ}$$

j. Thus, the heat pump cost is six times lower than the oil heater cost.
k. The COP may be calculated as well:

$$COP_H = \frac{Q_H}{W} = \frac{100,000 \ Btu/hr}{19,252 \ Btu/hr} = 5.194$$

Property values are taken from Holman (1980).

The following parameters are the values taken from the table:

a. $h_1 = 81.436$ Btu/lb$_m$ [189 kJ/kg]
b. $s_1 = 0.16586$ Btu/lb$_m$°R [0.69442 kJ/kg-K]
c. $h_3 = 36.013$ Btu/lb$_m$ [83.59 kJ/kg]
d. $h_4 = 36.013$ Btu/lb$_m$ [83.59 kJ/kg]
e. $p_2 = 172.35$ psia [1.2 MPa]
f. $p_3 = 172.35$ psia [1.2 MPa]

6.7 Geothermal Power Cycles

There are six general types of geothermal power cycles: (1) non-condensing cycle, (2) straight condensing cycle, (3) indirect condensing cycle, (4) single flash system, (5) double flash system, and (6) binary cycle. These systems will be discussed briefly. Figure 6.12 shows the schematic diagram of each system.

6.7.1 Non-Condensing Cycle

The non-condensing cycle is the simplest geothermal power cycle. The schematic diagram is shown in Figure 6.12a. Steam from geothermal wells is pumped out of the bore hole and directed into the turbine blades. Turbines usually have specifications for temperature, pressure, and steam rate. Engineers usually rely on ideal gas laws to adjust the requirements of the turbine. As pressure is increased, the temperature of the piping will increase as well. The steam introduced into the turbine blades comes from the superheated region, and an engineer can refer to steam tables, charts, nomographs, and diagrams. The spent steam is neither recycled nor condensed; it is simply exhausted to the atmosphere or outside environment. When the turbine is connected to a generator, power is produced. This is the oldest geothermal power system setup. However, with increasing demand for

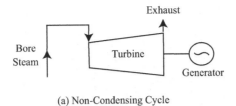

(a) Non-Condensing Cycle

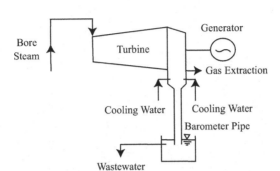

(b) Straight Condensing Cycle

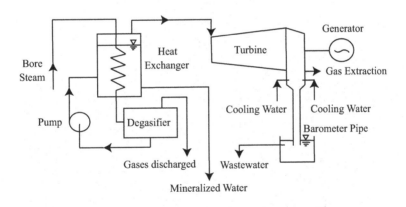

(c) Indirect Condensing Cycle

Geothermal Power Generation Cycles

FIGURE 6.12a–c
Basic geothermal power cycles (Bhattacharya, 1983).

cleaner exhaust product, the spent steam cannot simply be expended into the environment without prior treatment to the gas or the liquid product.

6.7.2 Straight Condensing Cycle

The straight condensing cycle is similar to the non-condensing cycle in the way steam is introduced into the turbine, as shown in Figure 6.12b. Spent steam is captured and

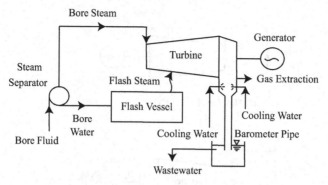

(d) Single Flash Power Generation Cycle

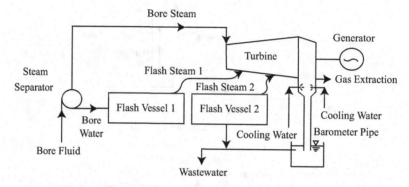

(e) Double Flash Power Generation Cycle

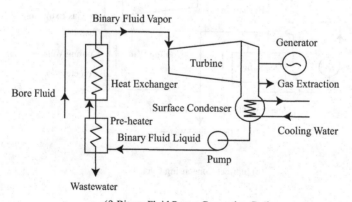

(f) Binary Fluid Power Generation Cycle

Geothermal Power Generation Cycles

FIGURE 6.12d–f
More geothermal power cycles (Bhattacharya, 1983).

condensed directly using cooling water. These systems would then require a wastewater treatment facility for the condensed steam. If there are toxic chemicals in the gas streams, they are also captured and directed to the wastewater treatment plant. Not all of the steam is condensed; otherwise, enormous amounts of cooling water would be required. Hence, excess vapors are usually present at the back end of the turbine and must be extracted or captured for other uses. The excess gas or steam is never used to generate additional power and is not used for other geothermal power applications. In many geothermal power plants, wastewater is quite difficult to maintain, especially if heavy metals such as arsenic are present. Sulfur is almost always present and would also have to be removed or captured and not exhausted into the waste streams or rivers. If one were to observe the homes near geothermal power plants, such as those in developing countries that have metal roofing, corrosion is a common problem. Smog is also an issue when the sulfur level is quite high. Therefore, non-condensing cycles are quite rare due to the requirement of capturing the pollutant elements in the gas streams.

6.7.3 Indirect Condensing Cycle

The indirect condensing cycle is shown in Figure 6.12c. In this system, which is common in many geothermal power plants, the steam coming from the bore hole is first processed through a heat exchanger. This pre-treatment process ensures that relatively dry super-heated steam is injected into the blades of the turbine. This device will remove some mineralized water to ensure that clean gas is introduced to the turbine blade section. A degasifier is also installed to remove or discharge properties not suitable for entry into the turbine blade section. The liquid portion, which is already demineralized, is introduced back into the heat exchanger, where some portion is heated further and allowed to enter the turbine blades. At the back end of the turbine, the used steam is condensed to capture harmful substances entrained through the gases. The product is collected in the wastewater treatment facility. Excess gases are also extracted for other uses not associated with the power plant. Cooling water is required in this section.

6.7.4 Single Flash System

The single flash system is shown in Figure 6.12d. The unique addition to the single flash system is the device called the flash vessel. As shown in the figure, this device re-evaporates some steam with enough energy to be reintroduced into the turbine blades, though at some point behind the entry section, as the quality of this steam is inferior to that of the steam introduced in the front-most section. Another unique feature is the steam separator, where superheated steam is separated from the liquid portion. The liquid portion is directed to the flash vessel. Flash (or sometimes called partial) evaporation occurs when a liquid compound undergoes reduced pressure. This usually occurs in a valve contraction, such as a throttling valve. This process is one of the simplest unit operation processes in the chemical engineering discipline. The flash vessel is also called a flash drum. A complication can occur when the liquid is not purified and contains numerous other compounds because compounds evaporate at different rates. Many industries deal with such vapor-liquid separation. This technique of water-vapor separation is not unique to geothermal systems. In oil extraction from the ground, the liquid portion is always desired but gaseous products, such as methane, must be separated or used properly elsewhere.

6.7.5 Double Flash System

The double flash system, as the name implies, has two flash vessels. Steam is injected into the steam turbine through various locations along the length of the turbine blades. A schematic is shown in Figure 6.12e. This system is very similar to the flash system, with an additional flash vessel. Efficiency of these systems is improved due to the recapture of steam that has already turned into its liquid state. Many newer geothermal power plants have double flash systems because of this efficiency improvement. The wastewater treatment facility is in place to clean up the spent water and steam after use. The other gases that are usually entrained with the steam include carbon dioxide, the odorous hydrogen sulfide, some methane, and ammonia. These pollutants contribute to global warming and must be handled properly. The amount of harmful emissions from a typical geothermal power plant is still lower than that of a conventional coal power plants. Nowadays, stringent air quality regulations require many new geothermal power plants to install emission control systems similar to those installed in coal power plants.

6.7.6 Binary Fluid Cycle

The more advanced geothermal power plants adopt the binary power cycle. The schematic is shown in Figure 12f. The advantage of this system is that it can even accept low temperature steam (as low as 57°C) [134.6°F] from the ground. This is possible through the use of a secondary fluid (hence the term "binary") that has a lower boiling point than water. This secondary fluid will flash vaporize at a lower temperature and can be used to drive the turbine (instead of steam from water). The typical enhancement to the binary fluid cycle is an additional heat exchanger that can vaporize the secondary liquid. The energy of the steam from the bore hole is used to transfer this energy to the secondary fluid, as seen in the schematic diagram. The binary fluid is condensed back for recycling and is pumped back to the heat exchanger where it is first preheated before being introduced into the main heat exchanger. Cooling water is then required to condense this secondary liquid. The spent steam must be processed before being exhausted into open water streams. Otherwise, the used steam must be reintroduced into the ground. Note that the thermal efficiencies of many geothermal power plants lie between 7% to 10% (Van del Sluis and Schavemaker, 2011).

The Philippines is second to the United States in geothermal power production worldwide. The Philippines has two large-scale geothermal power plants operated by Philippine Geothermal Incorporated, a U.S.-based company. These geothermal power stations are the Tiwi Geothermal Power Plant, located in Tiwi, Albay, and the Makban Geothermal Power plant, located behind the University of the Philippines at Los Baños in Laguna. A photo of one of these systems is shown in Figure 6.13. Figure 6.14 shows the pipeline delivering hot steam to the facility. The characteristics of these systems are shown in Table 6.2.

The Tiwi and Makban geothermal power plants were built in the Philippines in 1979. Each has a rated capacity of 330 MW [0.00986 Quad/yr] with capacity factor of 70.5% (Tiwi) to 85% (Makban). Their steam delivery capacity is similar at about a 6,380 k-lbs/hr rate [2,900 tonnes/hr].

The annual energy generation rate was reported at 1.9 M MWh [0.0065 Quad] for the Tiwi plant and 2.4 M MWh [0.0082 Quad] for the Makban plant, which is equivalent to displacing 3.1 M and 4 M barrels of crude oil, respectively. The Tiwi plant is of larger field size, taking up 1,800 hectares [4,446 acres], while the Makban plant has an effective area of 570 hectares [1,408 acres]. The Tiwi plant is near the majestic Mount Mayon, an almost perfectly conical volcano in the Philippines, and the Makban plant sits at the base of Mount Makiling, a dormant volcano in Los Baños, Laguna. Los Baños is a summer resort capital of the province of Laguna due to its famous hot springs and spas. These geothermal facilities employ close to a total of a thousand workers in the Philippines. The number of production wells are below 100 for both. The Makban plant reinjects its spent water/steam into the ground, while the Tiwi plant has wastewater treatment facilities and does not

FIGURE 6.13
The geothermal power plant in the Philippines at Tiwi, Albay.

reinject the liquid back to the ground. The well depth is about a mile (1.609 km) for the Tiwi plant and over a mile for the Makban plant.

Example 6.9: Estimating Number of Households Serviced by a Geothermal Facility

The average annual consumption of electricity in a developing country was reported at 3,235 kWh/yr. If the annual generation of power for a large geothermal power plant was reported at 2.4 M MWh [0.0082 Quad], how many households can this facility serve?

FIGURE 6.14
The delivery of steam through complicated pipelines.

TABLE 6.2

Characteristics of the Largest Geothermal Power Plant
in the Philippines

Characteristics	Tiwi	Mak-Ban
Installed capacity (MW)	330	330
First plant synchronized	1/79	4/79
1991 capacity factor	70.5	84.8
Steam delivery capacity (k-lb/hr)	6,382	6,812
Annual generation (MWh)	1.9 M	2.4 M
Oil equivalent (barrels)	3.1 M	4 M
Turbine inlet pressure (psig)	84.67	85.38
Field size (ha)	1,800	570
Length of pipelines (km)	72	62
No. of production wells	79	58
No. of injection wells.	10	21
Average well depth (ft)	5,010	6,643
Average steam flow rate (k-lb/hr)	80	118
Average brine flow rate (k-lb/hr)	55	130
Total field work force (NPC-PGI)	491	435

SOLUTION:

a. The number of households served by this geothermal facility will simply be the ratio of annual production divided by the yearly household consumption. However, the units must be the same. Hence, the production units must be converted into units of kWh/year as follows:

$$Production\left(\frac{kWh}{yr}\right) = \frac{2.4 \times 10^6 \ MWh}{yr} \times \frac{1,000 \ kW}{1 \ MW} = 2.4 \times 10^9 \ \frac{kWh}{yr}$$

b. The number of households served will be as follows:

$$Number \ of \ Households = 2.4 \times 10^9 \ \frac{kWh}{yr} \times \frac{HH - yr}{3,235 \ kWh} = 741,931 \ households$$

6.8 Geothermal Power Applications

There are many potential applications of geothermal hot water or steam besides electricity. These include the following:

a. Space heating and cooling

b. Industrial applications

c. Sale of by-products

d. Agricultural applications

Electrical power is still the foremost application of geothermal heat. However, the conversion efficiency of electrical power is low compared to other applications. The first is space heating and cooling.

Before discovering the great benefit of geothermal energy, Reykjavik, Iceland, was so dependent on fossil fuels that the town's environmental conditions were dismal. Today, the Reykjavik Municipal Heating Project, a geothermal energy–based project, serves 97% of a population of 113,000. Many countries or states should emulate this highly successful and environmentally positive renewable energy program.

There are also numerous industrial applications of geothermal heat. These include reheating, washing, cooking, blanching, peeling, evaporation, drying, and refrigeration. Industrial product processing requires enormous amounts of thermal energy, and a geothermal resource could be an economical option, depending on proximity.

The by-products of geothermal wells include minerals such as boron, calcium chloride, silica, and zinc. Many other compounds are found in geothermal wells, but most are considered too harmful to be recovered. These compounds are simply disposed of after separation from the liquid. Some are heavy metals that are carcinogenic, such as cadmium, arsenic, and even mercury in small amounts. In the future, some of these harmful metals may have applications and instead of being disposed of may become a source of additional revenue for some geothermal power plants.

Agriculture is perhaps the best application of geothermal energy. Greenhouses, including huge confined or concentrated animal house facilities, may be heated by geothermal heat. The aquaculture industry can also benefit from geothermal heat in many arid regions. Soil warming is a common practice of some agricultural farms, especially for growing mushrooms and other specialty crops. Anaerobic digestion or biogas production requires heated reactors, especially if using thermophilic organisms.

Still, the best direct use of geothermal energy is heating and cooling. A family in New Jersey (Jon and Kaley Heck: http://www.jonheck.com/Geothermal/Usage.htm) installed geothermal heating and cooling when they built their home in 2004. The basic concept was to use a constant temperature of the earth (about 53°C) [127.4°F] to temperature-control their house. The house has a total area of 2,350 ft² [218.4 m²] and has two zones, one on the first floor and one on the second floor. They initially spent $26,000 on this system, a price about $16,000 above the cost of a standard gas heating and air-conditioning unit). However, they also qualified for a rebate of $2,400 (net of $23,600). During the summer, they set their thermostat at 86°F [30°C] when they were not at home and at 78°F [25.6°C] when they were at home. In the winter, they set their thermostat at 64°F [17.8°C] when they were not at home and at 68°F [20°C] when they were at home. Every year, they reported their total electric usage as well as their geothermal usage and reported average savings of close to 30%. Example 6.7 shows the calculation for the return on their investment (ROI) by the offset in electricity costs alone. The ROI for this household application is very similar to that of solar energy for electricity consumption. The costs are reduced further if a rebate from the government is applied to the initial capital cost. In this example, a 15% rebate brings down the ROI by three years.

Example 6.10: Determining ROI of a Geothermal Project

Determine the ROI for a household that invested $26,000 on geothermal heating and cooling. Their average monthly geothermal usage was 416 kWh/month, and they paid an average of $0.138/kWh over the years. In how many years will they recover their initial investment for heating and cooling alone? If they have a rebate of $2,400, what is the ROI after the rebate?

SOLUTION:

a. The monthly savings in electricity is calculated as follows:
   Monthly Savings = $0.138/kWh × 416 kWh = $57.41

b. The yearly savings will be as follows:
   Yearly Savings = $51.41/month × 12 months = $688.90/year

c. Their ROI without the rebate is calculated as follows:
   ROI = $26,000/$688.90/yr = 37.74 years.

d. The ROI after applying the rebate is calculated as follows:
   ROI = $23,600/$688.90/yr = 34.26 years.

e. They have saved about three years after the rebate.

6.9 Levelized Cost of Selected Renewable Technologies

The levelized cost (in units of $/MWh) of selected renewable energy, including double flash geothermal and binary systems, is shown in Figure 6.15. Note that geothermal power is comparable with advanced nuclear systems and slightly better than wind. Levelized costs of small hydro power plants and solar concentrating photovoltaic systems are still high. The factors affecting the cost of geothermal power are enumerated below.

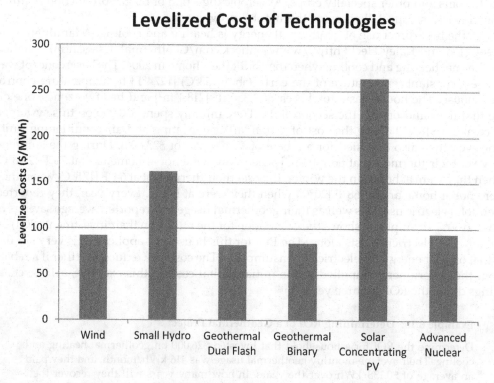

FIGURE 6.15
Levelized cost of renewable energy technologies (US EIA, 2018).

Factors Affecting Geothermal Power Plant Cost

a. Plant size

b. Technology used

c. Resource reliability

d. Nature of resource (temperature)

e. Geothermal water components

f. Depth and permeability

g. Tax incentives in state of country

h. Environmental issues

i. Market of end product

j. Financing options

The economic viability of a geothermal power system is a strong function of the choice of geothermal cycle. The binary system is perhaps the most cost effective of the systems due to its lower requirement for thermal heat. The capacity factor for geothermal systems is higher than wind energy, and its reliability is also very high. While geothermal steam temperature is a strong function of geothermal plant cost, the initial capital costs of advanced binary systems would offset the need for a higher steam temperature. The treatment of pollutants is another important issue to consider when planning a geothermal facility. Some areas have more pollutants, while others require minimal treatment. Almost always, a given site would have one particular pollutant to eliminate. The cost of drilling is another common factor of the initial capital cost. Almost always, a good temperature profile is desired, especially for those with shallow depths and high temperature gradients.

The incentives from the government count by way of tax rebates or other interest-payment offsets. All environmental issues will have to be addressed, and this will add the costs of obtaining permits. Environmental impact assessment is protocol for large-scale systems that are near urban areas. Many projects nowadays require equity and excellent financial options leading to profitable projects. Finally, some viable applications will have to be investigated for potential co-products of steam extraction. The environmental issues will be discussed thoroughly in the next section.

6.10 Environmental Effects of Geothermal Power Systems

The common environmental effects of installing a geothermal power system in a locality are as follows:

a. Land use

b. Water quality

c. Subsidence

d. Air pollution (noxious sulfuric gases, arsenic, hydrogen sulfide, etc.)

e. Heat rejection

f. Effect on hot springs

In any geothermal power project, the land will always be affected. The area where pipes are drilled covers several hectares, and conveying hot steam to the power plant requires several kilometers of pipeline, including the required insulation. For example, the total length of pipeline for one facility in the Philippines (Tiwi) is over 70 km [43.5 mi] with over 70 production wells and 10 injection wells. Another Philippine facility (Makban) had a total pipeline distance of only 62 km [38 mi] with 58 production wells and more injection wells of about 21. Water quality is also an issue with most geothermal power facilities. The average brine flow rate is about 130 k-lb/hr [59 tonnes/hr] for the Makban facility and 55 k-lbs/hr [25 tonnes/hr] for the Tiwi facility. Treating this much water can be an issue, especially if the brine is loaded with harmful pollutants. Because of continued depletion of water from the ground, there will be subsidence in the area where there are active production wells. Since the number of production wells far exceeds the number of injection wells, there will obviously be void space created by the water in the ground. Because of this, some areas require a greater number of injection wells. Areas within geothermal zones also have to deal with air pollution problems. Chronic bronchitis has been reported near some installations, and some villages or residential areas additionally contend with corrosion of metal roofs and changes in surrounding vegetation. It has been also reported that hot springs sites near geothermal plants decline in quality and number. Hence, tourist areas could also suffer from a decline in visitors after several years of geothermal heat usage. While heat rejection is not so widespread, it can affect areas very near the power plant. People would have to live with the constant emission of steam to the atmosphere and hope that these emissions are purely clean water vapor. In many instances, that is not the case. While the emissions can appear to be clean, the minute particles that the human eye cannot see, particularly $PM_{2.5}$, or particles less than 5 microns, have more damaging effects than visible particles.

6.11 Conclusion

Geothermal energy will continue to be a big portion of the energy mix in most countries that have ground heat at shallow depths. Not all countries will have this trait, but everywhere around the world, there is a gradient in temperature versus depth, and countries should take advantage of the natural phenomena when it is viable. While the overall conversion efficiency is rather low for most geothermal power applications, the environmental advantages of being free from fossil fuels can become the single most important impetus for geothermal energy adoption. For heating and cooling, we have seen that the payback period is longer and similar to that of solar and wind.

Geothermal power plants, due to the development of binary geothermal power cycles, are becoming advanced to the point that higher steam temperature is no longer a major requirement. Low-boiling-point refrigerant may be used instead of steam. Some means would still have to be arranged to rid of pollutants that come with the steam as it is pumped from the ground. These must be carefully dealt with since environmental issues could make or break a project. The economics of many renewables depend much on assistance from the government by way of subsidies or incentives to sell the renewable electrical power. In the future, many renewables such as geothermal power may add revenue through carbon credits. Carbon credit is a tradable commodity and is equivalent to one tonne of carbon dioxide reduced from the atmosphere, which a geothermal power has achieved for every power it produced. Many companies, for example in the European Union and in other developed countries, are buying the carbon credits (called carbon emission reduction [CER]) from other companies that have established that carbon dioxide reduction.

6.12 Problems

6.12.1 Well Selection

P6.1 Three wells were dug and the temperature profile is shown. Select which well is most likely to be a potential geothermal site.

Depth (km)	Well A (°C)	Well B (°C)	Well C (°C)
1	50	85	30
2	100	125	120
3	135	205	125
4	190	265	130
5	220	310	135

6.12.2 The Ideal Rankine Cycle

P6.2 Determine the first law efficiency (η_1) given the following data for the ideal Rankine Cycle:

a. The boiler pressure, $p_2 = 2.5$ Mpa $= 2.5 \times 10^6$ Pa [362.6 psi]

b. The boiler temperature, $T_3 = 280°C$ [536°F]

c. The condenser temperature, $T_1 = 60°C$ [140°F]

6.12.3 Efficiency of Ideal Geothermal Cycle

P6.3 Assume that an ideal power plant with 100% isentropic efficiency has a source with a temperature of 120°C [248°F] and a steam rate of 150 kg/s [19,800 lbs/min] of flow. The cooling water is assumed to enter the power plant at 10°C [50°F] and leave the plant at 20°C [68°F]. Assume further that the power plant is able to cool down the geothermal fluid to 75°C [167°F]. The resulting configuration is schematically shown in the figure. Determine the efficiency of this system.

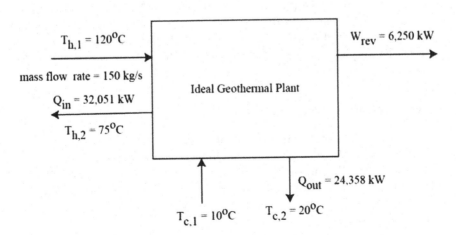

6.12.4 Changes in Efficiency and Power Output

P6.4 Assume the same configuration as Problem 6.3, with a 100% isentropic efficiency plant that has a source with a temperature of 120°C [248°F] and a steam rate of 150 kg/s [19,800 lbs/min] of flow. The cooling water is assumed to enter the power plant at 10°C [50°F] and leave the plant at 20°C [68°F]. However, assume that the power plant is able to cool the geothermal fluid down to 45°C [113°F] instead of 75°C [167°F]. The resulting configuration and new values for the energy output is schematically shown in the figure below. Determine the new efficiency of this system.

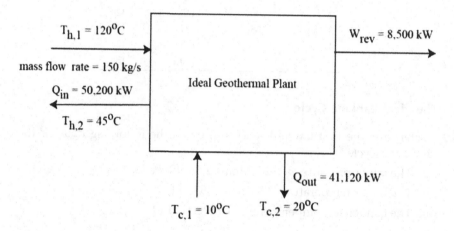

6.12.5 COP of Ideal Refrigeration Cycle

P6.5 A refrigerator is to be maintained at 4°C [41°F] by removing heat at a rate of 450 kJ/min [426.54 Btu/min]. If the required power to the refrigerator is 2.1 kW [2.82 hp], determine (a) the COP and (b) the rate of discharge of energy to the environment (i.e., the kitchen that houses the refrigerator).

6.12.6 Ideal Vapor Refrigeration System

P6.6 An ideal vapor refrigeration cycle uses Freon 12 as its working fluid. The saturation temperature in the condenser is 130°F [54.4°C], and the evaporator temperature is 40°F [4.4°C]. Calculate the COP and the horsepower (hp) input necessary to produce the cooling effect of 5 tons. Draw the diagram for this process.

6.12.7 Power Consumed in Heat Pump

P6.7 A heat pump is used to meet the heating requirements of a house and maintain its temperature at 25°C [77°F]. On a day when the outdoor temperature is 10°C [50°F], the house is estimated to lose heat at a rate of 50,000 kJ/hr [789.9 Btu/min]. If the heat pump has a COP of 3, determine (a) the power consumed by the heat pump and (b) the rate at which heat is extracted from the cold outdoor air.

6.12.8 Cost Comparison

P6.8 A heat pump using Freon 12 as its refrigerant was used to meet the heating requirements of a facility and maintain its temperature at 120°F [48.9°C]. The outdoor temperature was 40°F [4.44°C]. The compressor efficiency was 80%. The heating requirements are 90,000 Btu/hr [29.3 kW]. The electricity cost was 0.10/kWh. Compare the cost of operating this heat pump with that of a fuel oil heater that would cost $25 per 10^6 Btu [$23.70 per 10^6 kJ]. Assume a 70% efficient oil heater. Find the COP as well.

6.12.9 Number of Households Served by Geothermal Facility

P6.9 The average annual consumption of electricity in a developed country was reported at 12,000 kWh/yr. If the annual generation of power for a large geothermal power plant was reported at 2.4 M MWh [0.0082 Quad], how many households can this facility serve?

6.12.10 ROI of Geothermal Heating and Cooling

P6.10 Determine the ROI for a household that invested $18,000 into geothermal heating and cooling. Their average monthly geothermal usage was 650 kWh/month, and they paid an average of $0.12/kWh over the years. In how many years can they recover their initial investment for heating and cooling? If they have a rebate of $5,500, what is the ROI after the rebate?

References

Applegath, C. 2014. My future city. 2014 Symbiotic Cities International Design Ideas Competition: Urban Transformations: Designing the Symbiotic City. Symbiotic Cities Network. September 30, 2014 by Aclife. Available at: https://www.symbioticcities.net/index.cfm?id=64106&modex=blogid&modexval=18070&blogid=18070. Accessed April 20, 2018.

Bhattacharya, S. C. 1983. Lecture Notes in "Renewable Energy Conversion" Class. Asian Institute of Technology, Bangkok, Thailand.

Dickson, M. H. and M. Fanelli. 2004. What is geothermal energy? Geoscience and Georesource Institute, Pisa, Italy. February 2004. Available at: https://www.geothermal-energy.org/print/what_is_geothermal_energy.html. Accessed April 20, 2018.

EPRI. 1978. Geothermal energy prospects for the next 50 years. Special report ER-611-SR, February 1978. Electric Power Research Institute (EPRI), Palo Alto, California.

Exell, R. H. B. 1983. Thermodynamics and Heat Transfer Lecture Notes. Energy Technology Division. Asian Institute of Technology, Bangkok, Thailand.

Geothermal Education Office (US GEO). 2018. Geothermal Energy. Available at: http://www.geothermaleducation.org/. Accessed April 20, 2018.

Holman, J. P. 1980. Thermodynamics. 3rd Edition. McGraw-Hill, New York.

NREL, US DOE. 2018. Geothermal energy basics. Available at: https://www.nrel.gov/workingwithus/re-geothermal.html. Accessed April 20, 2018.

Texas State Energy Conservation Office (SECO). 2008. Texas renewable energy resource assessment. Prepared by Frontier Associates, LLC. December 2008. http://www.gov/todayinenergy/detail.php?id=17871. Accessed April 20, 2018.

US EIA. 2018. Levelized cost and levelized avoided cost of new generation resources in the annual energy outlook 2018. U.S. Energy Information Administration, Washington DC. Available at: https://www.eia.gov/outlooks/aeo/pdf/electricity_generation.pdf. Accessed April 20, 2018.

Van der Sluis, L. and P. Schavemaker. 2011. Electrical Power System Essentials. John Wiley & Sons, New York.

WEC. 2017. World Energy Resources: Geothermal 2016. World Energy Council. London Secretariat, London, UK. Available at: https://www.worldenergy.org/wp-content/uploads/2017/03/WEResources_Geothermal_2016.pdf. Accessed April 20, 2018.

Williams, M. 2016. What is the temperature of the earth's crust? Universe Today. Available at: https://www.universetoday.com/65631/what-is-the-temperature-of-the-earths-crust/. Accessed April 20, 2018.

US EIA. 2014. Geothermal resources used to produce renewable electricity in western states. U.S. Energy Information Administration, Washington DC. Issued September 8, 2014. Available at: https://www.eia.gov/todayinenergy/detail.php?id=17871. Accessed May 29, 2019.

7

Salinity Gradient

Learning Objectives

Upon completion of this chapter, one should be able to:

1. Describe the concept of generating energy and power from salinity gradients.
2. Enumerate the various ways of generating useful energy and power from salinity gradients.
3. Estimate available power from saline systems.
4. Describe the various applications of salinity gradients.
5. Describe the various conversion efficiencies in salinity gradient systems.
6. Relate the overall environmental and economic issues concerning salinity gradient power systems as well as the advantages and disadvantages of the system.

7.1 Introduction

Salinity gradient energy, sometimes referred to as osmotic energy, is technology that takes advantage of the osmotic pressure differences between salt water and fresh water. If a semi-permeable membrane (like that in a reverse osmosis filter) is placed between sealed bodies of salt water and fresh water, the fresh water will gradually travel through the filter by osmosis. By exploiting the pressure difference between these two bodies of water, energy can be extracted in proportion to the difference in osmotic pressure. There are two established practical methods currently being used—reverse electro-dialysis (RED) and pressure-retarded osmosis (PRO). The key process is the osmosis of ions in various membranes, which creates differential pressure. This phenomenon was first observed as early as 1954 when fresh water from inland was mixed with salt water along deltas (Pattle, 1954). However, the original observers did not come up with technology to harness the difference in osmotic pressure. In the 1970s, practical methods, using various impermeable membranes, were discovered by Professor Sidney Loeb at the Ben-Gurion University of the Negev, Beersheba in Israel (Weintraub, 2011). These methods were applied to an area along the Jordan River flowing into the Dead Sea. In 1977, Professor Loeb invented a method of producing power with a reverse osmosis electro-dialysis heat engine. The rest, as they say, is history. The technology has now evolved worldwide into commercial use, particularly in the Netherlands, which initiated the development of RED, and in Norway, the precursor of PRO. It is common knowledge in the energy community that membranes are quite costly. Now, the research is directed toward finding the cheapest membrane that can yield the most power. Salinity gradient or osmosis power is one of the renewable energy resources that does not contribute to an increase in carbon dioxide in the atmosphere; there is basically no associated fuel use, unlike some renewable technologies (such as fuel cells). Before the development of real osmotic power, the solar pond was thought to be the pathway for power generation, as a temperature gradient exists on open ponds, especially those with high concentrations of salt. The difference in temperature also gives rise to an increase in pressure, which may be enough to turn a low-power turbine. This phenomenon is the sole power source of a solar pond. This will be briefly discussed in the next section.

7.2 The Solar Pond

A solar pond is not quite a salinity gradient facility; rather, it is a facility that takes advantage of the thermal heat from the sun and generates energy or power for beneficial uses. Figure 7.1 shows a solar pond facility in El Paso, Texas. This facility has an area of 3,200 m² [34,427 ft²] in which the Bruce Foods Corporation operates (STM, 2017). This is the first-ever solar pond in the United States but only the second largest. The salt water along the pond forms a vertical salinity gradient that is called a "halocline." This demonstrates that low-salinity water floats on top of high-salinity water. In short, there is still a resulting gradient. To have a large vertical temperature gradient, a large area must be allocated. The primary cost is that of the impermeable liner, which is used to ensure that water will not percolate through the ground. The surface water is constantly evaporated due to the

FIGURE 7.1
A solar pond facility in Pecos, Texas (STM, 2017).

heat of the sun, causing the accumulation of salt, which would have to be removed. The energy obtained in this system is in the form of low-grade heat as a function of a temperature that is 70°C to 80°C [158°F to 176°F]. If the ambient temperature is at 20°C [68°F], there will be considerable heat difference (STM 2017). The heat energy absorbed is found with the sensible energy equation relating the mass, the specific heat, and the change in temperature presented in earlier chapters. This equation will be restated in this chapter as Equation 7.1:

$$Q(kJ) = mC_p\Delta T \tag{7.1}$$

where
Q = heat energy (kJ)
m = mass of body of fluid (kg)
C_p = specific heat of fluid (kJ/kg°C)
ΔT = change in temperature (°C)

Example 7.1 shows the estimated thermal energy of a hectare of seawater with a given depth as heated by the sun.

Example 7.1: Total Daily Energy Absorbed by a Solar Pond

A hectare of pond with a depth of 5 meters [16.4 ft] is being heated by the sun. The average temperature of the pond has risen from 20°C to 28°C [68°F to 82.4°F] during a hot day. Determine the amount of solar energy absorbed by this pond using the sensible heat Equation 7.1. Assume the density of sea water to be 1,027 kg/m³ [64 lbs/ft³] and its specific heat 3.93 kJ/kg°C [0.94 Btu/lb°F].

SOLUTION:

a. First, the mass of the body of water heated is calculated:

$$m(kg) = 1\ ha \times \frac{100\ m \times 100\ m}{1\ ha} \times \frac{3\ m}{1} \times \frac{1,027\ kg}{m^3} = 30,810,000\ kg$$

$$m(lbs) = 30,810,000\ kg \times \frac{2.2\ lbs}{1\ kg} = 67,782,000\ lbs$$

b. The heat energy equation is used:

$$Q(kJ) = mC_p\Delta T = 30,810,000\ kg \times 3.93 \frac{kJ}{kg°C} \times ((28-20)°C) = 968,666,400\ kJ$$

$$Q(Btu) = 968,666,400\ kJ \times \frac{1,000\ J}{1\ kJ} \times \frac{Btu}{1055\ J} = 918\ Million\ Btu$$

c. This is the energy absorbed by this vast area—more than 968 million kJ [918 million Btu].

Wide-scale generation of power from the salinity gradient/solar pond is shown in Figure 7.2. This system is similar to the El Paso solar pond facility in Texas. It was reported that the maximum Carnot efficiency of the solar pond (by taking advantage of this temperature gradient, usually at least 20°C [68°F] is around 17% [STM, 2017]). The Carnot Cycle efficiency

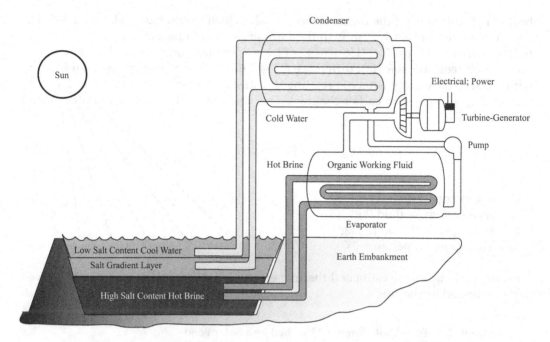

Typical Salinity Gradient Conversion System

FIGURE 7.2
Typical salinity gradient/solar pond conversion system.

is given by Equation 7.2. The efficiency may be calculated using the difference in high- and low-temperature regions. Example 7.2 shows how this calculation is made:

$$\eta_C\,(\%) = \left(\frac{W}{Q_H}\right) \times 100\% = \left(1 - \frac{T_C}{T_H}\right) \times 100\% \qquad (7.2)$$

where
W = work done by the system (kJ or KW)
Q_H = heat emitted by the system (kJ or kW)
T_C = absolute temperature of the cold reservoir (K)
T_H = absolute temperature of the hot reservoir (K)

Example 7.2: Theoretical Carnot Cycle Efficiency

Calculate the theoretical Carnot Cycle efficiency of a solar pond which has a high-temperature reservoir of 80°C [176°F] and a low-temperature reservoir of 20°C [68°F]. If the net output or work done in the system is 150 kW [201 hp], at this theoretical efficiency, how much heat was emitted by the system?

SOLUTION:

a. Equation 7.2 is simply used as follows:

$$\eta_C\,(\%) = \left(1 - \frac{T_C}{T_H}\right) \times 100\% = \left(1 - \frac{(273.15 + 20)\,K}{(273.15 + 80)\,K}\right) \times 100\% = 16.99\%$$

b. Using Equation 7.2, the heat emitted by the system is calculated as follows:

$$\eta_C(\%) = \left(\frac{W}{Q_H}\right) \times 100\%$$

$$Q_H(kW) = \left(\frac{150\,kW}{16.99\%}\right) \times 100\% = 882.87\ kW$$

$$Q_H = 882.87\ kW \times \frac{1,000\ W}{1\ kW} \times \frac{hp}{746\ W} = 1,183.5\ hp$$

c. About 883 kW [1,184 hp] of power is transferred to the reservoir.

The largest operating solar pond for electricity generation was the Belt HaArava pond in Israel, operated until 1988. It had an area of 210,000 square meters [688,800 ft²] and gave an electrical output of 5 MW [0.00015 Quad/yr] (STM, 2017). India was the first Asian country to establish a solar pond facility in Bhuj in Gujarat. This facility supplies 80,000 liters [21,136 gal] of hot water daily to the Gujarat Energy Development Agency and the Gujarat Dairy Development Corporation, Ltd (STM, 2017). An earlier project in Israel was a 150 kW [201 hp] solar pond built in 1980 (Mother Earth News, 1980).

The El Paso Solar Pond is a research, development, and demonstration project operated by the University of Texas at El Paso and funded by the U.S. Bureau of Reclamation and the state of Texas. The project, located on the property of Bruce Foods, Inc., a food canning company, was initiated in 1983. Since 1985, the El Paso Solar Pond has been operated continuously for seven years. The pond had been reconstructed with a geosynthetic liber (GCL) system, and operation resumed in the spring of 1995 (Lu and Swift, 2001). Current projects being undertaken at the El Paso Solar Pond include a biomass waste-to-energy project using heat from the pond, use of the pond for desalination and brine management, and its industrial application of sodium sulfate mining.

There are numerous advantages and disadvantages to a solar pond or salinity gradient facility. These are enumerated and discussed below.

7.2.1 Advantages

The advantages of salinity gradient facilities are as follows:

a. No CO_2 or other significant effluents or any global environmental effects
b. Completely renewable
c. Non-periodic (unlike wind or wave power)
d. Suitable for small- or large-scale plants (modular layout)

Solar ponds or salinity gradients should not contribute to global warming since no carbon dioxide is produced. No effluents are generated, and some products from the pond could be mined. This is a completely renewable energy technology with vast potential. It is not periodic like solar and wind, although the solar pond still needs the sun. Fully dedicated salinity gradients may not need the sun for their operation. There are small- and large-scale projects being initiated. Hence, the only hurdle is finding an available site for the facility. At the El Paso Solar Pond facility, the project halted because of a tear in its old, original XR-5 membrane liner. Newer and sturdier liners have since been developed (Lu and Swift, 2001).

7.2.2 Disadvantages

Overall drawbacks of salinity gradient facilities:

a. Plant equipment with the necessary efficiency has yet to be developed
b. High capital costs for plant construction, mostly of buildings and machinery
c. Energy cost is very sensitive to membrane cost and efficiency
d. Membranes used for plants are vulnerable to fouling

There are some minor drawbacks in the operation and management of salinity gradient projects or solar ponds. These are enumerated below.

Drawbacks in operation and management of salinity gradient projects:

a. Marine life is affected during construction and operation.
b. There will be massive geological changes on the flora.
c. There is a need for durable materials as liners.
d. Any materials used must be highly resilient.
e. The overall efficiency of the system is quite low.

The overall conversion efficiencies of salinity gradient systems are much lower than the typical efficiency of an internal combustion engine. However, newer dedicated salinity gradient projects have shown very high conversion efficiencies, with theoretical values reported at over 90% for the PRO system.

The current high cost of a plant is a hindrance in addition to the high cost of durable liners. Membranes used in salinity gradient systems are quite expensive, and this brings up the initial capital cost of the system. The membranes used are the same as those used in desalination plants, competing with that application.

Nevertheless, there are numerous valuable co-products generated. For example, the El Paso plant mines sodium sulfate for industrial applications.

7.3 Energy of Sea Water for Desalination

The osmotic pressure π from salt water is given in Equation 7.3. Example 7.3 shows how this equation is utilized to estimate the osmotic pressure as a function of molar concentrations and temperature. Example 7.4 shows how much work is done against pressure, which is also equal to the desalination work on salt water. Example 7.5 shows the sensible heat absorbed by a given amount of liquid:

$$\pi = cRT \tag{7.3}$$

where
  c = molar concentration
  R = universal gas constant = 0.082 L-bar/deg-mol
  T = temperature in absolute scale (K)

Example 7.3: Osmotic Pressure Calculations

Determine the osmotic pressure in units of kg/cm² [psi] for the following data. Temperature is 30 K, the molar concentration of salt ions is equal to 1.128 mol/L, and the universal gas constant is 0.082 L-bar/deg-mol.

SOLUTION:

a. Equation 7.1 is used directly to get the following:

$$\pi = cRT = 1.128\frac{mol}{L} \times 0.082\frac{L-bar}{degK-mol} \times 300\ K = 27.8\ bar = 27.8\frac{kg}{cm^2}$$

b. The osmotic pressure is around 27.8 bars [403.2 psi].

Example 7.4: Work Done in Osmotic Pressure Systems

Calculate the work done against pressure using the following data: pressure is equal to 27.8 kg/cm² [403.2 psi], and 1 liter [0.264 gal] of water is pushed against an area of 1 cm² [0.155 in²] over a distance of 10 meters [32.8 ft].

SOLUTION:

a. The work or energy is as follows (per liter of salt water):

$$W = \frac{27.8\ kg}{cm^2} \times 1\ cm^2 \times 10\ m = 278\ kg-m \times \frac{N-s^2}{kg-m} \times \frac{9.8\ m}{s^2} = 2,724\ Nm = 2.7\ kJ$$

b. The minimum energy required to desalinate 1 liter [0.264 gal] of seawater is about 2.7 kJ [2.582 Btu].

Example 7.5: Energy Required to Boil Salt Water

Determine the amount of energy required to boil 1 kg of salt water from 25°C to 100°C [77°F to 212°F]. Assume the specific heat of water to be 3.93 kJ/kg°C [0.9393 Btu/lb°F].

SOLUTION:

a. The heat energy equation is used:

$$Q(kJ) = mC_p\Delta T = 1\ kg \times 3.93\frac{kJ}{kg°C} \times \left((100-25)°C\right) = 294.75\ kJ$$

b. The energy required to heat up 1 kg [2.2 lbs] of salt water is around 295 kJ [279.62 Btu].

7.4 Pressure-Retarded Osmosis (PRO)

A pressure-retarded osmosis (PRO) system uses a membrane to separate a concentrated salt solution (like seawater) from fresh water. The fresh water flows through a semi-permeable membrane toward the sea water, which increases the pressure within the seawater chamber. A turbine is spun as the pressure is compensated, and electricity is generated. The world's first reported PRO plant was built by Statkraft, a Norwegian utility company (Patel, 2014). They estimated that in Norway, up to 2.85 GW [0.09 Quad/yr] of power would be available from this process. The plant, located in Oslo, opened in November 2009. Initial studies were made several years prior to its inauguration. The goal for this facility was to produce enough electrical power to provide light and heat energy to a small town near the Oslo fjord within a few years. The initial output was rather small—a 4 kW [5.36 hp] system (enough to heat a large electric kettle). As currently reported, the plant aims to increase the output to up to 25 MW [0.00075 Quad/yr], enough to equal the power-generating capacity of a small wind farm. The basic principle behind a PRO system is schematically shown in Figure 7.3.

In this system, fresh water is separated from the seawater through a semi-permeable membrane, creating a pressure differential that would be about 26 bars [2,600 kPa, or 377.1 psi]. This pressure is equivalent to a column of water with a hydraulic head of around 270 meters [82.3 ft] high in ideal conditions. In actual operation, the working pressure is only about half of this, or approximately 11 to 15 bars [1,100 to 1,500 kPa] (Williams, 2018).

The output of a PRO system is proportional to the salinity of the seawater. In 2014, researchers verified that 95% of a PRO system's theoretical power output can be produced with a membrane half the size needed for achieving 100% output. The system would yield very salty brine. The initial vision was to use this system in a wastewater treatment plant by mixing treated fresh water with seawater. However, for the size of a conventional wastewater treatment facility, this would require a membrane with an area of around 2.5 million square meters [26.9 million ft^2]. This would make the system quite expensive,

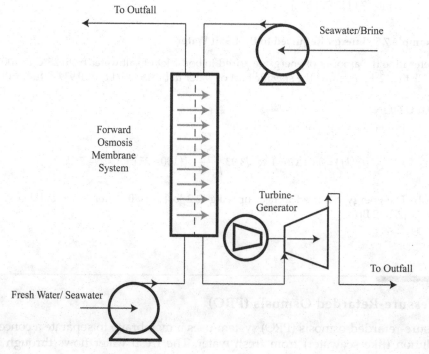

FIGURE 7.3
Basic PRO system schematic design.

and many scientists are currently developing newer membranes, such as the rolled system, which would not take up more space (Williams, 2018).

Studies in Norway showed that 12 TWh [0.041 Quad] of energy could be produced, which should meet about 10% of Norway's total demand for electricity. They have also estimated that each year, around 1600 TWh [5.46 Quad] could be generated worldwide.

7.4.1 PRO Standalone Power Plants (Statkraft, Netherlands, 2006)

The PRO standalone power plant developed in the Netherlands uses 2,000 m² [21,516.8 ft²] of flat sheet membranes. The original technical production was estimated to be 10 kW [13.4 hp], but the actual production was reported to be around 5 kW [6.7 hp]. The system uses hollow-fiber membranes (funded by Japan, the United States, and the Netherlands). A scaled-up design of a 2 MW [0.00006 Quad] power plant was envisioned in 2013 and planned for construction, subject to financing.

7.4.2 Statkraft Prototype (Norway, Co.)

Statkraft also developed a similar system in Norway. This system uses a polyamide membrane able to produce 1 W of power per square meter of surface area. This Statkraft plant has a capacity of 4 kW [5.36 hp] and employs a new way to harness osmotic power that would enable a 1 m² (10.7 ft²) membrane to have the same 4 kW [5.36 hp] capacity as the entire Statkraft plant. This performance was achieved using a 1 m² [10.76 ft²] boron nanotube membrane capable of generating 30 MWh/yr [3.4 kW, or 4.6 hp]. The next sets of studies included the development of boron nitride nanotubes. Example 7.6 shows the net power of this system and the requirement needed for a typical household. The PRO system envisions the use of these system types for communities near the shore. More recently, the Statkraft facility announced its closure (Patel, 2014). Example 7.7 shows the area needed to power a household if the performance is as recently reported. At present, these calculations may seem to be too good to be true.

Example 7.6: Sizing Pressure-Retarded Osmosis (PRO) Systems

The PRO salinity gradient prototype has a net power density of 14 W/m² [0.0017 hp/ft²]. Determine the size (in m²) of the polyamide membrane to power a household with a monthly requirement of 900 kWh.

SOLUTION:

a. This is a simple energy and power calculation as follows:

$$Area\left(m^2\right) = \frac{900\ kWh}{mo} \times \frac{m^2}{14\ W} \times \frac{1\ mo}{30\ days} \times \frac{1\ day}{24\ hrs} \times \frac{1,000\ W}{kW} = 89\ m^2$$

b. About 89 m² [957.5 ft²] of area is needed for this application.

Example 7.7: Calculating Required Membrane Area for Household Applications

If the average household in the United States uses around 10,908 kWh of electricity per year, how much membrane is needed to power 1 household if the yearly production of this membrane is 30 MWh/yr/m²?

SOLUTION:

a. This is a simple ratio of output to input as follows:

$$Area\left(m^2\right) = \frac{10,908 \; kWh}{yr} \times \frac{m^2 yr}{30,000 \; kWh} = 0.36 \; m^2$$

b. Thus, a nanotube with an area of 0.36 m² [3.9 ft²] is needed. The value seems too good to be true.

The power output of a PRO power plant depends on the salinity of the water bodies. The most common will be the mix of seawater along the shore and fresh water from the river along the deltas. These types of systems should generate around 25 bars [362.5 psi] of pressure and a power output of 5 to 10 W/m² [0.000623 to 0.00125 hp/ft²]. If the power requirement is quite high, large volumes of water may be needed.

If the project calls for a much higher power output, perhaps in the range of 10 to 20 W/m² [0.00125 to 0.00249 hp/ft²] a seawater source with extremely high salinity should be identified. For example, according to these criteria, the water from the Dead Sea or water from the Salt Lakes would qualify. In these systems, the power may be as high as 25 bars [362.5 psi]. The volume requirements for this system, despite the higher power output, may be significantly less.

Perhaps one option for many communities is a desalination treatment plant in the area. Desalination systems generate large volumes of brine, a by-product of the treatment facility and the nearby wastewater treatment plant. Brine has a pressure differential greater than that of fresh water from the river and salt water from the ocean but still less than water with extremely high salt content (like Dead Sea water). These systems could generate a pressure of around 50 bars and a power output of 10 to 20 W/m² [0.00125 to 0.00249 hp/ft²] (Williams, 2018).

Even if a desalination treatment facility is not available, a wastewater treatment facility and seawater may also be used. In this case, one has to find a wastewater treatment plant near the coastline. The wastewater treatment facility may save on electrical power if properly implemented. Engineers at the Massachusetts Institute of Technology (MIT) are envisioning this concept. The concept is shown in Figure 7.4. In this system, a long rectangular tank is built with a semi-permeable membrane in between the pressurized salty seawater and waste water. Through osmosis, the membrane lets water through but separates the salty water. Fresh water is drawn through the membrane to balance the saltier side. As the fresh water enters the saltier side, it becomes pressurized while increasing the flow rate of the stream on the salty side of the membrane. The pressurized mixture exits the tank, and a turbine recovers energy from this flow. This was a conceptualized project of Leonardo Banchik and John Lienhard at MIT; Abdul Latif Jameel, professor of water and food at MIT; and Mostafa Sharqawy of King Fahd University of Petroleum and Minerals in Saudi Arabia, reported in the *Journal of Membrane Science* (Chu, 2014).

7.5 Reverse Electro-Dialysis (RED)

Reverse electro-dialysis (RED) is also called reverse dialysis. RED uses the transport of salt ions through membranes. It consists of a stack of alternating cathode and anode exchanging perm-selective membranes. The compartments between the membranes are alternately filled with seawater and fresh water. The salinity gradient difference is the driving force in transporting ions, resulting in an electric potential that is then converted into electricity.

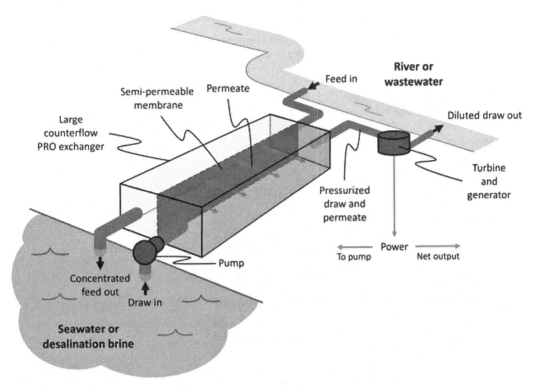

FIGURE 7.4
A simplified PRO system designed and envisioned by MIT engineers in the United States (Chu, 2014).

The RED technology is still quite new, even though the principles have been laid out since the 1950s. The main difference is the use of brackish water and seawater instead of fresh water. The important components of the RED system, as opposed to the PRO system, are electrodes and electrolytes—similar to fuel cell technology. Of course, the membrane is as important as these additional components. The spacers are also important such that the electrodes and the cathodes perform to their fullest. The basic premise from a reported study mixes 1 cubic meter of salt water and 1 cubic meter of fresh water while assuming a salinity gradient of 1:50 to generate energy of 1.4 MJ [1,327 Btu]. Using this assumption, numerous water bodies of known salinity gradient, when mixed with nearby brackish or fresh water, generate several MW of electrical power (IRENA, 2014). Example 7.7 illustrates these assumptions. The basic equation needed for calculation is the Nernst equation (Equation 7.4):

$$V(Volts) = \frac{RT}{zF} ln \frac{\left[high\ concentration \right]}{low\ concentration} \tag{7.4}$$

where
  V = cell potential (volts)
  R = universal gas constant = 8.314 J/mol-K
  T = absolute temperature (K)
  z = number of valence electrons per ion passed through a membrane (NaCl = 1)
  F = Faraday's constant = 96,485 C/mol

Example 7.8: Voltage Generation and Theoretical Power Calculations

The San Lorenzo River in California has an average annual discharge of 205 to 755 m³/s [3.5 to 11.96 million gpm]. The salinity at this location was measured at 0.593 mol/L [0.157 mol/gal]. The nearby fresh water salinity was 0.0033 mol/L [0.000872 mol/gal]; the salinity ratio was reported at 1:180. Use the Nernst equation (Equation 7.4) to calculate the voltage generated. Calculate also the theoretical power that may be generated from this facility if the amount of energy produced is on the order of 1.4 MJ/m³ [5 Btu/gal] of flow. If the overall efficiency is 1%, determine the estimated power output from this system.

SOLUTION:

a. The voltage is calculated from Equation 7.4 as follows:

$$V(Volts) = \frac{8.314\dfrac{J}{mol\,K} \times 298.15\,K}{1 \times \dfrac{96,485}{mol}} \times ln\frac{\left[0.593\dfrac{mol}{L}\right]}{\left[0.0033\dfrac{mol}{L}\right]} = 0.1334\ Volts$$

b. The power is estimated using the following relationship:

$$P_t(MW) = \frac{205\ m^3}{s} \times \frac{1.4\ MJ}{m^3} = 287\frac{MJ}{s} = 287\ MW$$

$$P_t(MW) = \frac{755\ m^3}{s} \times \frac{1.4\ MJ}{m^3} = 1,057\frac{MJ}{s} = 1,057\ MW$$

c. The actual power, using 1% conversion efficiency, will be as follows:

$$P_a(MW) = 287\ MW \times 0.01 = 2.87\ MW$$
$$P_a(MW) = 1,057\ MW \times 0.01 = 10.6\ MW$$

There are RED pilot plants that have been installed since 2005. The European Salt Company has a 5 kW [6.7 hp] RED pilot project initiated jointly by REDStack and Fuji Films. REDStack and Fuji Films started a follow-up project for a 50 kW [67 hp] pilot situated on a sea-defense site and major causeway called the "Afsluitdijk." This facility separates relatively clean fresh water on one side from relatively clean seawater present in the Wadden Sea/North Sea. Work on the 50 kW [67 hp] plant was completed in October 2013. A 200 MW facility is being planned for 2020 (Williams, 2018). Another innovation of the RED system is the improvement of the RED stack, employing the plug-flow tortuous flow path. This is called reverse electro-dialysis with a tortuous path flow spacer. By creating a unique path for fresh and salt water, the effectivity of the membrane is presumed enhanced, producing more power output or a higher pressure difference. Note that these systems rely on electro-dialysis and that voltage is generated without the use of turbines or moving parts. The beauty is in the combination of fuel cell–type systems and at the same time taking advantage of differences in osmotic pressure. Membrane use is still necessary, and again, its cost becomes an issue when commercialized. By increasing the membrane efficiency, the overall cost may be reduced. However, because of the additional necessity of electrodes and novel metals in the electrodes, another cost component is introduced. These components offset the cost for the turbines in PRO systems. More research is being directed toward using cheaper electrodes while maintaining overall system efficiency. The development of flat sheet membranes for RED systems is gaining traction. The power density of flat sheet membranes for RED systems is said to be lower, but they hold greater pressure (IRENA, 2014). These are different from the hollow fiber membranes being developed for the PRO system. Hollow fiber membranes are supposed to be of higher power density and have a reported range of 4.4 to 16 W/m² [0.00055 to 0.002 hp/ft²] (Kurihara, 2012; Kurihara and Hanakawa, 2013; and Han and Chung, 2014). The estimated current price of these membranes ranges from \$10.7/m² to \$32.1/m² [€10/m² to €30/m²] [\$1/ft² to \$3/ft²]. The ideal values should be \$2.14/m² to \$5.35/m² [€2/m² to €5/m²] [\$0.20/ft² to \$0.50/ft²]. The current capital cost for a RED power plant is around \$7.84 million/50 kW [\$116,973/hp], which is quite expensive by renewable energy standards, and this technology requires many years of research improvements. The bottom line is that salinity gradient power is viable and is a neat example of energy technology. Example 7.9 shows how much it would cost to build a 200 MW [0.006 Quad/yr] salinity power plant being planned in the Netherlands.

Example 7.9: Total Capital Cost Calculations for RED Systems

Determine the total capital cost for a 200 MW [0.006 Quad/yr] RED salinity gradient power plant if the reported unit cost is \$7.84 million/50 kW. Hack (2011) reported that the practical cost for a salinity gradient power plant is around \$3 million/MW. Determine the capital cost if this is the new estimate.

SOLUTION:

a. The unit cost in MW is simply calculated as follows:

$$\frac{Cost\,(\$)}{MW} = \frac{\$7.84\ M}{50\ kW} \times \frac{1,000\ kW}{MW} = \frac{\$156.8\ M}{MW}$$

b. The cost for a 200 MW system is as follows:

$$Cost\,(\$) = 200\ MW \times \frac{\$156.8\ M}{MW} = \$31,360\ M$$

c. This is quite high, and the more practical value for a 200 MW plant is \$600 million instead of \$31.36 billion.

7.6 Specific Applications or Locations

Numerous countries with vast ocean resources are embarking on aggressive osmotic power projects. Canada, leading in osmotic power, has identified at least 10 osmotic power projects along the St. Lawrence River based on preliminary resource assessments. These power plant outputs range from as low of 5 MW [0.00015 Quad/yr] to as high as 589 MW [0.0176 Quad/yr] (Laflamme, 2012). The total expected power potential along this river is around 805 MW [0.0024 Quad/yr].

The major areas with potential for salinity power are as follows:

a. Standalone plants in estuaries where fresh water runs into the sea
b. Energy generation processes recovering energy from high-salinity water streams (e.g., brine from salt mining or from desalination plants)
c. Industrial wastewater treatment plants with marked salt gradients of components
d. Land-based salt water lakes or other types of salt water reserves

Some research reports estimate the worldwide potential for salinity gradient at 3.1 TW (Stenzel, 2012). This value is obviously a small fraction of the world's power consumption of 17.7 TW (US EIA, 2017). There are also saline lakes and other saline water bodies inland where the same principle may be applied. In addition, there are underground reserves or aquifers that have various salinity gradients. These areas have not been included in the estimate for salinity gradient resource potential. Likewise, the estimates also exclude numerous sources of brine water from hundreds of desalination plants around the world. In this case, the overall potential estimate is likely to be conservative. Salinity gradient energy is perhaps the renewable resource that has the most potential over the other renewable energy sources already discussed in this book. It is also the one most considered as having no carbon footprint. The estimates may also expand to those areas that are very far from land and not currently included in estimates because of the long transmission lines required for power. At some point, these areas may become economical because of voltage drop losses. As some countries expand their sea fronts, there will be communities that are developed offshore, especially those with minimal land areas but vast ocean fronts.

7.7 Limitations and Factors Affecting Performance and Feasibility

Because salinity gradient is considered to be the renewable energy resource that will generate the most potential, success lies in research that will eliminate barriers against its adoption. Foremost are the expensive membranes as well as the balance of systems. This section discusses the limitations and factors affecting its feasibility:

a. Large-scale production of cheap membranes
b. Necessary location near sea areas where no human activities are present
c. Material durability (resistance to salt water)
d. Movement of large quantities of salt water
e. Salt water solids accumulation
f. Fouling of membranes
g. Water pretreatment when necessary

First of these issues is the selection of the proper membrane to use. The consistency of salt concentration at a given site as well as the purity of fresh water being used (or consistency of wastewater properties) must also be addressed. The higher the salt content, the better, though it calls for the proper selection of materials for the balance of systems—pumps and conveying devices to transport either salt water, brackish water, or fresh water. Each type of liquid has a different resistance to flowing through conveying parts.

Vast tracks of land may be needed, and the country planning for large-scale salinity gradient facilities should take this into account. Large ponds expectedly have large pumping requirements.

7.8 Performance and Costs

Different types of salinity gradient technology provide different performance values and initial capital costs. Performance reports vary in unit use. Usually, power density is reported in W/m^2, and the current maximum net power density maximum is reported at 2.7 W/m^2 [0.000336 hp/ft^2]. The current laboratory value is 14.4 W/m^2 [0.00179 hp/ft^2] for a PRO process. Higher net power density may be achieved by changing cell design (i.e., the membrane resistance, cell length, and use of nanotubes). Cell design change could bring net power to as high as 20 W/m^2 [0.0025 hp/ft^2]. Some research reports energy density in units of MJ/m^2. In one earlier example, we observed an energy density of 4.1 kJ/L [14.7 Btu/gal] of fresh water. Some reports place this number to around 2.2 kJ/L [7.9 Btu/gal] (IRENA, 2017)

The capital costs are also reported in various units. One data set shows the number between \$65 and \$125/MWh for the PRO system and around \$90/MWh for the RED design. The cost may also be reported in what is called levelized electricity cost. For the PRO system, this is reported at around \$0.15 to 0.30/kWh for the PRO and \$0.11 to 0.20/kWh for the RED (IRENA, 2014). Note that these values are simply estimates at this time and hardly reliable for commercial-scale systems. There are various uncertainties in these numbers, and therefore they should be treated appropriately (Vermaas, 2014).

As previously mentioned, the membrane is the crucial cost component. Some reports claimed that membrane cost comprised about 80% of the total capital cost of the salinity gradient system. The remaining 20% went toward the balance of power (mostly for water pretreatment). Regular membranes are two to three times less costly. These regular membranes are used in many desalination plants to minimize power production. The goal of a desalination plant is in contrast with the goal of a salinity gradient plant, that is, to increase the pressure difference between membranes. The added costs are due to having to make the membranes sturdier. The balance of power costs are comprised of costs of the installation, pumping energy, pressure vessels, and turbines or electrodes for the RED system (IRENA, 2014).

The operation and maintenance costs are made up primarily of the cost of energy utilized to pump the water bodies, both fresh water and seawater, as well as the brine that is produced.

The initial capital costs reported in literature also vary, and the widely accepted number is around \$3 million /MW. The demonstration costs are sometimes reported in the amount of billions per MW, but those quotes are impractical and are only true for research and demonstration systems. A complication occurs when energy cost is reported in units of cost per unit of energy, for example, \$65 to \$125/MWh. When converted into the usual \$/kWh unit, the values would make more sense. As such, electricity production costs are reported to be between \$0.06 and \$0.125/kWh, which lies within the range of most utility costs in the United States (IRENA, 2014). Example 7.10 shows the simple payback period example using some reported initial capital costs for salinity gradient power plant sizes. The cost for selling electricity when using hybrid systems is said to be lower. Hybrid systems combine various designs with other renewable energy resources, such as hydro power or even solar and wind power.

Example 7.10: Payback Period Calculations for Salinity Gradient Facility

Make a rough estimate of the payback period in years for a 200 MW salinity gradient facility that costs $600 million. The main revenue will come from selling electricity at $0.10/kWh. Assume that the payback period is simply based on the ratio of the initial capital cost and the initial capital investment. Assume also that the facility will operate for 365 days in a year with 24 hours in a day.

SOLUTION:

a. The number of kWh of electrical power produced in a year is calculated as follows:

$$\frac{kWh}{yr} = 200\ MW \times \frac{365\ days}{yr} \times \frac{24\ hrs}{day} \times \frac{1{,}000\ kW}{MW} = 1{,}752{,}000{,}000\frac{kWh}{yr}$$

b. The revenue from the sale of electricity is calculated from the product of energy output and the rate as shown below:

$$Revenue\left(\frac{\$}{yr}\right) = 1{,}752{,}000{,}000\frac{kWh}{yr} \times \frac{\$0.10}{kWh} = \$175{,}200{,}000$$

c. The simple payback period is the ratio of capital cost and revenue as shown below:

$$Pay\ Back\ Period\left(yrs\right) = \$600{,}000{,}000 \times \frac{yr}{\$175{,}200{,}000} = 3.42\ years$$

d. Hence, the payback period is less than five years. Of course, this could be greater if one considers the operating cost per year in the calculations.

7.9 Potential Energy and Barriers to Large-Scale Development

There are many estimates made on potential energy that may be derived from the salinity gradient. Alvarez-Silva, et al. (2016) estimated the practical potential to be around 625 TWh/acre based on factors such as suitability, sustainability, and reliability of exploitations. The article mentioned reports of theoretical calculations that reports the value to be around 15,102 TWh/a. The main economic barrier is the membrane cost, which accounts for 50% to 80% of the total capital cost. The current reported capital cost is roughly $3 million/MW as reported by the same article.

A summary of the theoretical and technical potential in many parts of the world is shown in Table 7.1. In this table, Asia and South America have the highest reported theoretical and technical potential with over 8,000 TWh/yr of energy and over 1,500 TWh/yr or technical potential. North America follows with 4,195 TWh/yr and 785 TWh/yr, respectively.

TABLE 7.1

Theoretical and Technical Potential of Salinity Gradient

Continent	Theoretical Potential		Technical Potential	
	[GWh gross]	[TWh gross/yr]	[GWh$_e$]	[TWh$_e$/yr]
Europe	241	2,109	49	395
Africa	307	2,690	63	503
Asia	1,015	8,890	208	1,663
North America	479	4,195	98	785
South America	969	8,492	199	1,589
Australia*	147	1,291	30	242
World	3,158	27,667	647	5,177

*Including Oceana.

Source: Stensel (2012).

Considerable research is still required to realize the economic and technical potential of numerous demonstration plants in development worldwide. Each country near the coast has the potential to generate enormous power from such systems, but the limited number of research resources hinders further development. Numerous other projects in some countries are still in the modeling stage. Newer technology will be developed, as with the surge of the newest capacitive method and the development of nanotubes, particularly the boron nitride version and other similarly novel materials. The use of some ideal thermodynamic cycles is also a welcome addition to the endless possibilities available for salinity gradient-related technologies. Both open-cycle and closed-cycle (using absorption refrigeration systems) technologies are projected to come into play in the near future. Solar ponds should also be maintained as an option and pursued as backup power systems in case major salinity gradient technology fails at some point. The thermal energy coming from the sun is constant and continuous. In the future, we will also see hybrid systems that combine solar energy, wind energy, and tidal power to amplify the effectiveness of existing systems.

7.10 Environmental and Ecological Barriers

There are obvious environmental and ecological barriers in the development of fully dedicated salinity gradient projects and solar ponds. The major ones are listed below:

a. Bio-fouling of membrane

b. Effect on nearby estuaries and rivers

c. Neither fresh water nor salt water is used up in the process

d. Some underground locations need to be built, which would require considerable amounts of digging

The main advantages are limited noise and minimal CO, CO_2, and NO_x emissions.

Marine life will be affected when large-scale salinity gradient ponds are developed. Aquatic species are either salt water tolerant or fresh water tolerant but not both. This forces an ecological adjustment upon some freshwater and marine species. The major waste product of a salinity gradient project is brackish water. When this wastewater is dumped in an area with a different salt concentration, there will be an ecological imbalance. The areas where these projects are located are not ideal grounds for aquatic animals or plants to thrive. One way to avoid this issue is to transport brackish waters into the deep ocean or to areas with larger tolerance to changes in salt water concentration. Of course, this would require a considerable amount of conveying pipes and transport devices.

Due to unpredictable surges in water caused by earthquakes, heavy storms, or typhoons, there is always the possibility of tearing the expensive membranes. The fouling of membranes is also an issue and may affect the aquatic environment. The mouths of rivers and sea fronts will have to be altered to give way to large tanks that will contain the membranes. Perhaps the only major advantages to the salinity gradient technology are its relatively quiet operation and the minimal amount of exhaust gases emitted. The development of new materials, which when disposed of are not harmful to the environment, is a crucial pursuit.

7.11 Conclusions

Salinity power is one of the largest sources of renewable energy that has not been fully exploited. The potential worldwide output is estimated to be 2,000 TWh annually (IRENA, 2014). However, the salinity gradient concept is not yet widely known. A considerable amount of technological development is necessary to fully utilize this renewable energy resource. The most important component of these systems, the impermeable membranes, are too costly at present. Also important is considering the harsh salt environment and the lack of efficient and suitable plant component parts. While the potential cost of energy from salinity gradient sources is higher than most traditional hydro power, it is comparable to other forms of renewable energy that are already being produced in full-scale plants.

There are numerous applications around the world for fully dedicated salinity gradient power plants. However, the combination of salinity gradient and solar power technology is still a practical application that takes advantage of combined energy from the sun and the pressure difference between fresh and salt water. Numerous wastewater treatment plants around the world close to areas with access to salt water as well as deltas where the river ends and the sea begins are ideal applications.

Future research on salinity gradient involves lowering the costs of membranes as well as improvement in overall efficiency of conversions, particularly power densities. The reader is warned that many costs and numbers reported in this chapter come from research undertaking. Only commercial undertaking based on actual situations will yield truly representative estimates on overall costs of salinity gradient systems for power generation.

7.12 Problems

7.12.1 Sensible Heat from Solar Pond

P7.1 Half a hectare [1.235 acres] of pond with a depth of 6.1 meters [20 ft] is heated by the sun. The average temperature of the pond has risen from 25°C to 30°C

[77°F to 86°F] on a hot day. Determine the amount of solar energy absorbed by this pond using the sensible heat Equation 7.1. Assume the density of sea water to be 1,027 kg/m³ [8.56 lbs/gal] and its specific heat equal to 3.93 kJ/kg°C [0.94 Btu/lb°F].

7.12.2 Theoretical Carnot Cycle Efficiency

P7.2 Calculate the theoretical Carnot Cycle efficiency of a solar pond which has a high reservoir temperature of 90°C [194°F] and a low reservoir temperature of 25°C [77°F]. If the net output or work done in the system is 200 kW [268 hp], at this theoretical efficiency, how much heat is put out of the system?

7.12.3 Osmotic Pressure Calculations

P7.3 Determine the osmotic pressure in units of kg/cm² [psi] for the following data. The temperature is 650 K, the molar concentration of salt ions is equal to 1.128 mol/L, and the universal gas constant is 0.82 L-bar/deg-mol.

7.12.4 Work Done against Pressure

P7.4 Calculate the work done against pressure using the following data: pressure equal to 60 kg/cm² [851.6 lbs/in²]; 1 liter of water pushed against an area of 2 cm² [0.31 in] over a distance of 20 meters [65.6 ft].

7.12.5 Energy Required to Boil Seawater

P7.5 Determine the amount of energy required to boil a kg of salt water from 30°C to 100°C [86°F to 212°F]. Assume the specific heat of water to be 3.93 kJ/kg°C [0.939 Btu/lb°F]

7.12.6 Size of PRO Unit to Generate Given Power

P7.6 The PRO salinity gradient prototype has a net power density of 15 W/m² [0.00187 hp/ft²]. Determine the size (in m²) of the polyamide membrane needed to power a household with a monthly requirement of 1,000 kWh.

7.12.7 Amount of Membrane to Use to Generate Power for a Household

P7.7 If the average household in the United States uses around 12,000 kWh of electricity per year, how much membrane is needed to power 1 household if the yearly production of this membrane is 30 MWh/yr per m²?

7.12.8 RED Salinity Gradient System

P7.8 The San Lorenzo River in California has an average annual discharge of 480 m³/s [7.6 million gpm]. The average salinity at this location is 0.6 mol/L [2.271 mol/gal]. The nearby fresh water salinity is 0.003 mol/L [0.0114 mol/gal]; the salinity ratio is reported at 1:200. The average temperature is 25°C [77°F]. Use the Nernst equation (Equation 7.4) to calculate the voltage generated.

Calculate the theoretical power that may be generated from this facility if the amount of energy produced is 0.74 MJ/m^3 [2.66 Btu/gal] of flow instead of the erroneously reported 1.4 MJ/m^3 [5.03 Btu/gal]. If the overall efficiency is 1%, determine the estimated power output from this system.

7.12.9 Cost of RED Power Plants

P7.9 Determine the total capital cost of a 500 MW RED salinity gradient power plant if the reported unit cost is $7.84/50 kW. Hack (2011) reported that the practical cost of salinity gradient plant is around $3 million/MW. Determine the capital cost if this is the new estimate.

7.12.10 Simple Payback Period for Salinity Gradient Power Plant

P7.10 Make a rough estimate of the payback period in years for a 125 MW salinity gradient facility that costs $375 million. The main revenue will come from selling electricity at $0.08/kWh. Assume the payback period is simply based on the ratio of the initial capital cost to the initial capital investment. Assume also that operation will be for 365 days in a year and 24 hours in a day.

References

Alvarez-Silva, O. A., A. F. Osorio and C. Winter. 2016. Practical global salinity gradient energy potential. Renewable and Sustainable energy Reviews 60 (2016): 1387–1395.

Banchik, L. D., M. H. Sharqawy and J. H. Lienhard. 2014. Limits of power production due to finite membrane area in pressure retarded osmosis. Journal of Membrane Science 468: 81–89. dx.doi.org/10.1016/j.memsci.2014.05.021.

Chu, J. 2014. The power of salt—power generation from the meeting of river water and sea water. Nanowerk Newsletter Email Digest. August 24. Available at: https://www.nanowerk.com/news2/green/newsid=37007.php. Accessed April 23, 2018.

Han, G. and T. S. Chung. 2014. Robust and high performance pressure retarded osmosis hollow fiber membranes for osmotic power generation. AIChE Journal 60, no. 3: 1107–1119.

IRENA. 2014. Salinity gradient technology brief. International Renewable Energy Agency (IRENA): Ocean Energy Technology Brief 2. June 2014. Available at: https://www.dutchmarineenergy.com/about-us/downloads/8%20-%20Salinity%20Gradient%20Energy_IRENA.pdf. Accessed April 23, 2018.

Kurihara, M. 2012. Toray Industries: Paper presented at the Fifth International Desalination Workshop, Jeju Island, 31 October 2012.

Kurihara, M. and M. Hanakawa. 2013. Mega-ton water system: Japanese National Research and Development Project on Seawater Desalination and Wastewater Reclamation. Desalination 308: 131–137.

Laflamme, C. 2012. Development of a salinity gradient power resource analysis by INES. International Workshop on Salinity Gradient Energy, 4–6 September 2012, Milan, Italy. Available at: www.energy-workshop.unimib.it/index.php?index=5. Accessed June 27, 2019.

Lu, H. and A. H. P. Swift. 2001. El Paso solar pond. Journal of Solar Energy Engineering 123: 178–180.

Mother Earth News. 1980. Israel's 150 kW solar pond. Mother Earth News: The Original Guide to Living Wisely. May/June. Available at: https://www.motherearthnews.com/renewable-energy/solar-pond-zmaz80mjzraw. Accessed April 23, 2018.

Patel, S. 2014. Statkraft shelves osmotic power project. Power Magazine. March 1. Business & Technology for the Global Generation Industry. Available at: http://www.powermag.com/statkraft-shelves-osmotic-power-project/. Accessed April 23, 2018.

Pattle, R. E. 1954. Production of electric power by mixing fresh and salt water in the hydroelectric pile. Nature 174 (1954): 660. doi:10.1038/174660a0. Available at: https://www.nature.com/articles/174660a0. Accessed April 25, 2018.

Post, J. 2009. Blue energy: Electricity production from salinity gradients by reverse electrodialysis. PhD Thesis, November 2009, Environmental Technology Department, Wageningen University, Wageningen, Netherlands.

Stenzel, P. and H. J. Wagner. 2010. Osmotic power plants: Potential analysis and site criteria. Paper presented at the 3rd International Conference on Ocean Energy, 6 October 2010, Bilbao, Spain.

STM. 2017. Solar thermal power and clean energy technology. Solar Thermal Magazine. Available at: https://solarthermalmagazine.com/learn-more/the-solar-pond/. Accessed April 23, 2018.

US EIA. 2017. International Energy Outlook 2017. #EIO2017. U.S. Energy Information Administration, Washington, DC. September 14. Available at: https://www.eia.gov/outlooks/ieo/pdf/0484(2017).pdf. Accessed April 23, 2018.

Vermaas, D. A. 2014. Energy generation from mixing salt water and fresh water, smart flow strategies for reverse electrodialysis. PhD Thesis, University of Twente. Available at: https://www.waddenacademie.nl/fileadmin/inhoud/pdf/06-wadweten/Proefschriften/thesis_D_Vermaas.pdf. Accessed April 23, 2018.

Weintraub, B. 2001. Sidney Loeb. Bulletin of the Israel Chemical Society 8: 8–9. Available at: https://drive.google.com/file/d/1hpgY6dd0Qtb4M6xnNXhutP4pMxidq_jqG962VzWt_W7-hssGnSxSzjTY8RvW/edit. Accessed April 23, 2018.

Williams, A. 2018. Norway's osmotic power a salty solution to the world's energy needs? Water and Wastewater International Magazine. Available at: http://www.waterworld.com/articles/wwi/print/volume-28/issue-2/regional-spotlight-europe/norway-s-osmotic-power-a-salty.html. Accessed April 23, 2018.

8

Fuel Cells

Learning Objectives

Upon completion of this chapter, one should be able to:

1. Describe the concept of generating energy and power from fuel cells.
2. Enumerate the various types of fuel cell systems.
3. Enumerate various input materials or feedstock that may be used for fuel cell systems.
4. Describe the various applications of fuel cell systems.
5. Describe the various conversion efficiencies of fuel cell systems.
6. Relate the future of fuel cell technology based on overall environmental and economic issues.

8.1 Introduction

A fuel cell is an electrochemical energy conversion device. It converts the chemical energy of a fuel (e.g., hydrogen) directly into electrical energy. The fuel and an oxidizing agent (usually oxygen from air) are continuously but separately supplied to the two electrodes of the cell, at which they undergo a reaction. Figure 8.1 shows the typical schematic and structure of a fuel cell that uses hydrogen gas and oxygen from the air. This is a typical proton exchange membrane fuel cell (PEM). The hydrogen fuel is channeled through the field flow plates to the anode on one side. The oxygen from the air is channeled to the cathode on the other side of the cell. A platinum catalyst is used at the anode, causing the hydrogen to split into positive hydrogen ions (protons) and negatively charged electrons. This PEM allows only the positively charged ions to pass through to the cathode. The negatively charged electrons must travel along the external circuit to the cathode, creating an electrical current. The electrons and positively charged hydrogen ions combine with oxygen at the cathode to form water, which then flows out of the cell.

The efficiency of electricity generation using a heat engine is limited by the Carnot Cycle efficiency, which is in the range of only 35% to 45%. Fuel electricity production is not subject to this limitation, and practical efficiencies of 60% may be achieved. The theoretical conversion efficiency of an ideal fuel cell of this type is a good 83% (Larminie and Dicks, 2013).

The main issue behind the use of fuel cells is the slow reaction rate, leading to low currents and power. Additionally, its source of input power is hydrogen gas, which nowadays

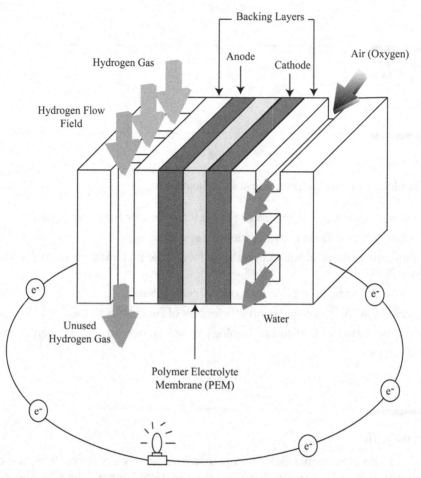

FIGURE 8.1
Typical design of a fuel cell (US DOE, 2018).

is produced from steam methane reforming due to the abundance of natural gas as a by-product of the oil industry.

The other principle behind the hydrogen-oxygen fuel cell, aside from the use of platinum material for the electrodes, is the use of a well-conducting electrolyte. Potassium hydroxide (KOH) or sulfuric acid (H_2SO_4) may be used (Brunton, et al., 2010). The schematic details are shown in Figure 8.2. In the figure, two platinum electrodes are immersed in the electrolyte. One electrode (the negative side) is supplied with hydrogen, while the other electrode is supplied with oxygen (the positive side). An electrical potential difference of 0.90 to 1.2 volts can be measured between the two electrodes—shown in Equation 8.1 for the negative electrode and Equation 8.2 for the positive electrode.

In the equations shown, for every molecule of hydrogen consumed, two electrons pass from the negative to the positive electrode, where they react with absorbed oxygen. Water is produced by this reaction:

$$2H_{ad} \rightarrow 2H^+ + 2e^- \tag{8.1}$$

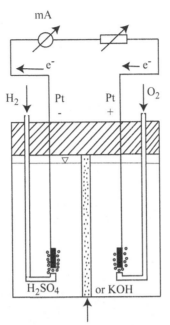

FIGURE 8.2
Basic schematic of a hydrogen-oxygen fuel cell and electrolytes used.

$$\frac{1}{2}O_2 + 2H^+ + 2e^- \rightarrow H_2O \tag{8.2}$$

Another way of looking at the fuel cell reaction is shown in Figure 8.3. Here, the overall reaction is shown:

$$H_2 + \frac{1}{2}O_2 \rightarrow H_2O \tag{8.3}$$

In Figure 8.3 and Equation 8.3, one will observe the "combustion" of hydrogen and oxygen, whereby heat energy is released. In the electrochemical reaction between hydrogen and oxygen in a fuel cell, electricity and heat are produced. Heat produced by combustion is not the same as electricity produced in a fuel cell. The difference is related to the Gibbs free energy.

The high heating value for the combustion of hydrogen is 285.8 kJ/kmol, but the Gibbs free energy for the reaction—and therefore the maximum electricity produced by the fuel cell—is only 237.2 kJ/kmol. The difference, which is 48.6, appears as heat produced in the fuel cell. The practical electricity produced is around 154 kJ/kmol (US DOE, 2013). Examples 8.1, 8.2, and 8.3 show how the calculations for efficiencies are made.

Example 8.1: Practical Fuel Cell Conversion Efficiency

Determine the practical conversion efficiency of a hydrogen fuel cell if the electricity produced is 154 kJ/kmol and the energy content of hydrogen is 285.8 kJ/kmol.

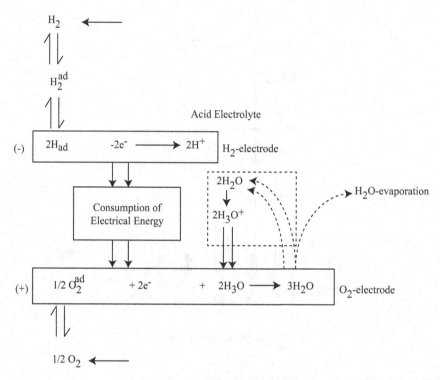

FIGURE 8.3
Another illustration of fuel cell reactions.

SOLUTION:

a. The conversion efficiency is simply the ratio of the practical electricity generated versus the heating value of hydrogen:

$$Practical\ Energy\ Conversion\ Efficiency(\%) = \frac{154\frac{kJ}{mol}}{285.8\frac{kJ}{mol}} \times 100\% = 54\%$$

b. Hence, the practical efficiency of a hydrogen oxygen fuel cell is about 54% much higher than the Carnot Cycle efficiency if hydrogen is burned in an engine.

Example 8.2: Theoretical Fuel Cell Conversion Efficiency

Determine the theoretical conversion efficiency for the hydrogen fuel cell if the electricity produced is 237 kJ/kmol and the energy content of hydrogen is 285.8 kJ/kmol.

SOLUTION:

a. The theoretical conversion efficiency is simply the ratio of the maximum electricity generated versus the heating value of hydrogen:

$$Theoretical\ ECE(\%) = \frac{237.2\frac{kJ}{mol}}{285.8\frac{kJ}{mol}} \times 100\% = 83\%$$

b. Hence, the theoretical efficiency of a hydrogen oxygen fuel cell is about 83%—much higher than the Carnot Cycle efficiency if hydrogen is burned in an engine.

Example 8.3: Heat Energy Losses in Fuel Cells

Determine the percentage of energy loss for the hydrogen fuel cell if the heat produced in a fuel cell is 48.6 kJ/kmol. The energy content of hydrogen is 285.8 kJ/kmol.

SOLUTION:

a. The percentage of energy losses is simply the ratio of the heat loss to the heating value of hydrogen:

$$Percentage\ Losses(\%) = \frac{48.6 \frac{kJ}{mol}}{285.8 \frac{kJ}{mol}} \times 100\% = 17\%$$

b. Hence, the theoretical efficiency of a hydrogen oxygen fuel cell is about 83%—much higher than the Carnot Cycle efficiency if hydrogen is burned in an engine.

8.2 The Various Types of Fuel Cells

There have been numerous types of fuel cells developed over the years (Larminie and Dicks, 2013). The basic premise is the same: to convert the chemical energy of a fuel into electricity. Hundreds of possibilities exist, but the economics of fuel cell technology relies on the chemicals or gaseous materials used for the reaction. The cheaper the input feedstock, the better.

8.2.1 Proton Exchange Membrane Fuel Cells

The most popular among fuel cell types is the proton exchange membrane (PEM) fuel cell, where the gaseous hydrogen is used. The selection of anode, cathode, and the electrolyte will also change, and this causes complications in design and its efficiencies. The membrane separates the anode and the electrolyte on one end and the cathode and the electrolyte on the other end. Other carbonaceous gases may also be used, such as methane and liquid compounds like methanol, diesel, and chemical hydrides. If hydrogen is used, the product of reaction will be water; if a carbonaceous compound is used, the product is mainly carbon dioxide. In the past, hydrogen was used and was produced first using steam reforming or methane reforming.

The same principle holds with the production of electricity. The hydrogen gas dissociates into its proton and electron components. On the anode side, hydrogen diffuses to the anode catalyst, where it later dissociates into protons and electrons. The protons react with oxidants, in this case, oxygen. The protons are conducted through the membrane to the cathode, but the electrons are forced to travel in an external circuit, supplying the desired output power. This is because the membrane is electrically insulating. On the cathode catalyst side, oxygen molecules react with the electrons and protons to produce water. Note that these electrons have traveled through the external circuit.

Cost is again the issue in why this system has not yet been fully commercialized. The best catalyst is platinum, a very expensive material. As a result, there is a move toward the use of inexpensive non-metal catalysts, such as carbon nanotubes. The use of water can also become an issue, as the process involves evaporation. This must be controlled; otherwise, the issue of water replenishment will be a problem for larger systems. The reaction is also exothermic, and a large quantity of heat is generated, especially by large systems.

The proton exchange fuel cells work using a polymer in the form of a thin, permeable sheet. The efficiency is about 40% to 50%, and the operating temperature is about 180°C [175°F]. The cell output ranges from 50 to 250 kW [67 to 335 hp] (US DOE, 2013). The solid flexible electrolyte will not leak or crack, and these cells operate at a low temperature, making them suitable for home use or car use. However, the fuel input must be purified, and the solid catalyst is usually platinum, making the unit rather costly.

8.2.2 High-Temperature Proton Exchange Membrane Fuel Cell

The high-temperature proton exchange fuel cell (HT-PEM; also called the Bacon cell) was also discovered in 1959 by Thomas Bacon and works in the same way as the PEM discussed above except that a higher temperature (343 K to 413 K) [617.4°R to 743.4°R] in which the cell operates most efficiently is needed. The electrodes used consist of porous carbon electrodes where novel metals such as platinum are impregnated. Some other catalysts have been tested, such as silver (AG) and cobalt-oxygen (CoO). The electrolytes used are simply concentrated solutions of potassium hydroxide (KOH) or sodium hydroxide (NaOH). The elements used are hydrogen and oxygen gas. The electrical potential output of this system is around 0.9 volts (Jungmann, 2010).

8.2.3 Direct Methanol Fuel Cell

The direct methanol fuel cell (DMFC) is in the family of proton-exchange fuel cells except that methanol is used as a fuel. The single advantage is that methanol is more easily transported than hydrogen gas. However, the efficiency for this system is quite low such that its targeted applications are low-power devices and portable units. There are groups that adhere to a hypothesized methanol economy, similar to hydrogen economy advocates. The word "direct" means that pure methanol is used instead of converting it to hydrogen:

$$Anode\ Reaction\ CH_3OH + H_2O \rightarrow 6H^+ + 6e^- + CO_2 \tag{8.4}$$

$$Cathode\ Reaction\ \frac{3}{2}O_2 + 6H^+ + 6e^- \rightarrow 3H_2O \tag{8.5}$$

$$Overall\ Reaction\ CH_3OH + \frac{3}{2}O_2 \rightarrow 2H_2O + CO_2 \tag{8.6}$$

Pure methanol is not used for the reaction. Only about 3% by mass is used (or 0.1 molar) to carry the reactant into the cell. The operating temperatures are also not high and range from 60°C to 130°C [140°F to 266°F] (FCT, 2014a). If a high temperature is desired, the system is usually put under pressure. Higher temperatures would sometimes be desired to take advantage of higher efficiencies at those conditions. The low concentration of methanol used brings down the overall efficiency of the system. One issue with the use of these compounds is the loss of some methanol that permeates through the membrane, and the development of better membrane is warranted. Methanol is also toxic and has the potential to cause blindness. Note also that carbon dioxide is produced, making the system susceptible to being categorized as a non-sustainable environmental greenhouse gas producer. As such, there is a benefit to using methanol if this compound is produced in a sustainable manner, like methanol derived from wood. If methanol is produced from fossil fuel–based feedstock, then it will cease to become a renewable technology.

Example 8.4: Water Production in Methanol Fuel Cells

Determine the amount of water (cubic meter and liters) produced for every tonne of methanol used if a methanol fuel cell system is used. Use the ideal Equation 8.6 for the calculation.

SOLUTION:

a. The amount of water is simply calculated as follows:

$$\frac{H_2O}{CH_4O} = \left[\frac{(1\times2)+(16\times1)}{(12\times1)+(1\times4)+(16\times1)}\right] = \frac{18}{32} = 0.5625\frac{kg}{kg}$$

$$\frac{H_2O}{CH_4O} = 0.562\frac{kg}{kg}\times\frac{1000\ kg\ H_2}{tonne}\times\frac{1,000\ L}{m^3}\times\frac{m^3}{1,000\ kg} = 562\frac{L}{tonne}$$

$$\frac{H_2O}{CH_4O} = 562\frac{L}{tonne}\times\frac{gallon}{3.785\ L}\times\frac{1\ tonne}{1.1\ ton} = 135\frac{gal}{ton}$$

b. About 562 liters [135 gal] of water is produced per tonne of methanol used in an ideal methanol fuel cell system.

8.2.4 Alkaline Electrolyte Fuel Cell

An alkaline electrolyte fuel cell (AEFC), as the name implies, uses an alkaline solution, such as potassium hydroxide, as an electrolyte. If air is used as one compound for the reaction, the aqueous alkaline does not reject carbon dioxide, and this would cause a problem within the system. The KOH will be converted into potassium bicarbonate and render the alkaline solution unusable or "poisoned." Hence, these systems are usually required to use pure oxygen as the oxidant instead of air or use purified air to scrub as much carbon dioxide as possible without increasing the expense in removing them totally. Alkaline was used as a fuel cell in the mid-1960s, and this type was also known as the Bacon fuel cell after its British inventor, Thomas Bacon. Many NASA shuttles in the 1960s used this type of fuel cell in their Apollo missions. They are among the first types of fuel cell ever developed with efficiencies reaching 70% at most times. In this original application, pure hydrogen and oxygen were used. Their use was justified then for their efficiency in spaceships. AEFCs are popular due to the fact that they are the cheapest to manufacture. The catalyst required could be any number of different chemicals that are inexpensive compared with those required of other fuel cell types. The commercial aspect of this type lies in the improvement of the use of multiple electrodes using a bi-polar plate version instead of mono-plate versions. The world's first fuel cell ship "HYDRA" used an AEFC system with 5 kW net output (US DOE, 2013).

Another recent development is the use of the solid-state alkaline fuel cell. This unit uses alkali anion exchange membranes rather than liquid. This resolves the problem of "poisoning" and allows the development of alkaline fuel cells capable of running on safer hydrogen-rich carriers, such as liquid urea solutions or metal amine complexes (Oshiba, et al., 2017).

The alkaline fuel cell has the general governing relationships shown in Equations 8.7 and 8.8 at the anode and the cathode, respectively:

At the anode

$$2H_2 + 4OH^- \rightarrow 4H_2O + 4e^- \qquad (8.7)$$

At the cathode

$$O_2 + 4e^- + 2H_2O \rightarrow 4OH^- \qquad (8.8)$$

The OH^- ions must be able to pass through the electrolyte, and there must be an electrical circuit for the electrons to go from the anode to the cathode. Twice as much hydrogen than oxygen is needed in this case.

The reactions occurring in the electrolytes are shown in Equations 8.9 and 8.10. Equation 8.9 shows the precipitation of K_2CO_3—this compound blocks the electrode power. This will lead to a decrease in the hydrophobicity of the electrode backing layer, leading to structural degradation and electrode flooding. On the other end, the hydroxyl ions in the electrolyte can also react with carbon dioxide, as shown in Equation 8.10, to form carbonate species. The amount of precipitates produced per weight of potassium hydroxide is given in Example 8.5:

$$CO_2 + 2KOH \rightarrow K_2CO_3 + H_2CO_3 \qquad (8.9)$$

$$2OH^- + CO_2 \rightarrow CO_3^{2-} + H_2O \qquad (8.10)$$

Example 8.5: Bicarbonate Production in Alkaline Fuel Cells

Determine the amount of bicarbonates (carbonic acid, H_2CO_3) (in kg or lbs) produced for every kg [lbs] of potassium hydroxide used in the alkaline fuel cell following Equation 8.9.

SOLUTION:

a. The amount of bicarbonate produced per weight of potassium hydroxide is simply calculated as follows:

$$\frac{H_2CO_3}{2KOH} = \left[\frac{\left((1 \times 2) + (12 \times 1) + (16 \times 1)\right)}{2 \times (39.0983 + 16 + 1)} \right] = \frac{30}{112.2} = 0.267 \frac{kg}{kg} \frac{H_2CO_3}{KOH}$$

b. About 26.7% of the weight of potassium hydroxide forms into carbonate.

8.2.5 Phosphoric Acid Fuel Cell

The phosphoric acid fuel cell (PAFC) was first designed and introduced in 1961 by G. V. Elmore and H. A. Tanner (FCT, 2014b). Phosphoric acid is used as a non-conductive electrolyte to pass hydrogen ions from the anode to the cathode. These cells commonly work in temperatures that range from 150°C to 200°C [302°F to 392°F]. This high temperature will cause heat and energy losses if the heat is not removed or used properly. Phosphoric acid is relatively cheap but is corrosive, and unit components must be able to withstand its harmful effects. What to do with the heat can vary from system to system. Some systems propose the use of heat for steam generation and ultimately power production, thereby increasing its overall efficiency. Others use this heat for cooling or air conditioning systems using absorption refrigeration systems. Others simply use the heat for various thermal applications. The increase in efficiency will bring the initial efficiency from 40% to 50% to as high as 80% (FCT, 2014b). Because phosphoric acid is nonconducting, the electrons are forced to go through the circuit producing power. Hydrogen and oxygen are still used as input gases. The hydrogen ion production rate is, however, quite small, and some other catalysts, like platinum, are still used to increase this ionization rate. The use of platinum would increase the initial cost of the system in this case.

8.2.6 Solid Oxide Fuel Cell, High Temperature

The solid oxide fuel cell (SOFC), as the name implies, uses solid materials such as ceramics (yttria-stabilized zirconia, or YSZ) as the main structural component. These systems require high operating temperatures of 100°C to 1,000°C [212°F to 1,832°F] (Singhal and Kendall, 2003). A positive draw is that these systems can run on fuels other than hydrogen gas. Ultimately, hydrogen is needed and must be produced internally by reforming other gases, such as methane, propane, or butane. SOFC systems are unique such that the negatively charged oxygen ions travel from the cathode (positive side of the fuel cell) to the anode (negative side of fuel cell) instead of positively charged ions traveling from the anode to the cathode. Oxygen gas is fed through the cathode, where it absorbs electrons to create oxygen ions. The oxygen ions then travel through the electrolyte to react with hydrogen gas at the anode. The reaction at the anode produces electricity and water as by-products. Carbon dioxide may also be a by-product depending on the fuel used, although carbon emissions are said to be significantly lower than those of fossil fuel combustion. The chemical reactions for an SOFC system are shown in Equations 8.11, 8.12, and 8.13:

$$\text{Anode Reaction } 2H_2 + 2O^{2-} \rightarrow 2H_2O + 4e^- \tag{8.11}$$

$$\text{Cathode Reaction } O_2 + 4e^- \rightarrow 2O^{2-} \tag{8.12}$$

$$\text{Overall Reaction } 2H_2 + O_2 \rightarrow 2H_2O \tag{8.13}$$

The main issue with SOFC is the high operating temperature coupled with the slow start-up time. New developments are trying to solve this issue. In the United Kingdom, a company called Ceres Power has developed a method that reduces the operating temperature to a lower range of 500°C to 620°C [932°F to 1,148°F] by replacing the YSZ structure with cerium gadolinium oxide (CGO) (Green Car Congress, 2016). The lower temperature has another advantage. They are able to use stainless steel instead of ceramics as support material. These replacements also improve the start-up time for the system. Example 8.6 shows how much water is produced per unit weight of hydrogen used in an SOFC system.

Solid oxide fuel cells may use other hard ceramic compounds made of metal oxides, such as calcium or zirconium, as electrolytes. In some applications, oxygen is used as electrolyte.

The efficiency is around 60%, and the operating temperatures are around 1,000°C [1,800°F], making this a high-temperature fuel cell. The fuel cell output could be about 100 kW [134 hp] (Singhal and Kendal, 2003). At this high temperature, a reformer to extract hydrogen is not required, and waste heat can be recycled to make additional electricity.

Example 8.6: Water Production in Solid Oxide Fuel Cell Systems

Determine the amount of water (cubic meter and liters) produced for every tonne of hydrogen used if an SOFC system is used. Use the ideal Equation 8.13 for the calculation.

SOLUTION:

a. The amount of water is simply calculated as follows:

$$\frac{2H_2O}{2H_2} = \left[\frac{(1\times2)+(16\times1)}{(1\times2)}\right] = 9\frac{kg}{kg}\times\frac{1,000\ kg\ H_2}{tonne\ H_2}\times\frac{1\ m^3\ H_2O}{1,000\ kg\ H_2O} = 9\frac{m^3\ H_2O}{tonne\ H_2}$$

$$\frac{H_2O}{H_2} = 9\frac{m^3}{tonne}\times\left(\frac{3.28\ ft}{1\ m}\right)^3\times\frac{1\ tonne}{1.1\ ton}\times\frac{7.48\ gal}{1\ ft^3} = 2,160\frac{gal}{ft^3}$$

b. About 9,000 liters [2,160 gal/ft³] of water is produced per tonne of hydrogen used in an ideal SOFC system.

8.2.7 Solid Acid Fuel Cell

The solid acid fuel cell (SAFC) uses solid acid sulfates and selenates such as $CsHSO_4$ or $CsHSeO_4$ (Haile, et al., 2001). These compounds belong to a class of compounds that exhibit super-protonic phase transitions at slightly elevated temperatures. This gives rise to very high conductivity and proton transport that does not rely on the presence of water and is therefore a very good electrolyte for fuel cell applications. These types are still under research and development. Research reports show that this type of fuel cell will remove complexity and, as a result, reduce the cost because of the impermeability of the electrolyte through membranes used. There are still some issues to be resolved. The sulfate and selenite–based acid compounds eventually react with hydrogen gas in the anode chamber in the presence of fuel cell catalysts. This generates a by-product such as hydrogen sulfide or H_2Se. These compounds then poison the fuel cell, especially with longer periods of operation. The report also shows power densities of over 400 mW/cm² [0.5 hp/ft²].

8.2.8 Molten Carbonate Fuel Cell, High Temperature

The molten carbonate fuel cell (MCFC) uses lithium potassium carbonate salt as an electrolyte. However, this salt requires a high temperature similar to that of the SOFC system. Its operating temperature is reported at 650°C [1,200°F] (Bischoff, 2006). Likewise, the high temperature is required to move the charge within the cell (the negative carbonate ions). Similar to SOFCs, MCFCs are capable of converting fossil fuels to hydrogen-rich gas in the anode, eliminating the need to generate hydrogen separately. The reforming processes generate CO_2 emissions. Candidate fuels for this type of process include methane, biogas from anaerobic digestion, or gaseous compounds generated from the conversion of coal. The hydrogen in the gas reacts with the carbonate ions from the electrolyte to produce water, carbon dioxide, electrons, and other chemical by-products. The electrons travel through an external circuit, creating electricity, and return to the cathode. There, oxygen from the air and carbon dioxide recycled from the anode react with the electrons to form carbonate ions that replenish the electrolyte, completing the circuit. The governing reactions are shown in Equations 8.14, 8.15, and 8.16:

$$Anode\ Reaction\ CO_3^{2-} + H_2 \rightarrow H_2O + CO_2 + 2e^- \tag{8.14}$$

$$Cathode\ Reaction\ CO_2 + \frac{1}{2}O_2 + 2e^- \rightarrow CO_3^{2-} \tag{8.15}$$

$$Overall\ Reaction\ H_2 + \frac{1}{2}O_2 \rightarrow H_2O \tag{8.16}$$

Note the similarities with Equations 8.9 and 8.6. The same amount of water is produced. Likewise, this type has slow start-up times and, with its corresponding high temperature, will not be suitable for vehicles or for standby power generation. There are reports that this system has a short shelf life as well. This is perhaps due to the corrosion of electrodes from high temperatures. The only reported advantage of this system is its resistance to impurities, which suggests that even low-quality fuel like coal could be a potential energy source. The other advantage is the reported high efficiency over SOFC. Efficiency was reported close to 50%, which is superior to the phosphoric acid system with low reported efficiency (37% to 42%). Using this system for combined heat and power (CHP) will further improve the efficiency to over 80% (Bischoff, 2006).

A company that sells and develops MCFCs is the Fuel Cell Energy Company based in Connecticut. They have reported power output range from 300 kW to 2.8 MW that can achieve 45% to 47% electrical efficiency and can improve upon efficiency if the system is used as a combined heat and power source (Bove and Moreno, 2008).

8.2.9 Regenerative Fuel Cell

The regenerative fuel cell is also called a reverse fuel cell (RFC). In short, they consume electricity as one chemical is converted (or dissociates) into another. Any of the above fuel cell types, when operated in reverse, fall under this category. As with the classic reaction of a hydrogen fuel cell combining with oxygen to produce water, the reverse could also be made to happen. In this case, water is dissociated into hydrogen gas and oxygen gas, which is a classic electrolysis reaction. The solid oxide fuel cell operation could also be performed in reverse, and this is called the solid oxide regenerative fuel cell. If applied to the SOFC system, it has been found that less electricity is required to reverse the process, perhaps due to the high required temperature. Note that in this application, the idea of power production becomes secondary and fuel production becomes the primary goal. Recalling the use of a solar photovoltaic cell to undergo electrolysis, fuel cell technology perhaps becomes a better option to utilize electrical power and produce combustible fuel (like hydrogen) from water. We are now seeing numerous research projects identifying novel compounds for the production of hydrogen fuel.

A recent study in Sweden (Zhu, et al., 2006) shows that ceria-based composite electrolytes can produce high current output for fuel cell operation and high hydrogen output for electrolysis operation. Zirconia doped with scandia and ceria was also investigated as a potential electrolyte in SORFC for hydrogen production at intermediate temperatures (500°C to 750°C) [932°F to 1,382°F]. The rationale behind these research projects is the possibility of recycling the water into fuel as well as using the fuel to produce electricity while generating water as a product. This would be a very good scenario for unmanned vehicles that rely on both electricity for their propulsion as well as fuel for other uses, or vice versa. This idea opens up new realms of applications and possibilities for many unmanned vehicles that will not require refueling. Ultimately, water losses through evaporation and leakage will cause the system to run out of this important fuel to produce the hydrogen gas. There are numerous other applications that can be enumerated for use in military operations, space travel, and even surveillance systems that are unmanned.

8.2.10 Solid Polymer Fuel Cell

The unique characteristic of a solid polymer fuel cell (SPFC) is the use of solid polymer as the electrolyte membrane. A fuel cell of this type is built around an ion-conducting solid membrane, such as Nafion material. This material is a trademark for a perfluorosulfonic acid membrane (Schumm, 2018). The electrodes are catalyzed carbon. Several construction alignments are feasible. Solid polymer electrolyte cells function well, but cost estimates are still rather high. The Gemini spacecraft, for example, has attested to the performance of this type of fuel cell. Likewise, this category also includes many other fuel cell types discussed above as long as they use solid membranes instead of liquid forms like phosphoric acid. System categorizations can overlap. We can also call an alkaline fuel cell a solid polymer fuel cell if it uses a solid polymer membrane. Most literature describing fuel cells can be so unclear in categorization that standardization may be warranted in the future to simplify the identification of various types. Some textbooks have attempted to start drawing the line between the many fuel cell types that are under development. It will take many more years before a majority will agree on delineation between similar types. Suffice it to say at this point that any researcher is welcome to differentiate his or her work from other technologies already developed.

8.2.11 Zinc-Air Fuel Cell

The zinc-air fuel cell (ZAFC) generates electricity by the reaction between oxygen and zinc pallets in a liquid alkaline electrolyte (Sapkota and Kim, 2009). This system was reported to be quite efficient and cheap since cheap zinc is required in place of precious metals. The only products produced are oxides of zinc, which are described to be benign without harmful gaseous emissions. These applications were reported in military operation, portable gadgets, and mobile and stationary systems. Research for this system is still in its infancy, and no commercial-scale systems have been developed. There are still numerous technical barriers to overcome. These include longer usage of the zinc-air modules to prolong the replacement of zinc anode cassettes. There need to be zinc anode regeneration facilities for centralized recycling as well. Many studies are aiming for the use of this system in vehicles. One such study combines 47 cells in a module with an open-circuit voltage of 67 volts and a capacity of 325 Ah (ampere-hours). The energy capacity was reported at 17.4 kWh with peak power at 8 kW [10.7 hp]. The unit weighs 88 kg [193.6 lbs] with a space volume of 79 liters [20.9 gal]. The cell has a central static replaceable anode cassette comprising a slurry of electrochemically generated zinc particles in a potassium hydroxide solution, compacted into a current collection frame, inserted into a separator envelope, and flanked on two sides by high-power air reduction cathodes (where oxygen is taken). This part extracts oxygen from the air for the zinc oxidation reaction. This design was developed in the electric vehicle division of the FTA Bus program in Germany (Goldstein, et al., 1999). The governing equation is shown in Equation 8.17:

$$2Zn + O_2 \rightarrow 2ZnO\,(E_o = 1.65\ Volts) \tag{8.17}$$

In the equation, note that zinc becomes oxidized with time and renders the electrode unusable at some point. Hence, it is important to recover the oxidized zinc and bring it back into its near pure form. The use of potassium hydroxide (KOH) has been proposed for the zinc oxide dissolution and for compacting the zinc after purification. The concept is to use a bus battery and after some time replace the cassette in a centrally located regeneration station. Example 8.7 shows the amount of oxides of zinc produced per unit weight of pure zinc used in a zinc-air fuel cell.

Example 8.7: Zinc Oxide Production from Zinc-Air Fuel Cell Systems

Determine the amount of zinc oxide produced (in kg) per tonne of pure zinc used in a zinc-air fuel cell. Use the ideal Equation 8.17 for the calculation. The molecular weight of zinc is 65.38 g/mol, and that of oxygen is 16 g/mol.

SOLUTION:

a. The amount of water is simply calculated using Equation 8.10 as follows:

$$\frac{2ZnO}{2Zn} = \left[\frac{(2\times(65.38+16))}{2\times(65.38)}\right] = 1.245\,\frac{kg\ ZnO}{kg\ Zn} \times \frac{1,000\ kg}{tonne} = 1,245\,\frac{kg}{tonne}$$

$$\frac{2ZnO}{2ZN} = 1,245\,\frac{kg}{tonne} \times \frac{2.2\ lbs}{kg} \times \frac{1\ tonne}{1.1\ ton} = 2,490\,\frac{lbs}{ton}$$

b. About 1.245 tonnes [2,490 lbs/ton] of the oxidized form of zinc is produced per tonne of pure zinc used in an ideal zinc-air fuel cell system.

8.2.12 Microbial Fuel Cell

The idea of using microbes to undergo electrochemical reactions is interesting. A microbial fuel cell (MFC), sometimes called a biological fuel cell, is a bio-electrochemical conversion system that drives a current using bacteria through mimicking bacterial interactions found in nature. These systems are not new. As early as the 20th century, scientists demonstrated that microbes transfer electrons in the cell to the anode using what is termed a mediator, or a special chemical that enhances the electrochemical process. In 1911, M. Potter, a botany professor at the University of Durham, managed to generate electricity from *E. coli*, although the work did not receive major publicity. Similar work was followed upon by Barnet Cohen when he created a number of microbial fuel cells that, when connected in series, were capable of producing over 35 volts though a current of only 2 milliamps (Jabeen and Farooq, 2017). Others used hydrogen produced by the fermentation of glucose using *Clostridium acetobutylicum* as the reactant at the anode of a hydrogen and air fuel cell (Finch, et al., 2011). It was only in the 1970s that these special chemicals were not needed for the reaction to proceed. This is because of the identification of bacteria that have electronically active redox proteins in cytochromes on their outer membranes that can transfer electrons directly to the anode. Several books have already been written on this, with a very exciting vision of producing electrical power from wastewater that has a microbial population doing the job, virtually freeing wastewater treatment plants of the need for external power for water treatment. There are also some groups that design soil-based microbial fuel cells. There are indeed interesting studies toward proving that desalination work in the future may be solely achieved by microbial populations. Some research studies claim that microbes produce more energy than is required for the desalination process from wastewaters (Clark, 2015). Likewise, there are still numerous technical issues to overcome, including scale-up of systems, contamination of other micro-types, as well as the problem of varied output from an open air system as seasons and weather change, including variations in loading day to day.

8.2.13 Other Fuel Cells: Biological, Formic Acid, Redox Flow and Metal/Air Fuel Cells

There are many more fuel cell types being developed, some using newer compounds such as formic acid or other metals besides zinc such as magnesium (for a magnesium-air fuel cell). The possibilities are vast until innovative researchers hone in on those most viable. It becomes complicated as each researcher finds another solid or liquid catalyst or another fuel to use. Some researchers still doubt the future of fuel cells due to their reported lower efficiencies compared to those of currently available battery technology. Fuel cells could be better than engines, but they would have to compete with the very rapid development of new batteries.

It will take more years of research and commercialization before the widespread use of fuel cell technology. The single most important issue is the type of fuel to use for the reactions. The use of hydrogen fuel in these systems has been universal, but the U.S. economy has abandoned the hydrogen economy at this point—hydrogen production in the cheapest possible way cannot be a reality until fossil fuel–based hydrogen sources are economically competitive. It will take a major breakthrough for this technology to become mainstream. For now, it is simply exciting to explore new concepts, new ways to improve efficiencies, and cheaper materials that may be easily regenerated or recycled to their original pure forms.

8.3 Data for the Different Major Types of Fuel Cells

Five different types of fuel cells have emerged, and the data are shown in Table 8.1 (US DOE, 2013). The alkaline fuel cells have been used in many space vehicles with power ranging between 10 to 100 kW. Some portable electronics have used direct methanol fuel cells with power ranging from a low of 100 mW to a high of only 1 kW. Molten carbonate fuel cell types have larger power capacity, potentially up to 100 MW. These units are appropriate for medium- to large-scale combined heat and power. The efficiencies of these systems could be as high as 75%.

TABLE 8.1

Data for Different Types of Fuel Cells (US DOE, 2013)

Fuel Cell Type	Electrolyte	Qualified Power	Applications
Alkaline FC	Aqueous alkaline solution	10–100 kW	Space vehicles
Proton exchange membrane FC	Polymer membrane (ionomer)	1 W–500 kW	Lower-power CHP
Direct methanol FC	Polymer membrane (ionomer)	100 mW–1 kW	Portable electronics
Phosphoric acid FC	Molten phosphoric acid	<10 MW	200 kW CHP systems
Molten carbonate FC	Molten alkaline carbonate	100 MW	Medium to large scale CHP
Solid oxide FC	O^{2-} conducting ceramic oxide	<100 MW	All sizes of CHP up to MW in range

8.4 Various Fuels Used for Fuel Cells and Issues

The versatility of this fuel system is that numerous other fuels may be used. Unfortunately, most compatible fuels are still derived from fossil fuels or from fossil fuel–derived final fuel. Enumerated below are the primary fuels already being used:

a. Petroleum products (gasoline, naptha, etc.)

b. Coal and coal gases (lignite, bituminous coal)

c. Natural gas (CH_4)

d. Propane (C_3H_8)

e. Methanol (CH_3OH)

f. Electrolysis of water (H_2O)

g. Biomass producer/synthesis gas from pyrolysis and gasification (CO and H_2)

The majority are still derived from fossil fuels and are therefore not sustainable. Only the production of gases such as hydrogen or methane from biomass resources can make these systems renewable. Electrolysis of water may also be considered sustainable and renewable if the power used comes from renewable resources such as wind or solar energy. All the others would have to be justified on the cost of the fossil fuels. At present, many applications use hydrogen derived from fossil fuels such as natural gas. If steam were to be used, the system may be considered renewable only if the steam was likewise produced from renewable sources. For a resource to be considered renewable, the fuel used must be replenished at the same rate it is used. Following this, fuel cell technologies cannot be considered renewable. The practical efficiency of using propane in a fuel cell is shown in Example 8.8.

Example 8.8: Practical Conversion Efficiency for Propane-Based Fuel Cells

Determine the conversion efficiency for when propane (C_3H_8) was used as fuel in a fuel cell. The electricity produced was 603 mWh over 3,900 hours. The energy content of propane was reported as 93 MJ/m³ [2,498 Btu/ft³]. During the test, some 51,870 m³

[13.7 million gal] of propane was used. Compare this efficiency with that of internal combustion engines.

SOLUTION:

a. The conversion efficiency equation as shown in an earlier chapter is given below:

$$Efficiency(\%) = \frac{Output}{Input} \times 100\%$$

b. The input and output must have the same units. The input is calculated first:

$$Input\ Energy\left(MJ\right) = 93\frac{MJ}{m^3} \times 51,870\ m^3 = 4,823,910\ MJ$$

$$Input\ Energy\left(MBtu\right) = 4,823,910\ MJ \times \frac{1 \times 10^6\ J}{1\ MJ} \times \frac{Btu}{1,055\ J} = 4,572\ Million\ Btu$$

c. Then the output energy is calculated:

$$Output\ Energy\left(MJ\right) = 603\ MWh \times \frac{1,000\ kWh}{1\ MWh} \times \frac{3.6\ MJ}{1\ kWh} = 2,170,800\ MJ$$

$$Input\ Energy\left(MBtu\right) = 2,170,800\ MJ \times \frac{1 \times 10^6\ J}{1\ MJ} \times \frac{Btu}{1,055\ J} = 2,058\ Million\ Btu$$

d. Finally, the efficiency is calculated as follows:

$$Efficiency(\%) = \frac{2,170,800\ MJ}{4,823,910\ MJ} \times 100\% = 45\%$$

e. The conversion efficiency is 45% higher than the Carnot Cycle efficiency if propane were burned in an engine.

8.5 Advantages and Disadvantages of Fuel Cells

There are many advantages in the production and adoption of fuel cell for power generation. These are as follows:

a. Greater efficiency than combustion engines
b. Small systems can be as efficient as large ones
c. Simplicity—very few moving parts, leading to high reliability and longer-lasting systems
d. Low emissions—the by-product is pure water
e. Silence—very quiet even with fuel processing unit

The first advantage of using a fuel cell, as illustrated in the first three example problems, is that its efficiency is by far greater than that of an internal combustion engine.

The second is that households may take advantage of these systems since many units can be designed for smaller systems for residential applications without penalizing overall efficiency. This is its simplicity. There are very few moving parts, leading to higher reliability and longer-lasting systems. Such a system has low emissions and only water as the by-product. As a plus, fuel systems are silent and unobtrusive, even with a fuel processing system in place.

The disadvantages in the use of fuel cell include the high cost of the unit and the eventual production of carbon dioxide, a greenhouse gas. This contribution of carbon dioxide can further heighten the carbon dioxide concentration in the atmosphere, which, as countless researchers demonstrate, increases average global temperature. The same reasoning applies to the widespread use of fossil fuels. The main disadvantage to this system is the source of fuel used in each fuel cell type. Hydrogen is a major fuel used but must ultimately be sourced cheaply to make the system economical. Hydrogen from biomass is a potential option, but the price of biomass feedstock may also be a barrier. In the future, the production of cheap hydrogen gas may be realized, perhaps by a thermal conversion system of biomass or through electrolysis of water using solar energy. Then the scarcity of water may also be at play.

8.6 Balance of Plant

The core components of fuel cells are the electrodes, electrolytes, and the bi-polar plate. The balance of the system—all the other components besides the fuel cell itself—are enumerated below:

a. Pumps or blowers, compressors, intercoolers

b. Power conditioning, DC/AC inverter

c. Electric motors, fuel storage, fuel processing systems, de-sulfurization systems

d. Control valves, pressure regulators, controller

e. Cooling systems, heat exchangers, pre-heaters

Pumps, blowers, and compressors are used to convey the liquid or gaseous materials to the fuel cell canister. Since most reactions occur at an elevated temperature and heat is always generated during the electrochemical reactions, intercoolers are almost always required. The power output of a fuel cell also needs conditioning due to slight changes in the output power as reactions occur in the fuel cell system. The primary output of a fuel cell system is also in direct current voltage (or DC), and this must be converted into alternating current (AC) such that appliances requiring alternating current may be used. Likewise, if the current is transported over longer distances, they must be upgraded to higher voltages, usually in alternating current form, to minimize transmission losses. In most instances, voltage transformers are required to either step up the voltage or bring it down for local applications to households or businesses. Control valves are also required in the fuel tanks and sources to control the flow of gases or liquids, and pressure regulators are needed to maintain system pressure. The heart of the electrical system is the controller. The controller usually monitors power output and monitors system conditions, such as temperature, pressure, and flow of fuels. Outside of the facility may be storage tanks and containers for the standby fuel or if pre-treatment or pre-conversion takes place. For example, some systems may not have a unit to convert higher-hydrocarbon fuels into the basic hydrogen gas, and steam reformers or natural gas reformers are needed. Many gases must be purified. For example, if anaerobic digestion gases are used, the hydrogen sulfide must be scrubbed, including other gaseous or liquid compounds that have sulfur. There will be numerous auxiliary systems in place, such as series of heat exchangers, pre-heaters, or other cooling systems.

8.7 Existing and Emerging Markets for Fuel Cells

There are numerous potential emerging markets for fuel cells. Some are enumerated below:

a. Aerospace applications
b. Auxiliary power units for light-duty vehicles, trucks, ships, and airplanes
c. Portable applications and light traction
d. Stationary applications
e. Fuel cells for buildings
f. Transportation applications
g. Trains and internal combustion engines

The foremost applications of fuel cell technology are in aerospace applications. NASA (2005) started most of the fuel cell research in the 1970s to address fuel and power needs in spaceships. The cost of this type of system is not economically viable for the general public. However, space exploration needs critical power in a small space and the use of lightweight fuel such as hydrogen for long-distance travel. Numerous projects will be discussed in this section, including some other systems that are starting to be commercialized. In addition to NASA, the U.S. military has also been experimenting with numerous fuel cell units for their light-duty vehicles and trucks as well as ships and airplanes. The same principle follows—these institutions want transportable power in a compact manner regardless of the cost. Thus, we see numerous portable applications for infantry power needs. One important application is standby or stationary power. In military installations, natural gas may be available, or propane tanks are loaded in some transport trucks, and fuel may be easily transported. When power is needed, these light gaseous fuels are used. Instead of noisy and large-scale internal combustion engines and generators, fuel cells would be more tactical to use.

There are numerous businesses and industrial companies that utilize cheap natural gas and have also adopted this technology, especially if they are in an area where higher noise levels are not allowed. Many companies use these fuel cells as standby power as well. When the price of natural gas is lowest, it is sometimes economical to use these fuel cell systems and generate lower-cost electricity despite the fact that the initial costs are still quite high.

One will also see several federal projects that use fuel cells for their bus fleets and to transport vehicles for goods. These bus fleets can take in heavy fuel cell systems while carrying light gaseous fuel. These systems are more popular in Europe than in the United States. Toyota and other major automobile companies have released some of their research vehicles powered by fuel cells. One will see more projects of this type released in the near future. There are reports that the railway systems can also begin to adopt these systems.

8.7.1 NASA Helios Unmanned Aviation Vehicle

NASA has also built an unmanned aviation vehicle (UAV) powered by solar energy and fuel cell technology. The solar array produces electricity to supply the electric motor and electrolyzer during the daytime. The electrolyzer produces hydrogen and oxygen, which are stored in tanks on board. During nighttime, the fuel cell is fed with H_2 and O_2 and delivers the power for the electric motor of the propulsion system. However, before this power system could be tested thoroughly, the aircraft crashed due to unexpected wind conditions during a test flight in 2003. Research is still in progress to improve the system, primarily because it would be convenient to fly these units without the need for refueling. Figure 8.4 shows the NASA Helios unmanned aviation vehicle powered by solar energy and fuel cell technology (NASA, 2003).

8.7.2 Naval Research Lab Spider Lion

The Naval Research Laboratory (NRL) has also developed a UAV fuel cell–powered vehicle named Spider Lion. This vehicle weighs 2.5 kg (5.6 lbs) and in November 2005 conducted

NASA Dryden Flight Research Center Photo Collection
http://www.dfrc.nasa.gov/gallery/photo/index.html
NASA Photo: ED01-0209-3   Date: July 14, 2001   Photo by: Nick Galante/PMRF
The Helios Prototype flying wing is shown over the Pacific Ocean during its first test flight on solar power from the U.S. Navy's Pacific Missile Range Facility in Hawaii.

FIGURE 8.4
The NASA Helios Unmanned Aviation Vehicle (UAV).

FIGURE 8.5
Photo of the Naval Research Laboratory's Spider Lion (photo courtesy of NRL).

a flight of 3 hours and 19 minutes, consuming 15 g [0.53 oz] of H_2. The 100 W [0.134 hp] fuel cell system was designed by NRL and used components developed by Protonex. The UAV was used to simulate tests to acquire data on performance and efficiency. The long-term goal is the development of an efficient fuel cell propulsion system for long-endurance (8 to 24 hrs) mini-UAV applications, something that cannot be achieved with current battery technology. Figure 8.5 shows the photo of the Spider Lion (Parsch, 2006). The efficiency of this system is shown in Example 8.9.

Example 8.9. Efficiency of Spider Lion Fuel Cell

The fuel cell on a Spider Lion vehicle has an output power of 100 watts or 100 J/s [0.134 hp]. The input energy was reported to be 15 grams [0.53 oz] of pure hydrogen consumed in 3 hours and 19 minutes. Determine the overall efficiency of this system assuming the energy content of pure hydrogen to be 142 MJ/kg [61,180.5 Btu/lb]. Compare the efficiency to an engine running on hydrogen gas.

SOLUTION:

a. The input energy in units of MJ is calculated:

$$Input\ Power(MJ) = 15\ g \times \frac{1\ kg}{1,000\ g} \times \frac{142\ MJ}{kg} = 2.13\ MJ\,[2,019\ Btu]$$

b. The output power, converted to the same units, will be calculated:

$$Output\ Power(MJ) = \frac{100\ J}{s} \times \frac{1\ MJ}{1,000,000\ J} \times \frac{3600\ s}{hr} \times 3.3167\ hrs$$
$$= 1.194\ MJ\,[1,132\ Btu]$$

c. The efficiency is calculated from the conversion efficiency equation:

$$Efficiency(\%) = \frac{Output}{Input} \times 100\% = \frac{1.194\ MJ}{2.130\ MJ} \times 100\% = 56.1\%$$

d. The efficiency of this test unit is around 56% greater than internal combustion engine efficiency.

8.7.3 The PEMFC Commercial Fuel Cell Module by Ballard (NEXA TM 1.2kW)

Ballard Power Systems (Burnaby, BC, Canada) is developing what they call state-of-the-art fuel cell systems for numerous applications. A Ballard company called Protonex claims to be the leading provider of advanced fuel cell technology for portable, remote, and mobile applications in the 100 to 1,000 watt range [0.134 to 1.34 hp] (https://www.protonex.com/). They have used their patented PEM and SOFC systems. The electrolyte of these fuel cells is a polymeric membrane with perfluorinated sulfonic groups that must be kept wet during operation to render the proton exchange. The catalysts used include platinum supported over carbon or bi-functional metallic electro-catalysts based on platinum or other metals, such as ruthenium. High-purity H_2 (99.9999%) is required at the anode, but at the cathode, air can be used instead of pure O_2. Any other fuel (ethanol, natural gas, gasoline, or derivatives) must be reformed into H_2 (Sevjidsuren, et al., 2012). A photo of the unit is shown in Figure 8.6, where they envision the system to be placed in transport vehicles. These systems are said to be among the industry's smallest, lightest, and highest-performing systems for many portable applications. The economics are justified for remote applications where grid power is not available. This company also works with the military, deploying some units to battlefields.

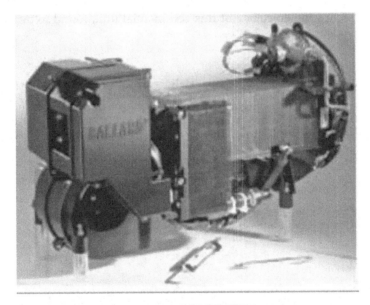

FIGURE 8.6
The PEMFC module by Ballard installed in a vehicle (Ballard, Burnaby, BC, Canada).

8.7.4 Heliocentris Fuel Cell System

Heliocentris (a company with its main headquarters in Germany) develops a wide range of products including batteries, solar panels, conventional power generators, and fuel cells. They have established satellite offices around the world. The Heliocentris Group based in California has reported to have developed a commercial-scale fuel cell system called NEXA TM 1200, also discussed above. This is a 1.2 kW [1.61 hp] fuel cell system with the following specifications:

a. The rated current is 52 amperes.

b. The rated output is 1,200 watts [1.61 hp].

c. The output voltage is 20 to 36 volts.

d. The reported hydrogen consumption is 15 L/min [3.96 gpm].

e. The air consumption is around 3,000 L/min [793 gpm].

f. The hydrogen purity must be 99.99% or better.

g. The main unit costs $9,000, while the controller costs $3,000.

Example 8.10 shows the efficiency of this commercial unit, found to be in the order of around 40% (Sevjidsuren, et al., 2012).

Example 8.10: Efficiency of Commercial Heliocentris Fuel Cell

The Heliocentris fuel cell system has an output power of 1,200 watts or 1,200 J/s [1.61 hp] The input energy was reported to be 15 liters [3.96 gpm] of pure hydrogen. Determine the overall efficiency of this system assuming the density of pure hydrogen to be around 0.0899 kg/m³ [0.00075 lbs/gal] with an energy content of 142 MJ/kg [61,180 Btu/lb].

SOLUTION:

a. The input energy in units of J/s is calculated:

$Input\ Power\,(Watts)$

$$= \frac{15\,L}{min} \times \frac{1\,m^3}{1,000\,L} \times \frac{142\,MJ}{kg} \times \frac{0.0899\,kg}{m^3} \times \frac{1 \times 10^6\,J}{MJ} \times \frac{1\,min}{60\,s} \times \frac{W-s}{J}$$

$$= 3,191\ Watts$$

$$Input\ Power\,(Hp) = 3,191\ W \times \frac{hp}{746\ W} = 4.28\ hp$$

b. The efficiency is calculated from the conversion efficiency equation:

$$Efficiency\,(\%) = \frac{Output}{Input} \times 100\% = \frac{1.200\ Watts}{3.191\ Watts} \times 100\% = 37.6\%$$

c. The efficiency of this commercial unit is around 37.6%.

8.8 The Future of the Fuel Cell

The future of the fuel cell will depend on the price of input fuel used (e.g., hydrogen, methane, methanol, or others). The high initial capital cost (in fact, not necessarily the cost of novel metals like platinum) will also affect its future application. The solid oxide fuel cell appears to be the most promising technology for small electric power plants over 1 kW [1.34 hp] (although high temperature can be an issue) (Wiens, 2010). The direct alcohol fuel cell (DAFC) seems to be the most promising battery material for portable applications, such as in cell phones and laptops. The PEFC is the most practical if the hydrogen (H_2) economy is to be supported by the government.

The Energy Policy Act of 2005 (EPACT, 2005) included the first tax incentive for fuel cell power plants (at least 30% conversion efficiency requirement). Take note that the cost during the 1960s–1970s space program was around $600,000/kW. The current cost is now around $4,500/kW (compared with the diesel generator of $800 1,500/kW and the natural gas turbine at $400/kW). The US DOE (2013) has also formed the Solid State Energy Conversion Alliance (SECA) with the goal of producing solid-state fuel cell modules that would cost no more than $400/kW (Surdoval, 2002). During the 3rd Annual SECA workshop in 2002, the group outlined current developments and initiatives across all U.S. DOE national labs on the following aspects:

a. Fuel processing technologies (e.g., high-temperature sulfur removal and contaminant-resistant electrodes)

b. Manufacturing (low-cost production of precursor materials)

c. Controls and diagnostics (development of sensors and active sealing systems)

d. Power electronics (interaction between fuel cells and its loads and DC-to-DC converters for SOFC)

e. Modeling and simulation (fuel cell failure analysis and manufacturing models)

f. Materials (improved cathodes, interconnecting devices, and innovative sealing concepts)

All U.S. DOE national labs (listed below) are involved in fuel cell development:

1. Pacific Northwest National Laboratory (PNNL) (SOFC component development and modeling)

2. National Energy Technology Laboratory (NETL) (SOFC electrochemical modeling)

3. Argonne National Laboratory (ANL) (SOFC materials research improvement)

4. Lawrence Berkeley National Laboratory (LBNL) (development of colloidal deposition techniques for SOFC)

5. Oak Ridge National Laboratory (ORNL) (SOFC material property and reliability evaluations including power electronics evaluation)

8.9 Conclusions

While fuel cell technology is quite an old energy technology, it has yet to take off commercially. We have observed several companies manufacturing commercial units with very few uses so far. The main reason is the fact that companies still rely on fossil fuel–based input materials for conversion directly into electricity. Large car manufacturers have embarked on ambitious fuel cell cars, although we will have to wait until they become mainstream. Until fuel cell technology breaks from the use of fossil fuel–based input materials and the associated fluctuations of crude oil prices, we may not reach the age of widespread adoption. Fuel cell technology may not be considered renewable energy if the primary fuel is sourced from fossil fuels. Of course, if the hydrogen comes from biomass, it may be considered a renewable technology.

In a study done to compare the use of fuel cell for domestic home use and vehicle use (Lipman, et al., 2004), it was found that domestic use is favored over vehicle use primarily because of the favorable natural gas cost at residences coupled with the potential low value of net metering to households. Commercial settings usually have higher natural gas costs used in the article.

Economic cost is again the deterrent for most units being sold in the market. If low-priced coal is used as primary fuel, there could be widespread adoption, suffering environmental consequences.

The list of fuel cell types discussed in this chapter still show widespread interest among research personnel, especially the promising microbial fuel cell or those that require sustainable and renewable feedstock. As always, the adoption of this technology hinges on the price of crude oil–based fossil fuels such as methane or natural gas or methanol or hydrogen gas from steam reforming or natural gas reforming.

8.10 Problems

8.10.1 Conversion Efficiency of a Direct Methane Fuel Cell

P8.1 Determine the conversion efficiency if one uses methane (CH_4) as fuel for a fuel cell system. The heating value of methane was reported at 37.7 MJ/m^3. The electricity produced was 150 kWh. The test was done in 100 hours, and 28.65 cubic meters of methane was consumed. Comment on the conversion efficiency compared to when methane is used in an engine. How much average power was produced during this test?

8.10.2 Maximum Conversion Efficiency for a Direct Methane Fuel Cell

P8.2 Determine the maximum conversion efficiency for the direct methane fuel cell if the maximum electricity produced is 8.17 kWh for every cubic meter of methane consumed. The energy content of methane is 37.7 MJ/m^3.

8.10.3 Heat Energy Losses in a Direct Methane Fuel Cell

P8.3 Determine the percentage of energy loss for the direct methane fuel cell if the heat produced in a fuel cell is 18.85 MJ for every cubic meter of methane consumed. The energy content of methane is approximately 50 MJ/kg, and its density was reported at 0.754 kg/m^3.

8.10.4 Hydrogen Needed (in kg) to Produce a Liter of Water

P8.4 Determine the theoretical amount of hydrogen needed (in kg) to produce a liter of water from the ideal reaction between hydrogen and oxygen (Equation 8.13 or 8.16).

8.10.5 Potassium Carbonate Produced in an Alkaline Fuel Cell

P8.5 Determine the amount of potassium carbonate produced for every kg of potassium hydroxide used in the alkaline fuel cell following Equation 8.9. The molecular weight of potassium is 39.0983 g/mol.

8.10.6 Ideal Water and Carbon Dioxide Produced for a Direct Methane Fuel Cell

P8.6 Determine the amount of water (kg or liters) and carbon dioxide produced (kg or cubic meter) for every cubic meter of methane used in an ideal direct methane fuel cell. Use a density of 0.754 kg/m^3. Also use the ideal combustion equation of methane in air for the calculation. Air has 79% oxygen and 21% nitrogen.

8.10.7 Zinc Needed for Every Tonne Zinc Oxide Produced in a Zinc-Air Fuel Cell

P8.7 Determine the amount of zinc needed (in kg) for every tonne of pure zinc oxide produced in zinc-air fuel cell. Use the ideal Equation 8.17 for the calculation. The molecular weight of zinc is 65.38 g/mol, and that of oxygen is 16 g/mol.

8.10.8 Practical Conversion Efficiency for a Direct Methanol Fuel Cell

P8.8 Determine the conversion efficiency when methanol (CH_3OH) was used as fuel for a fuel cell system. The electricity produced was measured to be 16.2 kWh. The energy content of methanol was reported to be 19.93 MJ/kg. During the test, some 10 liters of methanol was used. The density of methanol at standard conditions was 0.792 kg/L. Compare this efficiency with that of internal combustion engines.

8.10.9 Efficiency of a Spider Lion Fuel Cell

P8.9 The inaugural flight of the small Spider Lion vehicle lasted for 1 hour and 43 minutes using a half tank of hydrogen. The total weight of the hydrogen gas used was 8.7 g. The fuel cell used has an output power of 100 watts or 100 J/s. Determine the overall efficiency of this system assuming the energy content of pure hydrogen to be 142 MJ/kg. Compare the efficiency to an engine running on hydrogen gas.

8.10.10 Efficiency of a Commercial Fuel Cell

P8.10 A Ballard fuel cell has the following specifications:

    a. Average power produced during test = 5 kW

    b. Number of hours of test = 80 hrs

    c. Methanol consumption = 225 liters

    d. Fuel used = 62% methanol and 38% water

Determine the overall efficiency of this system assuming the methanol mixture has a density of 0.792 kg/L with the energy content of the methanol mixture at 14.3 MJ/kg.

References

Bischoff, M. 2006. Molten carbonate fuel cells: a high temperature fuel cell on the edge of commercialization. Journal of Power Sources 160: 842–845.

Bove, R. and A. Moreno. 2008. International status of molten carbonate fuel cell (MCFC) technology. JRC Scientific and Technical Reports. January. European Commission Joint Research Centre, Institute for Energy, Luxembourg, the Netherlands. Available at: http://publications.jrc.ec.europa.eu/repository/bitstream/JRC44203/mcfc_status.pdf. Accessed April 24, 2018.

Boysen, D. A., T. Uda, C. R. I. Chisholm and S. M. Haile. 2003. High performance solid acid fuel cells through humidity stabilization. Science Online Express, November 20, 2003, Science 303 68–70 (2004).

Boysen, D. A., C. R. I. Chisholm, S. M. Haile and S. R. Narayanan. 2000. Polymer solid acid composite membranes for fuel cell applications. Journal of the Electrochemical Society 147: 3610–3614.

Brunton, J., D. J. Kennedy, F. O'Rourke and E. Coyle. 2010. Design of a single alkaline fuel cell test bed. EEEIC, 9th International Conference on Environment and Electrical Engineering, Prague, Czech Republic, May 16–19. doi:10.1109/EEEIC.2010.5490009.

Chisholm, C. R. I., L. A. Cowan, S. M. Haile and W. T. Klooster. 2001. Synthesis, structure and properties of compounds in the $NaHSO_4$-$CsHSO_4$ system. 1. Crystal structures of $Cs_2Na(HSO_4)_3$ and $CsNa_2H(SO_4)_3$. Chemistry of Materials 13: 2574–2583.

Chisholm, C. R. I., L. A. Cowan and S. M. Haile. 2001. Synthesis, structure and properties of compounds in the $NaHSO_4$-$CsHSO_4$ system. 2. The absence of superprotonic transitions in $Cs_2Na(HSO_4)_3$ and $CsNa_2H(SO_4)_3$. Chemistry of Materials 13: 2909–2912.

Clark, H. 2015. Cleaning up wastewater from oil and gas operations using microbial-powered battery. New Atlas Online Newsletter. March 2. Available at: https://newatlas.com/microbe-powered-battery-purify-wastewater/36341/. Accessed April 24, 2018.

EPACT. 2005. Energy Policy Act of 2005. Public Law 109-58. August 8. Available at: https://www.gpo.gov/fdsys/pkg/PLAW-109publ58/pdf/PLAW-109publ58.pdf. Accessed April 24, 2018.

FCT. 2014a. Direct methanol fuel cell (DMFC). Fuel Cell Today. Available at: http://www.fuelcelltoday.com/technologies/dmfc. Accessed April 24, 2018.

FCT. 2014b. Phosphoric acid fuel cell (PAFC). Fuel Cell Today. Available at: http://www.fuelcelltoday.com/technologies/pafc. Accessed April 24, 2018.

Finch, A. S., T. D. Mackie, C. J. Sund and J. J. Sumner. 2011. Metabolite analysis of *Clostridium acetobutylicum*: Fermentation in a microbial fuel cells. Bioresource Technology 102: 312–315.

Goldstein, J., I. Brown and B. Koretz. 1999. New developments in the electric fuel zinc-air system. Journal of Power Sources 80, nos. 1–2 (July): 171–179.

Green Car Congress. 2016. Ceres Power to demonstrate SOFC stack technology for EV range extender with Nissan; light commercial vehicle. Green Car Congress. June 28. Available at: http://www .greencarcongress.com/2016/06/20160628-ceres-1.html. Accessed April 24, 2018.

Haile, S. M., D. A. Boysen, C. R. I. Chisholm and R. B. Merle. 2001. Solid acid as fuel cell electrolytes. Nature 410: 910–913.

Jabeen, G. and R. Farooq. 2017. Microbial fuel cells and their applications for cost effective water pollution remediation. Proceedings of the national Academy of Sciences, India Section B: Biological Sciences 87 (3): 625–635.

Jungmann, T. 2020. HT- and LT-PEM fuel cell stacks for portable applications. Fuel Cell Systems Group, Fraunhofer Institute for Solar Energy Systems (ISE). Paper presented at the Hannover Trade Fair 2010, April 23, 2013, Hannover, Germany.

Larminie, J. and A. Dicks. 2013. Fuel Cell Systems Explained. John Wiley and Sons, New York. doi:10.1002/9781118878330.

Lipman, T. E., J. L. Edwards and D. M. Kammen. 2004. Fuel cell systems economics: comparing the costs of generating power with stationary and motor vehicle PEM fuel cell systems. Energy Policy 32: 101–125.

Murray, E. Perry, T. Tsai and S. A. Barnett. 1999. A direct-methane fuel cell with ceria-based anode. Nature 400: 649–651.

NASA. 2003. Helios prototype. Dryden Flight Research Center. Available at: https://www.nasa.gov/ centers/dryden/news/ResearchUpdate/Helios/. Accessed April 24, 2018.

NASA. 2005. Fuel cells: a better energy source for earth and space. Glenn Research Center. February 11. Available at: https://www.nasa.gov/centers/glenn/technology/fuel_cells.html. Accessed April 24, 2018.

Oshiba, Y., J. Hiura, Y. Suzuki and T. Yamaguchi. 2017. Improvement in the solid-state alkaline fuel cell performance through efficient water management. Journal of Power Sources 345: 221–226.

Parsch, A. 2006. Naval Research Lab (NRL) Spider Lion. Directory of US Military Rockets and Missiles Appendix 4: Undesignated Vehicles. NRL, Washington, DC. Available at: http:// www.designation-systems.net/dusrm/app4/spider-lion.html. Accessed April 24, 2018.

Sapkota, P. and H. Kim. 2009. Zinc-air fuel cell, a potential candidate for alternative energy. Journal of Industrial and Engineering Chemistry 15 (4): 445–450.

Schumm, B. 2018. Fuel cell. Encyclopedia Britannica. Available at: https://www.britannica.com/ technology/fuel-cell#ref51258. Accessed April 24, 2018.

Sevjidsuren, G., E. Uyanga, B. Bumaa, E. Temujin, P, Altantsog and D. Sangaa. 2012. Exergy analysis of 1.2 kW NexaTM fuel cell module. Chapter 1. October 24. Intech Open Science Publishing, Croatia. Available at: https://www.intechopen.com/books/clean-energy-for-better-environment/ exergy-analysis-of-1-2-kw-nexatm-fuel-cell-module. Accessed April 24, 2018.

Singhal, S. C. and K. Kendall (Eds). 2003. High-Temperature Solid Oxide Fuel Cells: Fundamentals, Design and Applications. Elsevier Science, London.

Surdoval, W. A. 2002. SECA Core Technology program. Third Annual SECA Workshop. The Solid State Energy Conversion Alliance, Washington, DC. Available at: https://www.netl.doe.gov/ file%20library/events/2002/seca/SECA3Surdoval.pdf. Accessed April 24, 2018.

US DOE. 2013. Comparison of fuel cell technologies. U.S. Department of Energy, Office of Energy Efficiency and Renewable Energy. Available at: https://www1.eere.energy.gov/ hydrogenandfuelcells/fuelcells/pdfs/fc_comparison_chart.pdf. Accessed April 24, 2018.

US DOE. 2018. How fuel cells work. U.S. Department of Energy, Office of Energy Efficiency and Renewable Energy and US. Environmental Protection Agency (EPA). Available at: https:// fueleconomy.gov/feg/fcv_pem.shtml. Accessed April 24, 2018. Also available at www .fueleconomy.gov, the Official Government Source for Fuel Economy Information.

Wiens, B. 2010. The future of fuel cells. Ben Wiens Energy Science Site. Available at: http://www .benwiens.com/energy4.html. Accessed April 24, 2018.

Zhu, B., I. Albinsson, C. Andersson, K. Borsand, M. Nilsson, and B.-E. Mellander. 2006. Electrolysis studies based on ceria-based composites. Electrochemistry Communications 8 (3): 495–498. doi:10.1016/j.elecom.2006.01.011. Accessed February 20, 2006.

9

Tidal Energy

Learning Objectives

Upon completion of this chapter, one should be able to:

1. Describe the principles of harnessing tidal energy to generate useful power.
2. Classify the various schemes of power generation from tidal energy.
3. Describe the potential tidal energy available in the United States and worldwide.
4. Compare the estimated cost of electrical energy produced from tidal energy from various countries.
5. Relate the environmental and economic issues concerning tidal energy conversion.

9.1 Introduction

Tidal energy results from the forces of gravitational attraction between the earth, the sun, and the moon. The moon contributes to these gravitational forces and either creates additional tidal energy or subtracts from the sun's forces depending on its physical orientation. Because of gravity, the parts of the earth closest to the moon have higher tides, while those in the middle parts have the lowest. These "humps" occur twice every 24 hours and 50 minutes, which is also the time of the moon's apparent rotation around the earth. Tides due to the attraction of the moon, called semi-diurnal tides, occur every 12 hours and 25 minutes. Figure 9.1 shows the primary areas in the world with high potential for tidal energy. Figure 9.2 shows the variation of the height of the wave deviation as a function of time based on mean sea level. The way to take advantage of these tidal energies is to find a place on the shores of islands or continents where the tide elevations are at maximum. The energy derived comes from the potential and kinetic energy of water bodies similar to those used in hydro power plants.

The sun has an understandably strong effect on the variations of these tides. During a full moon or new moon, the sun's attractive forces add to the pull of the tidal "humps," and this makes the tidal heights higher than normal (see illustrations in Figure 9.1 on forces associated with tidal energy). These events are called spring tides. Of course, there

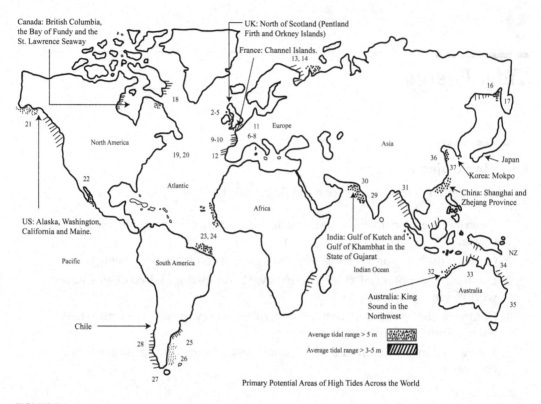

FIGURE 9.1
Primary areas around the world with high potential for tidal energy.

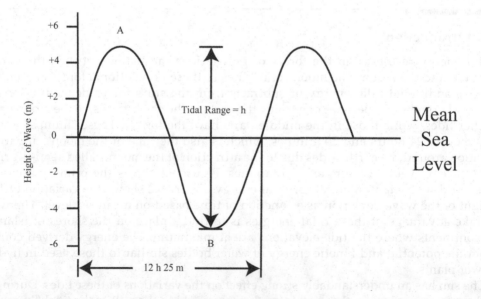

FIGURE 9.2
Variations in height of time as a function of time.

are instances where the attractive forces are in the opposite direction and one gravitational attraction cancels the effect of the other, resulting in lower tides. These events are called neap tides. As mentioned earlier, the variation in mean tidal range will vary from one location to another depending upon the unique characteristics of each coastal area. The other effect is due to the rotational action of the earth's movement around the sun, or what is called the "Coriolis effect." These forces create some form of angular displacement and create a three-dimensional effect on tidal forces and tidal heights. The orientation of the coastal areas is also affected by the variations of these tides as a result of earth's rotational movement.

The variations in height or tidal range may be approximated by a sine curve as shown in Equation 9.1. Example 9.1 shows one how this equation may be used.

$$Tide\ Height(m) = Maximum\ Tide(m) \times sin\Phi \tag{9.1}$$

where
tide height = in meters
Φ = angle from zero to 90°
maximum tide = highest tide achieved in a given location (meters)

Example 9.1: Tidal Height Variation with Time Using a Sine Curve

The variations in height of the seawater tide may be described using a sine curve. Use Equation 9.1 to determine tide variations with time and make a plot of height as a function of time. The maximum tidal height or the tidal range shown in Figure 9.2 is 12 meters [39.36 ft] (peak height from the mean is 6 meters) [19.68 ft]. Assume that the total time for this cycle to occur is 12 hours and 25 minutes (or 12.4167 hrs).

SOLUTION:

a. The table below shows variations in height at various sine angles at increments of ±15° angles for the range stated. If one were to divide the whole cycle (12 hours and 25 minutes) into various segments, then the time for tide levels achieved may also be given in the table (i.e., a time difference of approximately 31 minutes for each 15° sine angle used. Note that the value of sine 90° is 1, corresponding to the maximum height, and sine 0°, 0 being the lowest value of the sine function.

Time	Height (m)	Time	Height (m)	Time	Height (m)	Time	Height (m)
6:00	0.0000	9:37	5.7956	13:14	−3.0000	16:51	−4.2426
6:31	1.5529	10:08	5.1962	13:45	−4.2426	14:22	−3.0000
7:02	3.0000	10:39	4.2426	14:16	−5.1962	17:53	−1.5529
7:33	4.2426	11:10	3.0000	14:47	−5.7956	18:24	0.0000
8:04	5.1962	11:41	1.5529	15:18	−6.0000		
8:35	5.7956	12:12	0.0000	15:49	−5.7956		
9:06	6.0000	12:43	−1.5529	16:20	−5.1962		

b. The data may be plotted to reflect time variations in height of tide level as a
   function of time.

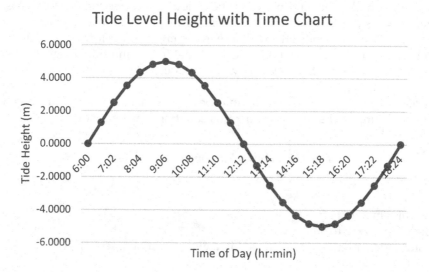

9.2 Worldwide Potential of Tidal Energy

Figure 9.3 shows the possible sites for tidal power in many parts of the earth. The
average tidal range could be from 3 to 5 meters [9.84 to 16.4 ft] in most areas, while
there are some areas with greater than 5 meters [16.4 ft] of variations in tidal head.
Tidal Energy Today (2015) estimates the global potential of tidal energy exceeds
120 GW [3.59 Quad/yr] and could supply more than 150 TWh per annum. The report
also enumerated the potential power for numerous countries as well as ranges of
head available. Table 9.1 shows the potential power from tidal energy across major
countries.

The tidal potential of the United Kingdom is estimated to be about 50% of Europe's
tidal energy capacity. About 25% of Europe's tidal energy potential resources comes from
Scotland. The Pentland Firth, widely considered to be one of the world's best sites for tidal
power, could provide half of Scotland's electricity, according to the study recently completed by Oxford University. (Adcock, et al., 2014). The water flow is said to be rapid in
that area because the tide shifting from the Atlantic to the North Sea is forced through a
narrow eight-mile channel.

Oxford University engineers calculated that underwater turbines strung across the
entire width of the Pentland Firth could generate a maximum of 1.9 GW of power, averaged across the fortnightly tidal cycle. That is equivalent to 16.5 TWh of electricity a
year, almost half of Scotland's entire annual electricity consumption in 2011. As Scotland
already produces 14.6 TWh a year of renewable energy, a fully exploited Pentland Firth
would bring Scotland close to meeting its aim of 100% renewable electricity by 2020
(Adcock, et al., 2014).

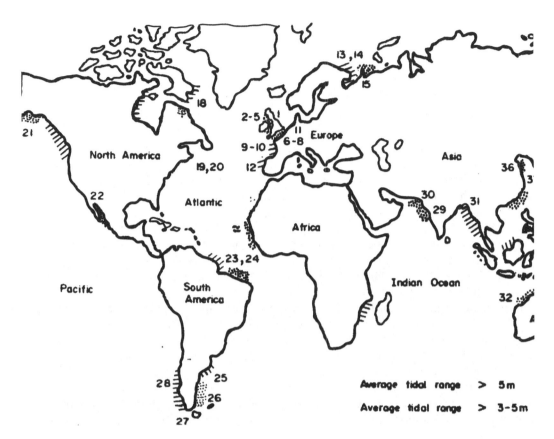

FIGURE 9.3
Possible sites for tidal power stations worldwide.

TABLE 9.1

Worldwide Power Potential from Tidal Energy (US EIA, 2018)

Country	Depth (m)	Power Potential (MW)
United States	200–100	350
Mexico	20–50	100
Chile	20–50	100
Brazil	30	200
Canada	30–80	2,000
United Kingdom	30–50	11,400
France	30 – 50	1,000
Russia	100	350
Japan	20–80	2,200
China	100	2,200
Philippines	50	500
South Korea	50	1,000
Papua New Guinea	30	200
New Zealand	30	200
Australia	100	500
India	500	700

There are three categories of tidal energy technologies (IRENA, 2014). They are enumerated and explained below:

a. Tidal range technologies
b. Tidal current or tidal stream technologies
c. Hybrid applications

Tidal range technologies use a barrage, dam, or other physical barrier to harvest power from the height difference between low and high tides. Tidal current (more than 40 devices introduced from 2006 to 2013) takes advantage of tidal flow and by placing turbines along the flow path. Hybrid systems take advantage of both systems described above with additional unique features for infrastructures along the shores.

One of the best places to witness high tides is the Bay of Fundy. The Bay of Fundy, between New Brunswick and Nova Scotia, is also the most promising location in Canada for tidal energy and could potentially produce as much as 30,000 MW [0.897 Quad/yr] of energy (Karsten, et al., 2008). Figure 9.4 shows the areas within the Bay of Fundy where potential tidal power plants may be constructed.

Other countries are beginning to realize the potential of tidal energy and have started conducting feasibility studies. For example, China has abundant resources of tidal power with more than 18,000 km [11,187 mi] of mainland coastline and more than 14,000 km [8,701 mi] of island coastline. They have an estimated tidal power capacity of 3.5 GW [0.10 Quad/yr], according to China's Ocean Energy Resources Division (Tweed, 2014). Australia and New Zealand have large ocean energy resources but do not yet generate any power from them. Other territories with significant tidal power potential include parts of North America, Argentina, Russia, France, India, the Philippines, and South Korea (IRENA, 2014).

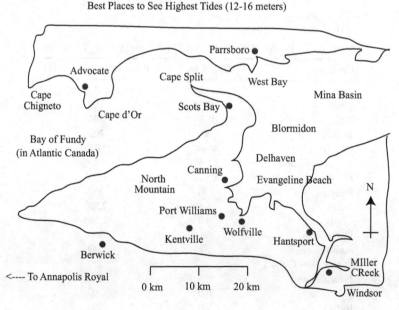

Best Places to See Highest Tides (12-16 meters)

FIGURE 9.4
Best places to observe high tides that vary in height from 12 to 16 meters [40 to 52.5 ft] in a day.

9.3 How Tidal Energy Works

Tidal energy works very simply, as shown in Figure 9.5. A barrier or dam is usually built to divide the sea and an estuary. This barrier is provided with a turbine and generator where the water from the sea may be directed as the tide comes in. As the tide lowers, the water from the estuary is then returned back to the sea, generating power as well. The rise and fall of the tide would simply actuate the blades of the turbine, which produces power when attached to a generator. This simple system will have some lag when the seawater level and the estuary are about the same dynamic head or when the head is too small and the turbine blades are not able to move. Examples 9.2 to 9.6 show how much power may be produced given a certain volume of water stored in a reservoir as well as the time it takes to empty this reservoir. One will realize after this series of examples that power may be produced if the dynamic head is high or when the volume of water increases. The limitation in a tidal system is the dynamic head. Unlike hydro power plants, in which the dynamic head could be several hundred meters, tidal dynamic heads are just a fraction of the land-based water heads. Large tracts of land or large reservoirs are required for tidal power systems.

How Tidal Energy Works

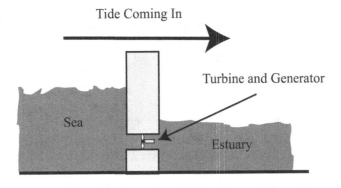

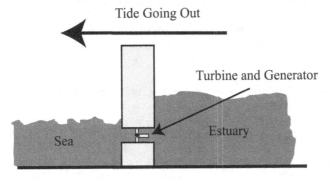

FIGURE 9.5
Simple schematic of tidal power generation system (Bhattacharya, 1983).

Example 9.2: Water Storage Volume Calculations

Determine the volume of water contained in a 100 acre [40.5 hectare] plot filled with water with a depth of 32.8 feet [10 m]. Report your answers in gallons and cubic meters.

SOLUTION:

a. This is a simple volumetric calculation:

$$Volume\left(m^3\right) = \text{Area} \times \text{Depth} = 100 \ acres \times \frac{1 \ ha}{2.47 \ acres} \times \frac{10,000 \ m^2}{1 \ hectare} \times 10 \ m = 4,048,583 \ m^3$$

b. Hence, the total volume of this container is over 4 million cubic meters. In the English system, the volume will be as follows:

$$Volume\left(gallons\right) = 4,048,583 \ m^3 \times \frac{1,000 \ L}{m^3} \times \frac{1 \ gallon}{3.785 \ L} = 1,069,638,836 \ gallons$$

c. Over a billion gallon of water can be contained in a 100 acre plot with a water depth of 10 meters [32.8 ft].

Example 9.3: Release Time for Stored Water

If the volume of water in a 100-acre [40.5-hectare] plot with a depth of 32.8 feet [10 m] reservoir is released at a rate of 4.5 million gallons per minute [17 million liters/min], how long (in hrs) will the reservoir become empty?

SOLUTION:

a. This is a simple volume over volumetric flow rate calculation:

$$Time\left(hrs\right) = \frac{Volume\left(m^3\right)}{\left(\dfrac{m^3}{hr}\right)} = \frac{1,069,638,836 \ gal}{} \times \frac{min}{4.5 \times 10^6} \times \frac{hr}{60 \ min} = 3.96 \ hrs$$

$$Time\left(hrs\right) = \frac{4,048,583 \ m^3}{} \times \frac{1,000 \ L}{1 \ m^3} \times \frac{min}{17,000,000 \ L} \times \frac{1 \ hr}{60 \ min} = 3.96 \ hrs \ \text{ would}$$

take over 17,827 hours, or 742 days, or over 2 years to empty this reservoir.

Example 9.4: Power Generated from Elevated Stored Water

Determine the power generated for a water storage facility having an area of 100 acres [40.5 hectares] with a depth of 32.8 feet [10 m], if water is released at a rate of 4.5 million gallons per minute [17 million L/min] at a height of only 32.8 feet [10 m]. Use an overall conversion efficiency of 85%. Report your answer in kW and hp.

SOLUTION:

a. The solution simply makes use of the hydro power equation previously presented (Equation 5.3):

$$Actual \ Power\left(kW\right) = \frac{QHE_o}{102} = \frac{17,000,000 \ L}{min} \times \frac{1 \ kg}{1 \ L} \times \frac{1 \ min}{60 \ s} \times \frac{10 \ m}{102} \times 0.85 = 23,611 \ kW$$

$$Actual \ Power\left(hp\right) = 23,611 \ kW \times \frac{1,000 \ W}{1 \ kW} \times \frac{hp}{746 \ W} = 31,650 \ hp$$

b. This is quite an appreciable power or around 23.6 MW.

Example 9.5: Energy Produced in Elevated Stored Water

Determine the energy produced (in kWh) by emptying a 100 acre [40.5 hectare] plot filled with water having a depth of 32.8 feet [10 m] at a rate of 4.5 million gal/min [17 million L/min]. Determine the number of households that will benefit from this facility if the average daily household energy requirement is around 35 kWh/day in the United States. Use an overall conversion efficiency of 85%.

SOLUTION:

a. The energy is simply the product of power (from previous example problem) and time and calculated as such:

$$Energy\,(kWh) = 23,611\ kW \times 3.96\ hrs = 93,500\ kWh$$

b. This energy should satisfy the energy requirements of a little over 2,670 of households in the United States.

$$Number\ of\ Households = 93,500\ kWh \times \frac{HH}{35\ kWh} = 2,671\ households$$

Example 9.6: Controlled Release of Elevated Reservoir Water

At what rate would you release water (in kg/s) for a reservoir measuring 100 acres [40.5 hectares] in area and 32.8 feet [10 m] in depth if you simply want to generate 100 kW [134 hp] of electrical power? Assume a conversion efficiency of 85%. Convert the units to gpm.

SOLUTION:

a. The power equation may also be used as follows:

$$100\,kW = \frac{QHE_o}{102}$$

$$Q\left(\frac{kg}{s}\right) = \frac{100\,kW \times 102}{H \times E_o} = \frac{100 \times 102}{10 \times 0.85} = 1,200\,\frac{kg}{s}$$

b. In English units, this is equivalent to 9,511 gpm as shown:

$$Q(gpm) = 1,200\,\frac{kg}{s} \times \frac{60\ s}{min} \times \frac{1\ L}{1\ kg} \times \frac{1\ gallon}{3.785\ L} = 19,022\ gpm$$

9.4 Tidal Power Generation Schemes

There are various basin configurations that could be designed to generate electrical power from tidal energy. The simplest is a single basin system, where only one basin is filled with water during high tide and power is generated during low tide and vice versa. These systems would then be intermittent in nature, and there would be times when power is not generated at all. The other system designs include multiple basins to ensure that continuous operation and power generation are achieved. This section discusses the various ways of generating electrical power from tides based upon the number of basins available as well as whether the tide cycle or the ebb cycle is favored (Bhattacharya, 1983).

9.4.1 Single-Basin Ebb Cycle Power Generation

In the single-basin ebb cycle generation system, the basin is filled during high tide, and power is generated during low tide. As illustrated in Figure 9.6, the tide cycle is illustrated by the broken arrow.

The events occurring in a single-basin ebb-cycle system are enumerated below. Refer to Figure 9.7 for the events and the condition of the tide in this scheme.

Events:

1. T0 = Tide is up; the basin is open and filled with seawater.
2. T1 = The basin has reached maximum peak, and the gates are closed.
3. T2 = Minimum head (basin and sea head) is achieved, and power production starts.
4. T3 = Minimum head is reached again, and power production ceases.
5. T4 = The sluice gates are opened again to fill up the basin.

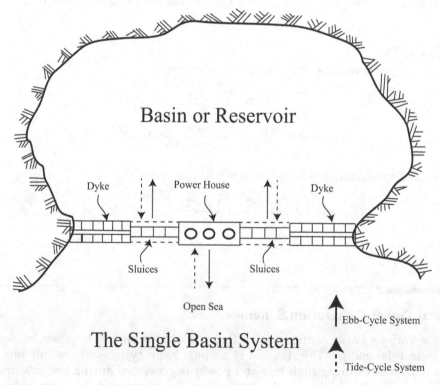

FIGURE 9.6
Diagram of the single-basin system (Bhattacharya, 1983).

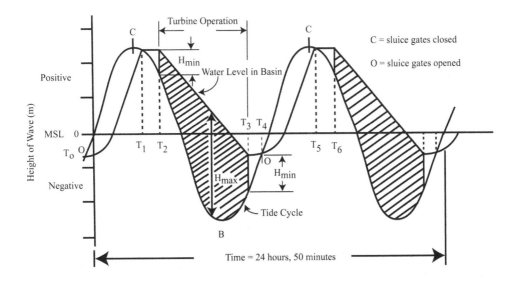

Single Basin Ebb-Cycle System

FIGURE 9.7
Diagram of the events in a single-basin ebb cycle system (Bhattacharya, 1983).

In tidal power operations, there is usually a minimum head where the turbine cannot operate. As in wind turbines when the blades cannot move at lower wind speed, the blades of a tidal wave system will have to be moved by a minimum head. This usually measures about a meter. For example, as the water level in the basin is at its maximum at time T1, power cannot yet be generated since the basin level is the same as the tide level. A short period of time between T1 and T2 is needed for the turbine to start the operation. At time T2, the turbine will start to generate power since the minimum head is reached. Note from the water level profile that the release of water from the basin to the ocean is made constant (linear line). This, however, will generate a different power level due to increasing dynamic head and is at its maximum when the tide level is at its minimum. The tide level will reach its minimum, but power is still produced. Only at time T3 will power production stop since the minimum head is achieved again. In this instance, since the water level in the basin is still higher than that of the ocean, the sluice gate would still be closed. It is at time T4 when the sluice gate opens. At this condition, the water level from the basin is the same as that of the ocean. The sluice gates are then opened to allow the ocean water to refill the basin. Note that there is also a lag in the water level between the basin and the ocean tide due to the fact that a large volume of water is being transferred. The sluice gates are then closed when the tide is at its maximum and the reservoir level is also at its maximum. The cycle then begins again as in time T1.

Example 9.7 illustrates how fast the decrease is in the level of an ideal square basin if the rates of water release are increased. Engineers must carefully analyze this relationship such that power is prolonged. However, if the basin volume is quite small, the same rate of release will make the rate of decline in the level increase. Example 9.8 shows how much power is produced from the basin given in Example 9.7.

Example 9.7: Water Level Variation on Timed Release in a Reservoir

Compare the level of the water from two basins, one with a cross-sectional area of $100 \times 100 \times 10$ m height [$328 \times 328 \times 32.8$ ft] to another basin with a cross section of $10 \times 10 \times 10$ m height [$32.8 \times 32.8 \times 32.8$ ft]. The rate of release is 500 L/min [132.1 gpm]. Estimate the water level after 10 hours of release in both systems.

SOLUTION:

a. The volume of each basin is first calculated:
   $V_{small} = 10 \times 10 \times 10$ m = 1,000 m^3 = 1,000,000 L [264,201 gal]
   $V_{large} = 100 \times 100 \times 10$ m = 100,000 m^3 = 100,000,000 L [26,420,079 gal]

b. After 10 hours, the volume released will be 500 L/min $\times$ 60 min/hr $\times$ 10 hrs = 300,000 liters [79,260 gal].

c. The level in the small basin will be calculated as follows:
   1,000,000 L – 300,000 L = 700,000 L = 700 m^3/(10 $\times$ 10 m) =7 meters [22.96 ft]
   The small basin has decreased by 3 meters [9.84 ft].

d. The level decrease in the large basin is calculated the same way:
   100,000,000 L – 300,000 L = 99,700,000 L = 99,700 m^3/(100 $\times$ 100 m) = 9.97 m [32.7 ft].

The large basin only decreased by a fraction of a meter, which is quite minimal.

Example 9.8: Power Generated in a Tidal Storage Facility

Determine the power generated when water is released at a rate of 1.8 million L/min [475,561 gpm] with 1 meter [3.28 ft] of head. Assume an overall efficiency of 75%.

SOLUTION:

a. Use the classical hydro power equation, converting units properly:

$$Q\left(\frac{kg}{s}\right) = \frac{1,800,000\,L}{min} \times \frac{1\,kg}{1\,L} \times \frac{1\,min}{60\,s} = 30,000\frac{kg}{s}$$

$$Power\,(kW) = \frac{QHE_o}{102} = \frac{30,000\,kg}{s} \times \frac{1\,m}{102} \times 0.75 = 220.6\,kW\left[295.7\,hp\right]$$

b. Releasing a million liters of water per minute will only generate a low power output of 220 kW.

9.4.2 Single-Basin Tide Cycle Power Generation

In the single-basin tide cycle generation system, the power is generated and the basin is filled during high tides. This is also illustrated in Figure 9.6 using solid arrows.

The events occurring in a single-basin tide cycle system is enumerated below. Please refer to Figure 9.8 for the events and how the tides vary with time.

Events:

1. T0 = Power is produced when seawater is dumped into an empty basin.
2. T1 = Minimum head is reached, and thus power production stops. The sluice gates are closed, and as soon as the tide level is below the water level in the basin, water is drained from the basin into the sea.
3. T2 = The sluice gates are closed to allow the water level in the ocean to rise.

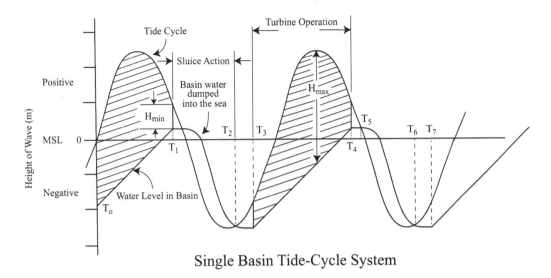

Single Basin Tide-Cycle System

FIGURE 9.8
Diagram of the single-basin tide cycle system and the corresponding events (Bhattacharya, 1983).

4. T3 = Minimum head is reached, and power production starts by dumping water from the ocean into the basin (same point as T0). Water level is controlled at greater than minimum head.

5. T4 = Minimum head is reached again (same as T1), and power production stops. The cycle repeats.

Note that the tide cycle system is very similar to the ebb cycle system. The only difference is that water is dumped from the ocean to the storage system. The closing and opening of the sluice gates are performed at opposite times. The process usually starts with an empty basin. For example, at point T1 in Figure 9.8, the sluice gates are closed since the dynamic head is quite low and no power is produced. The seawater level is allowed to go down, and since this level is lower than that of the basin, water is dumped into the sea. The goal is to empty the basin as the sea level goes down. At point T2, when the seawater level and the basin level is the same, the sluice gates are closed. The seawater level will reach its minimum depth and is allowed to rise again. As the dynamic head between the sea level and the basin is the minimum head where the turbine would move, power is generated. This is point T3 in the diagram of Figure 9.8. Water from the ocean is dumped into the empty basin, and power is produced. The volume of water released from the ocean is controlled such that the minimum head is maintained. Note again the constant amount of water being dumped into the ocean. This will give rise to increasing dynamic head and increased power generation as the seawater level gets higher. The sluice gates will be closed again at point T4, where the seawater level and basin level are so low that the turbine blades are not able to move appropriately. Point T5 is when the seawater level is the same as the basin level. Beyond point T5 is a condition where basin water is dumped into the ocean to keep the basin level head as low as possible. If one were to compare the ebb cycle and tide cycle systems, one could observe that similar power production output is achieved. Power is simply produced during opposite periods. That is, power is produced either when water from the ocean is dumped into the sea or the other way around. The main advantage of the tide cycle system is the fact that the basin is always kept empty and maintenance procedures, such as siltation activities and pond improvement, may be performed on the basin while water is being dumped into the ocean. Besides that, the energy output in each system will be the same. In the ebb cycle system, the pond maintenance procedure is done during power production. The advantage of the ebb cycle system is that if power is required for the maintenance procedures, power is available. In contrast, for the tide cycle system, external power must be sourced elsewhere during maintenance procedures.

9.4.3 Single-Basin Two-Way Power Generation

There is a way to combine the ebb cycle and the tide cycle system. It is possible in a single-basin system to generate more power by taking advantage of both cycle systems described above. The events in single-basin two-way power generation is enumerated below. If one is familiar with either the ebb cycle or the tide cycle system, this power generation scheme simply combines the two. There is still a lag in power production but it is minimal. Refer to Figure 9.9 for the events.

Events:

1. T1 = The sluice gates are closed, and as soon as the tide level reaches minimum head (between the basin and the sea), the sluice gates are opened again.

2. T2 = The sluice gates are opened to dump water to the sea to generate power.

3. T3 = Minimum head is reached, and power production stops. More water is dumped into the sea but not generating power.

4. T4 = The sluice gates are closed. The ocean water level and sea water level are the same.

5. T5 = Water is dumped from the sea to the basin as soon as minimum head is reached.

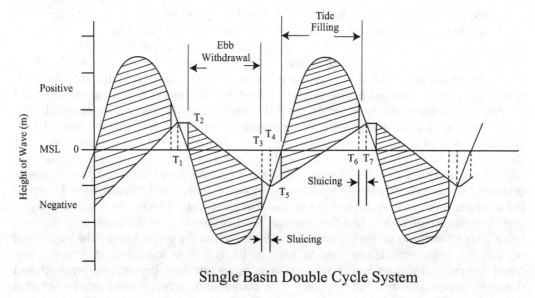

Single Basin Double Cycle System

FIGURE 9.9
Diagram of the single-basin double-cycle system and the corresponding events (Bhattacharya, 1983).

6. T6 = Power production stops. The sluice gates remain open to achieve higher basin head.

7. T7 = This is the same as T1, and the cycle repeats.

In the single-basin double-cycle system illustrated in Figure 9.9, the cycle could start at point T1 when the sluice gate from the basin is closed and the basin level is at its maximum. At this point, the seawater level is allowed to go down until the minimum dynamic head is reached. This should happen at point T2 when water from the basin is being released to the ocean. This event is called ebb withdrawal. Likewise, water release is controlled constantly even if the power production is increased with time, with the maximum at the time when the ocean level is at its minimum. By controlling the rate of release, power is still produced even when the seawater level has risen. However, there will be a point where the minimum dynamic head between the seawater level and the basin level is achieved and the turbine is not able to move and generate power, and in this case, the sluice gate is closed to allow the seawater level to rise. At point T4, the seawater level and the basin level are the same. The seawater level is allowed to increase its dynamic head, and at the point where the dynamic head is at its minimum to generate power, the water from the ocean is dumped into the basin. This is called tide filling. Likewise, water release is controlled such that power is maximized for longer periods. As the seawater level and the basin level are at the minimum for power generation, the operation is halted. The seawater level is allowed to go down. Note that the basin level is made full by filling in more water from the sea even though power is not generated. When the basin level is full, the sluice gates are then closed. When the seawater level and the basin level difference is at its minimum to generate power, water is dumped from the basin to the ocean in a controlled manner to generate the power, and the cycle is repeated. Note that the amount of time to generate power is twice that of either the ebb cycle or the tide cycle alone. Power is produced either by withdrawing water from the basin to the ocean or likewise by dumping water from the ocean to the basin, combining the ebb cycle and the tide cycle systems. However, there is still a time when no power is generated at all. This is the primary disadvantage of the single-basin system. Power generation may be extended, but it cannot be made continuous throughout the day; there must be a downtime period. The only advantage of this period is that maintenance procedures may be accomplished during the periods when power is not being generated. If power is needed during downtime and repair, it should be sourced externally or come from standby systems.

9.4.4 Double-Basin Systems

The only way that continuous power can be generated in a tidal power system is with two basins. The schematic arrangement of this system is shown in Figure 9.10. In this system, turbines are strategically placed between Basin A and Basin B as shown. Sluice gates are found at the entrances of both these basins as illustrated. Ideally, Basin A and Basin B should be of about the same reservoir volume. There is no advantage to having one basin larger than the other. The schematic diagram for all the processes involved in this system is shown in Figure 9.11.

The events occurring in this system are enumerated below.

Events

1. T1 = Basin A dumps seawater to Basin B, producing power.

2. T3 = Basin B dumps water to the sea (since the sea level is lower than Basin B).

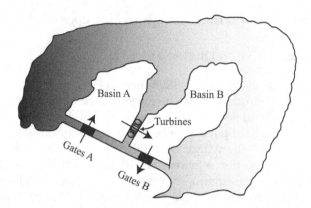

Typical Layout for Two-Basin Scheme for Continuous Power Generation

FIGURE 9.10
Diagram of the double-basin system for continuous power generation (Bhattacharya, 1983).

3. T4 = Basin B stops dumping water to the sea but still receives water from Basin A.
4. T2 = Basin A is filled up with seawater while producing power.
5. C and E = Basin A closes its gates and stops receiving water from the sea.
6. D and F = Basin A opens its gates to receive water from the sea.
7. J and L = Basin B opens its gates to dump water to the sea.
8. K and M = Basin B closes its gates since the seawater level is higher.
9. In all cases, power is produced continuously by dumping water from Basin A to Basin B.

The process may start at point T1, where Basin A is dumping seawater to Basin B and generates power. Basin A is at its maximum dynamic head. Since the level of water from the sea is going down, the sluice gate from the sea to Basin A is closed. The level in Basin B increases because of the water coming from Basin A. There is a point T3, where the level of water in Basin B is the same as the seawater level. In this instance, water from Basin B

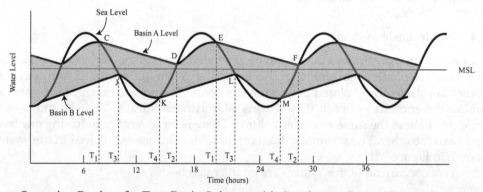

Operating Regime for Two-Basin Scheme with Continuous Power Generation

FIGURE 9.11
Operating regime for double-basin continuous power generation system (Bhattacharya, 1983).

may be dumped to the ocean as the water level there decreases. Hence, the dynamic head between Basin A and Basin B is about constant, and power is continuously being generated. At point T4, the water level in Basin B and the seawater level are the same again, and the seawater gate must be closed to keep the water level in Basin B low to generate higher power. Point D is when the water level in Basin A and the seawater are the same. In this case, the gates in Basin A are opened to allow water from the sea to contribute to the water level in Basin A, generating more power as water is continuously being dumped into Basin B. The gate from Basin A remains open until point E, where the water levels in Basin A and the sea are the same again. Then the Basin A gate is closed again and stops receiving water from the sea. Basin A continues to dump water to Basin B. The cycle is repeated, power is continuously generated, and, more importantly, instead of switching water between the two basins, there is a steady dump of water from Basin A to Basin B [Figure 9.11] and only one turbine system is needed. Basin A is maintained at a constantly higher water level, and Basin B is kept at a lower water level. The seawater level simply does its own routine diurnally while power is generated throughout the day. This system necessitates the proper rate of release of water from Basin A to Basin B such that continuous power is produced by ensuring that the minimum dynamic head is achieved in all instances.

Note that the water being released from Basin A to Basin B is also constant. This is depicted by the relative slope of the Basin A and Basin B line as shown in Figure 9.11. In this system, when the tide variation is established throughout the year, only the amount of power is calculated, and its magnitude is a strong function of the size of the storage facilities. In most examples presented in this chapter, power is increased when the rate of release of water is increased. This means that higher power simply means increasing the reservoir size because the dynamic head does not vary so much. Example 9.9 shows a simple example of power generation from a double-basin system.

Example 9.9: Power and Energy from Double-Basin System

A double-basin system has two reservoirs, Basin A and Basin B, with similar dimensions of about $100 \times 100 \times 12$ m depth [$328 \times 328 \times 39.39$ ft]. Basin A is the higher-level basin, and Basin B is the lower-level basin. The minimum dynamic head of power generation is 10 meters [32.8 ft]. (a) Determine the amount of water to be released such that 2 meters [6.56 ft] of head remains in Basin A as it releases its water to Basin B. In this case, Basin A will use up about 10 meters [32.8 ft] of its dynamic head continuously (i.e., 24 hrs). (b) Determine the minimum power (kW) generated based on the appropriate release rate from Basin A. Assume an overall conversion efficiency of 75%. (c) Determine the energy produced (kWh) if this system operates 24 hours per day, 365 days in a year. Report this in kWh/yr. (d) Determine the number of households that may be served by this facility if one household uses 12,000 kWh/yr. Note that since this is a double basis system and may be operated year round, it assumed to be operating continuously in a year.

SOLUTION:

a. If Basin A is to release water to Basin B in 8 hours with 2 meters of depth remaining, the volumetric rate of release will be calculated as follows:

$$Q\left(\frac{kg}{s}\right) = \frac{100\ m \times 100\ m \times 10\ m}{24\ hrs} = \frac{4,167\ m^3}{hr} = 4,167\frac{m^3}{hr} \times \frac{1,000\ kg}{m^3} \times \frac{1\ hr}{3600\ s} = 1,158\frac{kg}{s}$$

$$Q\left(\frac{lbs}{min}\right) = 1,158\frac{kg}{s} \times \frac{2.2\ lbs}{1\ kg} \times \frac{60\ s}{min} = 152,856\frac{lbs}{min}$$

b. The power is calculated as follows:

$$P(kW) = 1,158\frac{kg}{s} \times \frac{10\ m}{102} \times 0.75 = 85\ kW\left[114\ hp\right]$$

c. The energy is simply the product of power and time:

$$Energy\,(kWh) = 85\ kW \times \frac{24\ hrs}{day} \times \frac{365\ days}{yr} = 744,600\frac{kWh}{yr}$$

d. If a household consumes 12,000 kWh/yr, then the number of households served by this facility is given below:

$$Number\ of\ Households = 744,600\frac{kWh}{yr} \times \frac{yr - HH}{12,000\ kWh} = 62\ households$$

e. Close to 62 households may be served by this facility.

A 100 × 100 m area is about a hectare [2.47 acres]. Following the above example, a hectare of land may be used to serve over 60 households if the average dynamic head for a tidal power plant is about 10 meters [32.8 ft]. A thousand households would need approximately 16 hectares [~40 acres] and so on. One would do this by back calculation or simply ratio and proportion if given values are the same. The size of the land required for this system increases substantially. In conclusion, a combination of higher tidal range and large tracts of land are needed for tidal power systems.

9.5 Other Tidal Power Generating Methods

It has been reported that about 2 to 3 TW of energy are dissipated through tides. This amount is one-third of today's world consumption. However, only a small fraction (1 TW) could be derived due to a limited number of locations (IRENA, 2014). There are numerous installations already in place, and the first such tidal power plant, in La Rance, France, was built and operated for a period of only six years (1960–1966). The installed capacity for this plant was 240 MW [0.007 Quad/yr]. Generating this much power with a 10 meter [32.8 ft] dynamic head and 75% overall conversion efficiency would require a volumetric flow rate of 3,264,000 kg/s [7.2 million lbs/s] of seawater or over 3,264 cubic meters per second of water. Thus, power generation using this potential energy may not be feasible if land is not available. Fortunately, there are newer ways to generate power from tidal energy:

a. Tidal stream generator
b. Tidal barrage
c. Dynamic tidal power
d. Tidal lagoon
e. Hybrid systems

Tidal stream generators (TSGs) make use of kinetic energy instead of potential energy to move the turbine blades. This is very similar to wind power generation technology and does not require large tracts of land along the shores. It only requires movement of water streams along deltas or bridges. Land constructions, such as riverways that direct water to the ocean, create substantial flow of water, and turbines may be placed underneath to harness the water flow. High velocities of water are usually created along these structures. Turbines are usually oriented horizontally but with some ingenuity; with the use of a special duct or vertical turbine blades, the water can be diverted in a specific stream to the ocean.

Tidal barrages make use of the potential energy between high and low tides by creating a very long dam (a barrage) along the full width of a tidal estuary. This would not make use of the inland and may be placed a few kilometers away from the shore. The Philippine government recently created a plan to build a tidal barrage (also called tidal fence) that would generate 2,200 MW [0.066 Quad/yr] of electrical power. The plan had an estimated budget of $3 billion (IRENA, 2014).

Example 9.10: Cost to Recover Initial Investment in Tidal Power Systems

Determine the return to investment for a tidal barrage project similar to the one in the Philippines with a rated power output of 2,200 MW [0.066 Quad/yr]. Assume that the yearly rated power output is 60% of the rated output (i.e., 1,320 MW) [0.039 Quad/yr] and the power is to be sold for $0.10/kWh. The plant operates for 360 days in a year and 24 hours per day. Base the return on investment (ROI) on just the initial capital cost of $3 billion and the sale of electricity.

SOLUTION:

a. The gross sale of electricity is calculated by first calculating energy output in a year:

$$E\left(\frac{kWh}{yr}\right) = 2,200\ MW \times 0.60 \times \frac{360\ days}{yr} \times \frac{24\ hrs}{day} \times \frac{1,000\ kW}{1\ MW} = 11.4 \times 10^9\ \frac{kWh}{yr}$$

b. The gross receipt after a year will be as follows:

$$Gross\ Income(\$) = 11.4 \times 10^9\ \frac{kWh}{yr} \times \frac{\$0.10}{kWh} = \frac{\$1.14 \times 10^9}{yr}$$

c. The ROI is the ratio of the initial capital cost and the yearly gross revenue:

$$ROI(years) = \frac{\$3 \times 10^9}{\$1.14 \times 10^9/yr} = 2.63 years$$

d. Hence, the ROI could be less than 5 years based simply on the gross receipts.

Dynamic tidal power (DTP) is yet untried but a promising technology that would use both the kinetic energy and the potential energy of water bodies. This concept proposes a very long dam of 30 to 50 km [18.65 to 31 mi] long built along the coastline and straight out to the sea. Tidal phase differences are introduced across the dam, leading to various water head differences in shallow coastal areas. These coastal areas must have strong oscillating tidal currents. Potential applications include those near the coasts of China, the United Kingdom, and South Korea. The dams are usually constructed perpendicular to the coastline, and another dam parallel to the coastline acts as a barrier between the high-tide side and the low-tide side. This will form a structure shaped like the letter T. This concept was invented and patented by Kees Huldbergen and Rob Steijn, two Dutch engineers. A short video explaining the concept was released and completed in October 2013. It was esti-mated that some of the largest dams could accommodate 15,000 MW [0.448 Quad/yr] of installed capacity. A DTP dam with 8 GW installed capacity and a capacity factor of about 30% could generate about 21 TWh of energy per year. For example, an average European person consumes about 6,800 kWh of energy per year, and one DTP of this size could sup-ply energy to about 3.09 million Europeans (Steijn, 2015).

For tidal lagoon designs, circular retaining walls are built with integral turbines under-neath. They are also designed to capture the potential energy of water. The concept is simi-lar to that of tidal barrages except it does not contain a pre-existing ecosystem. The design encloses a certain region within the shore. The tidal lagoon being planned in Swansea Bay in Wales will be the first of its kind. The country plans to build six units of this type of facility with a capacity of 320 MW [0.01 Quad/yr] each (Gardiner, 2014).

Hybrid tidal systems are those that combine the other types previously discussed. In the future, we will see numerous projects that would build large reservoirs along the shores while creating dams and barrages outside of the shoreline. Potentially, further past the shorelines are lagoons that could capture both kinetic and potential energy if the topo-graphic location is ideal (Electric Light and Power, 2013).

9.6 Cost of Tidal Energy Systems

Tidal energy systems have been in place since the 1960s in Canada, China, and France. More recently, Korea has also initiated tidal power technology. The initial capital cost is still quite high, but the payback period is excellent considering the high electricity cost worldwide. Many of these installations in the 1960s and 1970s are still operational without many problems. Very little general economic data are available at present due to the fact that each project is unique and has specific topography. As in many examples presented in this chapter, the cost of tidal systems is a strong function of tidal dynamic height as well as the cost of the reservoirs. For tidal barrages, the size of the dams determines cost.

The La Rance facility in France has roughly $500/kW of installed capacity and has a cost of electricity of 0.0026/kWh. The Bay of Fundy system has a reported electricity production cost of $0.18 to 0.30/kWh, which is quite high. A project in Sunderbans, India, had a reported electricity cost of $0.60 to 0.90/kWh (Boyle, 2004). The Sihwa power plant in South Korea has the largest reported tidal range (or tidal barrage) installation in the world. The estimated cost is around $300 million, and the plant produces electricity for about $0.024/kWh (Patel, 2015).

A comparison of existing and proposed tidal barrages is shown in Table 9.2. Many of the reported systems and their costs are still in their demonstration phases and are not repre-sentative of long-term commercial costs. The capacity factors are also quite high, while initial capital costs are low. For tidal systems, like wind power generation systems, the capacity fac-tor may be between 25% and 40%, while availability factors range from 70% to 90%. Hence, the overall multiplier for installed capacity may range between 20% and 40% (IRENA, 2014).

One main advantage of tidal energy is its possible proximity to urban areas and resulting use of electrical power where it is most needed. While reservoirs and dams create changes in the environment and topography of places, it should not contribute to widespread dam-aging environmental impact. At worst, these facilities may become tourist spots. There are, however, a few environmental issues that should not be neglected, and these will be discussed in the next section.

TABLE 9.2

Estimated Costs for Existing and Proposed Tidal Barrages

Barrage	Country	Capacity (MW)	Power Generation (GWh)	Construction Costs (Million $)	Construction Cost/kW (US$/kW)
Operating					
La Rance	France	240	540	817	340
Sihwa Lake	Korea	254	552	298	117
Proposed					
Gulf of Kutch	India	50	100	162	324
Wyre Barrage	United Kingdom	61.4	131	328	534
Garorim Bay	Korea	520	950	800	154
Mersey Barrage	United Kingdom	7,000	1,340	5,741	820
Incheon	Korea	1,320	2,410	3,772	286
Dalupiri Blue	Philippines	2,200	4,000	3,034	138
Severn Barrage	United Kingdom	8,640	15,600	36,085	418
Penzhina Bay	Russia	87,000	200,000	328,066	377

Source: IRENA, 2014.

9.7 Environmental Concerns

There are beneficial and non-beneficial environmental impacts along with the installation of tidal energy systems. These are enumerated below.

9.7.1 Beneficial

a. Use of road across the barrage/dam structure

b. Land reclamation

c. Tourism (in some areas)

9.7.2 Non-Beneficial

a. Land drainage problems

b. Affects wildlife and fisheries

c. Flooding in some areas

One of the most beneficial environmental contributions of tidal energy is that these huge structures, in the form of dams or barriers, may also act as passageways for the population as roads or bridges. Numerous shorelines are now being converted into habitation for the growing population, and the land reclamation introduced as part of reservoir creation could be a positive infrastructure development. Tourism could be enhanced while generating renewable power for a large portion of the population. There will be minimal moving parts outside of the power generation system. The life spans of these projects are longer compared to that of other renewables. Operation and maintenance costs are also minimal.

On the other hand, land drainage can be a problem, especially if the natural construc-tions near deltas are affected. The sea creatures and wildlife in the area may also be affected. The requirement for new infrastructure to transport electricity (i.e., new grid) will also become a nuisance for large-scale systems.

In developed countries like the United States, acquiring permits may be an issue as coastal communities become affected. Environmental impact assessment is a necessity for all these projects, and the effect to the normal ecosystem will have to be analyzed.

9.8 Conclusions

Tidal power is definitely a key player in the renewable energy of the future and can con-tribute significantly to the world's energy mix. The main advantage of this type of system is its predictability. In any given place around the earth, tidal variations have already been established. Most calendars even used to have high-tide and low-tide indicators as well as reports of the moon's location and fullness. Tidal power is thus more predictable than wind or solar energy. The only problem with tidal power is that there are limited sites with potential to generate significant amounts of power. In addition, the initial capital cost for any of these systems is still quite expensive. Compared with hydro power, tidal energy is inferior due to its limited dynamic head. It would take enormous tracts of land for a reservoir to be built that would equal the capacity of a large-scale hydro power plant like the Hoover Dam between Nevada and Arizona. Like wind power, tidal energy's capacity factor is also quite low even if its availability is significantly high.

There is also a need to develop new turbine technologies and material types that can withstand the harshness of seawater and its saltiness. Nonetheless, the power from the tide is inexhaustible as long as the sun and the moon maintain their gravitational attrac-tion. The cost barrier for tidal power generation systems will be overcome quite easily, primarily because of the longevity of the tidal power systems compared to solar or wind energy systems. While it was reported that the movement of tides causes a loss in mechan-ical energy, the losses are low compared to the magnitude of tidal resources. Hence, for practical purposes, these losses are almost negligible in the long run and should not be felt by the human population.

Many countries have recently revived their renewable energy programs, and the tidal power systems are always included, especially at sites that have the obvious potential. Capital cost will continue to decline as new materials for manufacturing are discovered. Projects are reportedly growing at a significant rate, and this is an excellent sign for this inexhaustible renewable energy.

9.9 Problems

9.9.1 Variation of Tide Level with Time Using Sine Curve

P9.1 The variations in height of the seawater tide may be described by a sine curve. The equation for height is simply as follows: Tide level = Maximum tide × sin Φ, where Φ is an angle between $0°$ and $90°$ with sine considered. Plot and make a table of the height as a function of time for a tide that has a maximum height of 10 meters [32.8 ft]. Assume that the total amount of time for this cycle to occur is 12 hours and 25 minutes (or 12.4167 hrs).

9.9.2 Reservoir Volume Calculation

P9.2 Determine the volume of water contained in a 1,000 hectare [2,470 acres] plot filled with water with a depth of 30 meters [98.4 ft]. Report your answers in cubic meters and in gallons.

9.9.3 Time to Release Water from Reservoir

P9.3 If the volume of water in a 1,000 hectare [2,470 acres] plot with a depth of 30 meters [98.4 ft] is released at a rate of 800 million liters per minute [211.4 million gpm], how long does it take (in hrs) for this reservoir to run out of water?

9.9.4 Power from Tidal Reservoir

P9.4 Determine the power generated for a water storage facility with an area of 1,000 hectares [2,470 acres] and a depth of 30 meters [98.4 ft] if water is released at a rate of 800 million liters per minute [211.4 million gpm] at a height of 30 meters [65.6 ft]. Use an overall conversion efficiency of 75%. Report your answer in kW.

9.9.5 Energy from Tidal Reservoir

P9.5 Determine the energy produced (in kWh) by emptying 1,000 hectares [2,470 acres] of land filled with water having a depth of 30 meters [65.6 ft] at a rate of 800 million L/min [211.4 million gpm]. Determine the number of households that will benefit from this facility if the average annual household energy requirement is around 12,000 kWh/yr in the United States. Use an overall conversion efficiency of 75%.

9.9.6 Matching Household Energy Requirements

P9.6 At what rate would you release water (in kg/s) for a reservoir measuring 1,000 hectares [2,470 acres] in area and 30 meters [98.4 feet] in depth if you simply want to generate 300 kW [402 hp] of electrical power? Assume an overall conversion efficiency of 75%. Convert the units to gpm.

9.9.7 Water Level Decline with Time for a Given Basin

P9.7 Compare the level of water from two basins, one with a cross sectional area of $50 \times 50 \times 5$ meters [$164 \times 164 \times 16.4$ ft] height to another basin with a cross section of $5 \times 5 \times 5$ meters [$16.4 \times 16.4 \times 16.4$ ft] height. The rate of release is 325 L/min [85.9 gpm]. Estimate the water level after 5 hours of release in both systems.

9.9.8 Power Generated from Small Basin

P9.8   Determine the power generated when water is released at a rate of 5 million L/min [1.321 million gpm] with 1 meter [3.28 ft] of head. Assume an overall efficiency of 75%.

9.9.9 Power and Energy from Double-Basin System

P9.9   A double-basin system has two reservoirs, Basin A and Basin B, with similar dimensions of about 50 × 50 × 12 meters [164 × 164 × 39.36 ft] in depth (half a hectare). Basin A is the higher-level basin, and Basin B is the lower-level basin. The minimum dynamic head of power generation is 10 meters [32.8 ft]. (a) Determine the amount of water (kg/s) to be released such that 2 meters [6.56 ft] of head remains in Basin A as it releases its water to Basin B. In this case, Basin A will use up about 10 meters [32.8 ft] of its dynamic head in 24 hrs. (b) Determine the minimum power (kW) generated based on the appropriate release rate from Basin A. Assume an overall conversion efficiency of 75%. (c) Determine the energy produced (kWh) if this system operates 24 hours per day and 365 days in a year. Report this in kWh/yr. (d) Determine the number of households that may be served by this facility if one household uses 12,000 kWh/yr.

9.9.10 Cost to Recover Initial Investment

P9.10   Determine the return to investment for a tidal barrage project in Korea (Sihwa Lake) with a rated power capacity of 254 MW (considered power input). Assume the yearly rated power output is 80% of the rated output [i.e., 203 MW] and the power is to be sold for $0.08/kWh. The plant operates for 360 days in a year and 24 hours per day. Base the return on investment (ROI) on just the initial capital cost of $298 million and the sale of electricity.

References

Adcock, T. A. A., S. Draper, G. T. Houlsby, A. G. L. Borthwick and S. Serhadlioglu. 2014. Tidal stream power in the Pentland Firth—long-term variability, multiple constituents and capacity factor. Proceedings of the Institute of Mechanical Engineers, Part A: Journal of Power and Energy 228 (8): 854–861. doi:10.117/0957650914544347.

Bhattacharya, S. C. 1983. Lecture Notes in "Renewable Energy Conversion" Class. Asian Institute of Technology, Bangkok, Thailand.

Boyle, G. (Ed.). 2004. Renewable Energy: Power for a Sustainable Future. 2nd Edition. Oxford University Press, The Open University, Oxford, UK.

Electric Light & Power. 2013. Hybrid wind, tidal power turbine to be installed off Japanese coast. Power Grid International. June 11. Available at: https://www.elp.com/articles/2013/06/hybrid -wind–tidal-power-turbine-to-be-installed-off-japanese-co.html. Accessed April 24, 2018.

Gardiner, B. 2014. Generating power from tidal lagoons. Energy and Environment Newsletter. Available at: https://www.nytimes.com/2014/10/29/business/energy-environment/swansea -bay-generating-power-from-tidal-lagoons.html. Accessed April 24, 2018.

IRENA. 2014. Tidal Energy Technology Brief. International Renewable Energy Agency (IRENA). Ocean Energy Technology Brief 3. June. Available at: http://www.irena.org/DocumentDownloads/Publications/Tidal_Energy_V4_WEB.pdf. Also available at: http://www.irena.org. Accessed April 24, 2018.

Karsten, R., J. M. McMillian, M. J. Lickley and R. Haynes. 2008. Assessment of tidal current energy for the Minas Passage, Bay of Fundy. Proceedings of the Institute of Mechanical Engineers, Part A: Journal of Power and Energy 222 (2008): 493–507. doi:10.1243/09576509JPE555.

Patel, S. 2015. Sihwa Lake tidal power plant, Gyeonggi Province, South Korea. Power Magazine. Available at: http://www.powermag.com/sihwa-lake-tidal-power-plant-gyeonggi-province-south-korea/. Accessed April 24, 2018.

Steijn, R. 2015. Dynamic tidal power technology advances. Renewable Energy World Newsletter. January. 13. Available at: https://www.renewableenergyworld.com/articles/2015/01/dynamic-tidal-power-technology-advances.html. Accessed April 24, 2018.

Tweed, K. 2014. China wants to make a splash in ocean energy. IEEE Spectrum, Energywise Newsletter. April 1. Available at: https://spectrum.ieee.org/energywise/green-tech/geothermal-and-tidal/china-wants-to-make-a-splash-in-the-ocean-energy. Accessed April 24, 2018.

10

Wave Energy

Learning Objectives

Upon completion of this chapter, one should be able to:

1. Describe the concept of generating energy and power from waves.
2. Enumerate the various types of energy generation systems from waves.
3. Enumerate various potential sites for wave power generation around the world.
4. Describe the various applications of wave energy.
5. Describe the various conversion efficiencies of wave energy systems.
6. Relate the overall environmental and economic issues concerning wave power systems.

10.1 Introduction

Wave energy is derived from wind energy, which in turn comes from solar energy. The wind passing over the surface of the ocean creates a propagation as the water level rises and subsequently lowers in a nearly synchronized fashion. This power source can only be harnessed from oceans or vast seas. The extraction equipment must be able to operate in a marine environment. This necessitates proper maintenance and includes various construction costs and lifetime reliability concerns. The energy converters must be capable of withstanding very severe peak stresses in storms. A machine that is used to capture energy from waves is called a wave energy converter (WEC). Waves generated in the ocean have characteristic heights, speeds, and cycle periods. Water density is another factor to consider. Large waves are more powerful than small waves, and the bigger the wave, the more energy may be produced or extracted. It should also be noted that oscillation is highest at the ocean surface and diminishes with ocean depth. If one observes a wave, it rises and falls as well as moves toward the seashore.

There are numerous wave energy developers around the world (over 100 in a recent report), and each company may be classified by their location (shoreline, nearshore, and offshore) and by the type of power takeoff (hydraulic ram type, hose pump, or air turbine). Wave power may also be differentiated according to the four most common approaches: point absorber buoys, surface attenuators, oscillating water columns, and overtopping devices (Bhattacharya, 1983). Each of these systems will be discussed through examples in this chapter.

10.2 Power from Wave

The power generated from a wave has three very distinct applications: power production, water desalination, or water pumping to a reservoir. The governing equation for wave power is shown in Equation 10.1. This formula is applicable in deep-water regions, where the water depth is greater than half the wavelength. This equation is also called the wave energy flux equation. Example 10.1 shows how to derive the approximate constant to make the relationship in this equation quite simple:

$$Power\left(\frac{kW}{m}\right) = \frac{\rho g^2}{64\pi}H^2T \approx \left(0.50\frac{kW}{m^3s}\right)H^2T \tag{10.1}$$

where
ρ = density (approximately 1,000 kg/m³)
g = acceleration due to gravity = 9.8 m/s²
H = wave height, (m)
T = wave period (s)

Example 10.1: Calculating Estimated Value of Constant in Wave Energy Equation

Determine the estimated value of constant such that g and ρ do not appear in the equation (i.e., value of $\rho g^2/64\pi$ or units of kW/m³s). Assume the density of water to be 1,000 kg/m³ [8.33 lbs/gal].

SOLUTION:

a. The value of constant is approximately 0.5 as shown below:

$$\frac{\rho g^2}{64\pi} = \frac{1,000\ kg}{m^3} \times \left(\frac{9.8\ m}{s^2}\right)^2 \frac{1}{64 \times 3.1416} \times \frac{N \times s^2}{kg \times m} \times \frac{J}{N \times m} \times \frac{W \times s}{J} \times \frac{kW}{1,000\ W}$$

$$= 0.478\frac{kW}{m^3s} \cong 0.5\frac{kW}{m^3s}$$

$$\frac{\rho g^2}{64\pi} = 0.478\frac{kW}{m^3s} \times \frac{1,000\ W}{1\ kW} \times \frac{hp}{746\ W} \times \frac{1\ m^3}{35.29\ ft^3} \times \frac{60\ s}{min} = 1.1\frac{hp}{ft^2min}$$

b. The power constant is about half a kW per cubic meter per second [1.1 hp/ft³min].

Wave power is very different from tidal power or power harnessed from tidal currents. However, there is yet no large-scale project for wave converters even though hundreds of companies have started to develop their unique wave converted designs. Offshore wind power systems still dominate power generation from the oceans. Example 10.2 shows the magnitude of wave power for a given set of wave heights and duration. In major storms, the power could be greater than a megawatt as shown in Examples 10.3 and 10.4.

Example 10.2: Calculating Power from Wave per Meter of Crest Head

The wave height for a particular deep-sea wave was measured to be about 4 meters [13.12 ft] with a wave period of 6 seconds. Estimate the power per meter of crest head.

SOLUTION:

a. Use Equation 10.1 and simply substitute the values as follows:

$$Power\left(\frac{kW}{m}\right) = \frac{\rho g^2}{64\pi}H^2T \approx \left(0.50\frac{kW}{m^3s}\right)H^2T$$

$$Power\left(\frac{kW}{m}\right) = \left(0.50\frac{kW}{m^3s}\right)\times(4\ m)^2 \times 6\ s = 48\frac{kW}{m}$$

b. There is about 48 kW [64.3 hp] of power potential per meter of wave crest.

c. In major storms, the largest waves offshore may be as much as 15 meters [49.2 ft] high for a period of about 15 seconds.

Example 10.3: Estimating Power for a Given Length of Wave

In a major storm, the wave height could be 15 meters [49.2 ft] for a period of 15 seconds. How much is the total power available if the wave is 100 meters [328 ft] long?

SOLUTION:

a. Use Equation 10.1 and simply substitute the values as follows:

$$Power\left(\frac{kW}{m}\right) = \frac{\rho g^2}{64\pi}H^2T \approx \left(0.50\frac{kW}{m^3s}\right)H^2T$$

$$Power\left(\frac{kW}{m}\right) = \left(0.50\frac{kW}{m^3s}\right)\times(15\ m)^2 \times 100\ m \times 15\ s = 168,750\ kW\left[168.75\ MW\right]$$

b. Several hundreds of MW are available from this wave.

Example 10.4: Estimating Power per Given Longer Distance of Wave

Given a wave height of 4 meters [13.12 ft] and a wave period of 6 seconds, estimate the kW power per meter [hp/ft] of crest head. Determine the power for every km of wave crest [MW/km or MW/mi].

SOLUTION:

a. Use the formula as follows:

$$Power\left(\frac{kW}{m}\right) = \frac{\rho g^2}{64\pi}H^2T \approx \left(0.50\frac{kW}{m^3s}\right)H^2T$$

$$Power\left(\frac{kW}{m}\right) = \left(0.50\frac{kW}{m^3s}\right)\times(4\ m)^2 \times 6\ s = 48\frac{kW}{m}$$

$$Power\left(\frac{hp}{ft}\right) = 48\frac{kW}{m}\times\frac{hp}{0.746\ kW}\times\frac{1\ m}{3.28\ ft} = 19.62\frac{hp}{ft}$$

b. There is about 48 kW of power potential per meter of wave crest and in 1 km, or 1,000 meters, the total power in MW will be as follows:

$$Power(MW) = 48\frac{kW}{m}\times 1,000\ m = 48,000\ kW \times\frac{MW}{1,000\ kW} = 48\ MW$$

c. The total power per mile of wave is given as follows:

$$Power\ per\ mile\left(\frac{MW}{mi}\right) = 48\frac{MW}{km} \times \frac{1.609\ km}{mi} = 77.2\frac{MW}{mi}$$

10.3 World's Wave Power Resource

The amount of wave energy available worldwide was estimated to be around 2 TW (IRENA, 2014). The locations around the world with the most potential include western Europe, the northern United Kingdom, the Pacific coastlines of North America and South America, Australia, and New Zealand. The world's wave power resources are illustrated in Figure 10.1. The numbers shown in the figure correspond to wave power levels in units of kWh/m of crest length. The highest value shown in the figure is around 100 kWh/m crest length and is located southwest of Australia and in the Atlantic Ocean west of the United Kingdom. As depicted in the figure, most applications are near the shore and are also in many islands and archipelagoes.

Deep-sea or deep-ocean applications are also possible, and these are not shown in the figure. Note that science has not yet tapped the power from storms. There is difficulty due to the fact that these powerful storms can destroy even the sturdiest of structures. Earlier examples show how powerful storm waves are in the order of hundreds of MW.

Of the current projects on wave power, most are along the coastlines enumerated above. The first known patent of wave power dates back to 1799 and was filed in Paris, France,

World's Wave Power Resource

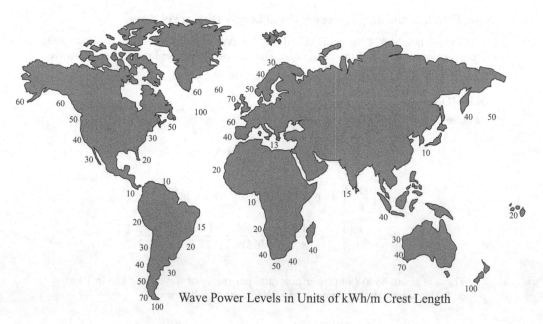

Wave Power Levels in Units of kWh/m Crest Length

FIGURE 10.1
World's wave power potential (kWh/m).

by Monsieur Girard (Ocean Energy Council, 2018). In 1910, Bochaux Praceique built a wave power system to light and power his house at Royan, near the Bordeaux in France (Renewable Solar Energy, 2017). This was the first oscillating water column ever built. From 1855 to 1973, there were 340 patents filed in the United Kingdom for wave energy converters. The oil crisis in the 1970s also brought forth newer designs and newer patents. The more popular researchers include the developers of the Masuda buoy from Japan, the Salter's duck from Edinburgh University in Scotland, and, more recently, the Pelamis wave energy converter (Boyle, 2004).

10.4 Various Generic Wave Energy Converter Concepts

There are various wave energy converter concepts that have been developed through the years (Figure 10.2). Six types are discussed below.

10.4.1 Point Absorber Buoy

Point absorber wave converters float on the surface of the water and are anchored on the sea ground surface. These buoys or floats use the rise and fall of swells or waves

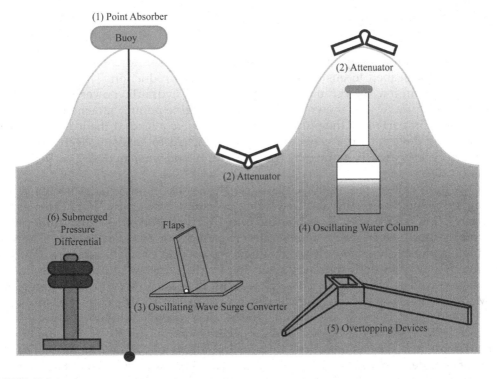

FIGURE 10.2
Various wave energy converter concepts (1 = point absorber, 2 = attenuator, 3 = oscillating wave surge converter, 4 = oscillating water column, 5 = overtopping devices, 6 = submerged pressure differential) (Bhattacharya, 1983).

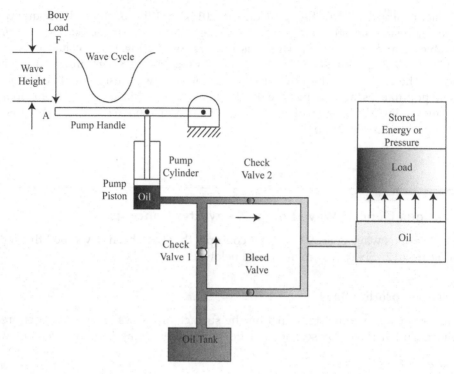

FIGURE 10.3
Hydraulic power from a wave point absorber.

to drive a hydraulic pump and generate electricity. These devices must be properly secured on the seabed to ensure reliability of power production. The power from a hydraulic system may be estimated using the simple hydraulic power Equation 10.2, similar to a hydraulic jack shown in Figure 10.3. Imagine point A attached to the buoy and pressure imparted on the pump cylinder to actuate a load. If the load is replaced by a rotating device, then a generator may be attached to generate electrical power. The wave cycle is usually defined by wave height, H, and wave period, T, as defined in Equation 10.1.

As the buoy is lowered as the wave lowers, pressure is imparted onto the piston cylinder, moving the piston and displacing a volume of working fluid. This action generates flow and power. Equations 10.2 to 10.6 define all these hydraulic power parameters and can be used to generate mechanical or electrical power. The succeeding examples show how these equations are used to estimate pressure and flow as well as power in both the metric and the English system of units. Note that in the metric system, the unit of pressure (Pascal) is equal to 1 N/m^2. Newton's Law of Motion (Force = mass × acceleration) is a very useful relationship that simplifies most calculations. In these calculations, the acceleration due to gravity value used is 9.8 m/s^2 [32.2 ft/s^2]. In the English system of units, lbs/in^2 (or psi) is a common unit for pressure, and flow is measured in units of gallons per minute (gpm). Since the units of cylinders are usually in cm, the unit kPa (kiloPascal), which is 1,000 N/m^2, is more convenient to use. Note that converting psi to kPa is done by simply multiplying psi units by 6.89.

The pressure created when the buoy is lowered is given in Equation 10.2. In metric units, the weight unit is in Newton (N) (lbs in the English system):

$$p = \frac{F_b(N)}{A_p(cm^2)} = \frac{F_b(lbs)}{A_p(in^2)} \tag{10.2}$$

where
p = pressure created as buoy load drops (N/cm² or psi)
F_b = buoy load (N or lbs)
A_p = piston area (cm² or in²)

Liquid flow is created when the buoy is also lowered as the piston moves a given distance—in this case, the wave height, H, over a given period of time. In this case, the element of time is the wave period. These relationships are shown in Equations 10.3 and 10.4:

$$Q\left(\frac{L}{s}\right) = \frac{H(m)}{T(s)} \times A_p(m^2) \times \frac{1,000\ L}{m^3} \tag{10.3}$$

$$Q\left(\frac{gallons}{min}\right) = \frac{H(ft)}{T(s)} \times A_p(in^2) \times \left(\frac{ft}{12\ in}\right)^2 \times \frac{60\ s}{min} \times \frac{7.48\ gal}{ft^3} \tag{10.4}$$

where
Q = fluid flow created (L/s, gal/min)
H = wave height (m or ft)
T = wave period (s)

The power generated during this activity is simply the product of volumetric flow rate created and the pressure as shown in Equations 10.5 and 10.6 for metric and English systems, respectively:

$$Power(kW) = \frac{Q_{in}\left(\frac{L}{s}\right) \times p\left(\frac{N}{m^2}\right)}{1,000,000} \tag{10.5}$$

$$Power(hp) = \frac{Q_{in}(gpm) \times p(psi)}{1714} \tag{10.6}$$

Example 10.5 shows how much pressure is generated for every 445.5 N [100 lbs] of buoy load with a given cylinder size and wave period.

Example 10.5: Theoretical Hydraulic Power from Waves

Determine the pressure generated if the buoy load is 445.5 N [100 lbs] with a wave height of 4 meters [13.12 ft] and a wave period of 6 seconds. Estimate also the volumetric flow rate generated assuming the diameter of the small piston cylinder is 5 cm [1.97 inch]. Please refer to Figure 10.3 for this arrangement. Finally, estimate the theoretical power generated for this action.

SOLUTION:

a. Using Equation 10.2, the pressure generated is as follows for both the English and metric system of units:

$$p = \frac{F_b\,(N)}{A_p\,(cm^2)} = \frac{445.5\ N}{\frac{\pi}{4}(5\ cm)^2} = 22.7\frac{N}{cm^2} \times \left(\frac{100\ cm}{m}\right)^2 = 226,891\frac{N}{m^2} = 226.89\ kPa$$

$$p = \frac{F_b\,(lbs)}{A_p\,(in^2)} = \frac{100\ lbs}{\frac{\pi}{4}(1.97\ in)^2} = 32.81\ psi$$

b. The volumetric flow rate is calculated using Equations 10.3 and 10.4:

$$Q\left(\frac{L}{s}\right) = \frac{H(m)}{T(s)} \times A_p\,(m^2) \times \frac{1,000\ L}{m^3}$$

$$Q\left(\frac{L}{s}\right) = \frac{4(m)}{6(s)} \times \frac{3.1416}{4} \times \left(\frac{5\ cm}{100\ cm/m}\right)^2 \times \frac{1,000\ L}{m^3} = 1.31\ \frac{L}{s}$$

$$Q\left(\frac{gal}{min}\right) = \frac{H(ft)}{T(s)} \times A_p\,(in^2) \times \left(\frac{ft}{12\ in}\right)^2 \times \frac{60\ s}{min} \times \frac{7.48\ gal}{ft^3}$$

$$Q\left(\frac{gal}{min}\right) = \frac{13.12(ft)}{6(s)} \times \frac{3.1416}{4} \times (1.97\ in^2)^2 \times \left(\frac{ft}{12\ in}\right)^2 \times \frac{60\ s}{min} \times \frac{7.48\ gal}{ft^3} = 20.77\ gpm$$

c. The power is then calculated using Equations 10.5 and 10.6:

$$Power\,(kW) = \frac{Q_{in}\left(\frac{L}{s}\right) \times p\left(\frac{N}{m^2}\right)}{1,000,000}$$

$$Power\,(kW) = \frac{1.31\ L}{s} \times \frac{226,892\ N}{m^2} \times \frac{m^3}{1,000\ L} \times \frac{J}{Nm} \times \frac{Ws}{J} \times \frac{kW}{1,000\ W} = 0.297\ kW$$

$$Power\,(hp) = \frac{Q_{in}\,(gpm) \times p\,(psi)}{1714}$$

$$Power\,(hp) = \frac{20.77\ gpm \times 32.81\ psi}{1714} = 0.398\ hp$$

Example 10.6 shows how much power is generated when pressure generated from one cylinder is translated to another with a different piston diameter. The idea is to translate one small force and convert this to a higher level of force via Pascal's Law. This law is also called the principle of transmission of fluid pressure and is used to generate a much higher actuating load from a smaller load through a combination of cylinder dimensions. Imagine the hydraulic jack shown in Figure 10.3.

Example 10.6: Power Output from Piston-Type Wave Systems (Metric)

Determine the output power or power stored (in kW) for a piston-type system assuming the stroke is 4 meters [13.12 ft] at a rate of one cycle per 6 seconds, matching the wave cycle in the previous example. The buoy load force is around 445.5 N [100 lbs].

Use the dimensions in Figure 10.3 for calculations. Assume that the handle's total length is 3 m [9.84 ft] and the distance from the handle pivot is 1 m (the short side distance to anchor, or 3.28 ft). The pump has a 5 cm [1.97 in] diameter piston, and the load cylinder has a 12 cm [4.72 in] diameter piston. The load will be raised 10 m [32.8 ft]. Determine the number of cycles required to lift this load at this height. Each complete cycle consists of two pump strokes (intake and power).

SOLUTION:

a. The buoy load, F, is first calculated:

$$F_{rod} = \frac{445.5 \ N \times 3 \ m}{1 \ m} = 1,336.5 \ N$$

b. The pressure created out of this action will be as follows:

$$p = \frac{F}{A} = \frac{1,336.5 \ N}{\frac{3.1416}{4} \times (5 \ cm)^2} = 68.1 \frac{N}{cm^2} = 681,000 \frac{N}{m^2} = 681 \ kPa$$

c. The load on the other piston cylinder is then calculated:

$$F_{load} = pA = 68.1 \frac{N}{cm^2} \times 3.1416 \times \frac{(12 \ cm)^2}{4} = 7,702 \ N$$

d. The force load balance for each stroke of movement of the wave cycle is given by the following relationship:

$$(A \times S)_{pump-piston} \times \# \ of \ cycles = (A \times S)_{load-piston}$$

$$\left(\frac{3.1416 \times (5 \ cm)^2}{4} \times (4 \ m) \right)_{pump-piston} \times \# \ of \ cycles = \left(\frac{3.1416 \times (12 \ cm)^2}{4} \times (10 \ m) \right)_{load-piston}$$

$$Number \ of \ Cycles = \left[\frac{(12 \ cm)^2}{(5 \ cm)^2} \times \frac{10 \ m}{4 \ m} \right] = 14.4 \ cycles$$

e. The power is simply the product of the distance to raise the load 12 inches [1 ft = 0.3048 m] and the force associated with this load:

$$Power(hp) = \frac{Fs}{t}$$

$$Power(W) = \frac{7,702 \ N \times 10 \ m}{(14.4 \times 6)s} = 891.4 \ W \left[1.2 \ hp \right]$$

Example 10.7: Power Output from Piston-Type Wave Systems (English System)

Determine the output power or power stored (in hp) for this system assuming that the stroke is 4 meters [13.12 ft] at a rate of one cycle per 6 seconds, matching the wave cycle in the previous example. The buoy load force is around 445.5 N [100 lbs]. Use the dimensions in Figure 10.3 for calculations. Assume the handle's total length is 3 m [9.84 ft] and the distance from the handle pivot is 1 m (the short side distance to anchor or 3.28 ft). The pump has a

5 cm [1.97 in] diameter piston, and the load cylinder has a 12 cm [4.72 in] diameter piston. The load will be raised 10 m [32.8 ft]. Determine the number of cycles required to lift this load at this height. Each complete cycle consists of two pump strokes (intake and power).

SOLUTION:

a. The buoy load, F, is first calculated:

$$F_{rod} = \frac{100 \; lbs \times 9.84 \; ft}{3.28 \; ft} = 300 \; lbs$$

b. The pressure created out of this action will be as follows:

$$p = \frac{F}{A} = \frac{300 \; lbs}{\dfrac{3.1416 \times (1.97 \; in)^2}{4}} = 98.4 \; psi$$

c. The load on the piston cylinder is then calculated:

$$F_{load} = pA = 98.4 \frac{lbs}{in^2} \times \frac{3.1416 \times (4.72 \; in)^2}{4} = 1{,}721.7 \; lbs$$

d. The force load balance for each stroke of movement of the wave cycle is given by the following relationship:

$$(A \times S)_{pump-piston} \times \# \; of \; cycles = (A \times S)_{load-piston}$$

$$\left(\frac{3.1416 \times (1.97 \; in)^2}{4} \times (13.12 \; ft) \right)_{pump-piston} \times \# \; of \; cycles = \left(\frac{3.1416 \times (4.72 \; in)^2}{4} \times (32.8 \; ft) \right)_{load-piston}$$

$$Number \; of \; Cycles = \left[\frac{(4.71 \; in)^2}{(1.97 \; in)^2} \times \frac{32.8 \; ft}{13.12 \; ft} \right] = 14.4 \; cycles$$

e. The power is simply the product of the distance to raise the load 12 inches and the force associated with this load:

$$Power(hp) = \frac{Fs}{t}$$

$$Power(hp) = \frac{1{,}721.7 \; lbs \times 32.8 \; ft}{(14.4 \times 6)s} \times \frac{hp - min}{33{,}000 \; ft - lb} \times \frac{60 \; s}{min} = 1.18 \; hp$$

10.4.2 Surface Attenuator

Surface attenuators are similar to single-point absorbers except that there are numerous floating segments that are connected to one another and are oriented perpendicular to the incoming waves. Imagine a book being opened and closed as a wave propagates. Imagine also that two piston pumps are inserted between the leaves of this book. This will create a reciprocating in-and-out motion. When the cylinders are on top of the wave, the piston is extended, and when the pistons are at the bottom of the wave, the pistons are in contact. This in-and-out motion then creates reciprocating motion and generates pressure and

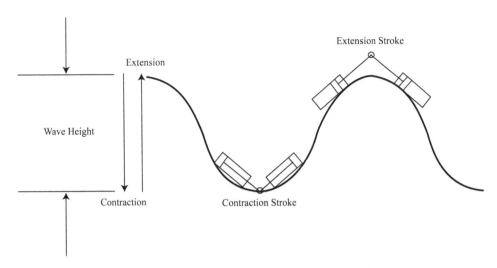

FIGURE 10.4
Piston hydraulic motion as the wave changes.

power. Figure 10.4 shows the piston hydraulic motion as the wave changes. Example 10.6 shows how much pressure or power is developed for every cycle for a given set of piston sizes. The power associated with this action is given simply by Equation 10.7:

$$Power\,(Watts) = \frac{Energy\,(Joules)}{time\,(sec)} = \frac{(pA) \times S}{t} = p(Av) \tag{10.7}$$

where
$Power$ = watts (N-m/s, J/s, watts)
$Pressure$ = N/cm² (or N/m² or Pascal)
A = area of piston (cm²)
S = stroke or displacement (m)
t = time per cycle (s)

Example 10.8: Power Generated in a Piston per Stroke

Determine the power generated in each piston for every complete stroke if the piston diameter is 0.5 meters and the stroke is 1 meter. The wave motion completes its cycle every 5 seconds (i.e., 5 seconds per stroke). Assume that the pressure was created by a 200 liter [52.84 gal] volume of water creating 1,925 N of load onto the piston rod, which drives the piston in and out due to the wave movement. The pressure created by this force was approximately 9,800 N/m² [2,203 lb$_f$].

SOLUTION:

a. This is a simple power equation (given in Equation 10.7):

$$Power\,(Watts) = \frac{9,800\ N}{m^2} \times \frac{3.1416 \times (0.5\ m)^2}{4} \times \frac{1\ m}{5\ s} = 385\ Watts\,[0.52\ hp]$$

b. This is the theoretical power generated by each piston of given dimensions:

The Cockerell Wave Contouring Raft

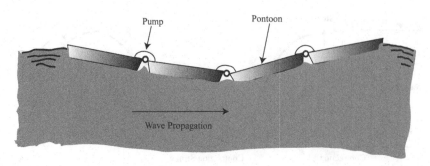

FIGURE 10.5
The Cockerell wave contouring raft (Bhattacharya, 1983).

10.4.2.1 Wave Contouring Rafts (Cockerell Rafts)

Wave contouring rafts are examples of attenuators. The system is comprised of hydraulic pumps placed between each raft (the pumps act as double-acting pistons). Power is generated by the action of two piston cylinders opposed to each other. As seen in Figure 10.4, when the piston pairs are at the top of the wave, the cylinders are on the extension stroke and are on the compressed stroke when the piston is at the bottom of the wave. The uniqueness of the Cockerell wave contouring raft is that there is an endless number of rafts joined together using pontoons (see Figure 10.5). The hydraulic pumps are fitted between each raft pair. This arrangement converts the energy derived from the motion of the raft into high pressure in a fluid that can be used to run a turbine.

10.4.3 Oscillating Wave Surge Converter

An oscillating wave surge converter usually has one end fixed to a structure or at the bottom of the seabed while another end is free to move. Since there is a fixed point, energy is generated from the relative motion between the fixed component and the movable component. These movable components may be flaps, floats, or membranes. Many possibilities are available, including the development of parabolic reflectors to increase the wave energy at the point of capture and transport the energy of power to some mechanisms (electrical power cables).

The Pendulor shown in Figure 10.6 is an example of an oscillating wave surge converter. The Pendulor was invented in Japan (Gunawaradane, et al., 2010). The commonly used hydraulic pump is a rotary vane type attached to a hydraulic motor that in turn runs a generator to produce electrical power. The proven efficiency is around 40% to 50%.

10.4.4 Oscillating Water Column

Oscillating water columns are devices that oscillate as waves pass through. An oscillating water column could be anchored to the bottom bed of the ocean and will move back and forth as the waves propagate, or a system could be located on the surface, where back-and-forth movement creates compressed air and forces a turbine blade to move and generate power. These devices will create some noise when air is pushed through the turbine chambers. An example is shown schematically in Figure 10.7. In this figure, a certain volume of

The Pendulor (Pivoting Flap Device)

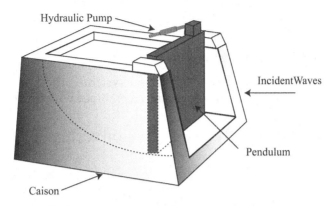

FIGURE 10.6
The Pendulor is an example of an oscillating wave surge converter (Gunawaradane, et al., 2010).

air is displaced by the water column in a collecting chamber and is pushed back and forth past a power-takeoff unit, which is a turbine and generator. The turbine should be bidirectional such that it will spin even if the direction of water is reversed. Power is continuously generated throughout this process.

10.4.5 Overtopping Device

Wave energy converters of the overtopping device type have long structures that use wave velocity to fill a reservoir to a greater water level. Simple hydro power schemes are utilized, similarly to tidal or hydro power plants. The innovations in these systems include the type of collector for the water directed to a reservoir. There are numerous designs utilizing parabolic structures for easy collection of water to a higher elevation. The most popular design is the Wave Dragon designed in Denmark (http://www.wavedragon.net).

Oscillating Water Column (OWC)

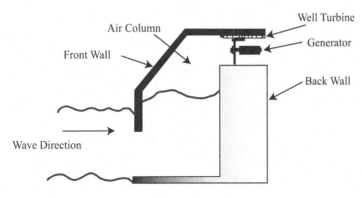

FIGURE 10.7
The oscillating water column (US DOE, 2018).

This patented technology claims to be a pioneering large-scale ocean energy solution for bulk electricity generation (Kofoed, 2006). The European Commission Project is the demonstration unit in Wales, estimated to be about 7 MW, and there is a project in Portugal that is aiming toward 50 MW wave power from Portuguese waters (Henriques, et al., 2016).

10.4.6 Submerged Pressure Differential

An example of a submerged pressure differential system is the Isaac-Seymour system shown in Figure 10.8. In this system, air pressure, P, combined with hydrostatic head, forces

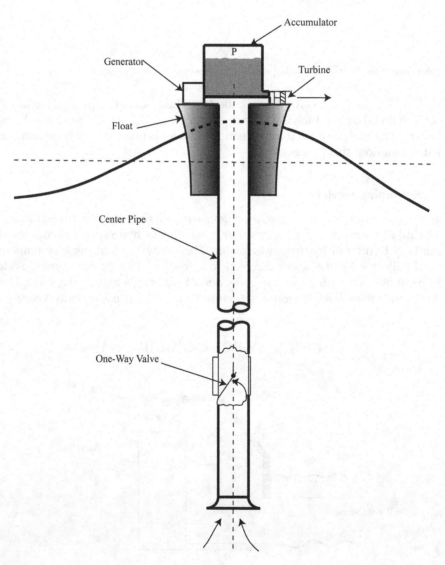

FIGURE 10.8
The Isaac wave energy converter (Bhattacharya, 1983).

water into the turbine. When the turbine is attached to a generator, power is immediately produced. Likewise, the turbine should be double-acting so that it works continuously even with varying surge pressures in the system. The principle is based on Archimedes' wave swing (Damen, et al., 2006). The system also operates like a hydraulic system in many tractors or vehicles.

The Isaac-Seymour device consists of a float through which a long, vertical center pipe passes (Bhattacharya, 1983). The bottom of the center pipe is in free communication with the sea. When the float heaves downward, the valve within the center pipe opens, allowing water to flow upward. When the float heaves upward, the valve is closed, and the internal water level remains at a constant height relative to the pipe. Over each wave cycle, more water accumulates above the valve until air above the water in the accumulator is so compressed that no more water is accepted. The air pressure, combined with the hydrostatic head of the water, forces the water through the turbine.

10.5 Other Common Types of Currently Deployed Wave Energy Converters

There are other types of wave energy converters that fall under a combination of concepts introduced in a previous chapter. These include the hose pump, the Salter's duck system, and the Masuda buoy (Bhattacharya, 1983). These systems will be discussed in this section.

10.5.1 Hose Pump

The hose pump is shown in Figure 10.9. The system has a float that follows the propagation of the wave. As the float goes up and down, compressed seawater is generated,

The Hose Pump (Heaving Bouy Device)

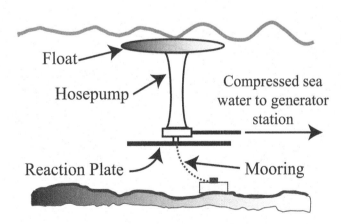

FIGURE 10.9
The hose pump (a heaving buoy device).

and this is sent to a generator station. The unit may be placed offshore or near the shore such that mooring is required. This system is also called a heaving buoy device. A heavy reaction plate is usually needed at the base of the series of hoses to ensure stability of the system.

10.5.2 Salter's Duck

The Salter's duck system is another invention by S. H. Salter of Edinburgh University in Scotland (McGrath, 2008). The design is shown in Figure 10.10. The system includes four double-acting pumps with non-return valves in the ridged wall of the cylinder. The float has a front surface that moves with the water from the incoming wave as shown. It has a back surface that does not disturb the water behind. The float rocks about an axis through a center pivot unit. Nearly 80% of the wave energy was reported to be absorbed by the duck. The actual power-producing device consists of a string of ducks floating on the sea, with each set rotating around a common backbone in the form of a hollow cylinder with parallel ridges on its surface. The pumped water can be used to generate electricity. Note the similarity of these actions with that of a typical hydraulic pump of the internal gear type.

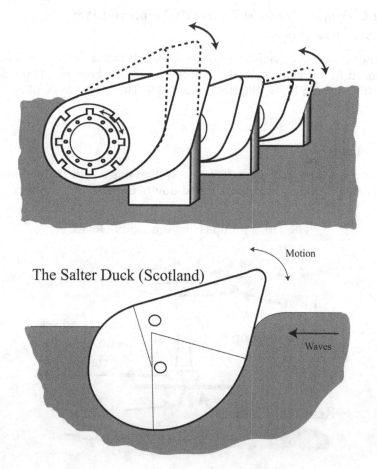

FIGURE 10.10
The Salter's duck (Bhattacharya, 1983).

Basic Principles of the Masuda Bouy

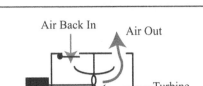

FIGURE 10.11
The mechanism of the Masuda buoy.

10.5.3 Masuda Buoy

The principles of the Masuda buoy are shown in Figure 10.11. Yoshio Masuda, a former Japanese naval commander, was regarded as the father of modern wave power technology. The principle used is that of an oscillating water column but with the unique action of compressed air to run a turbine and generator.

Imagine the Masuda buoy to be an inverted floating box. Air inside the buoy is compressed with a rising wave and expanded with a falling wave. A turbine-generator unit is driven by air flowing due to the presence of a non-returning valve. It has been claimed that the small-scale air turbines have very long operational lives with relatively good efficiencies. The output of the device has both short-term and long-term fluctuations. Short-term fluctuations will be lessened in a relatively big device. Note that Masuda started his research in the 1940s. He has tested various concepts of wave energy conversion devices at sea with several hundred units that are used to power navigational devices and lights, primarily because of his naval experience as a commander. One of his inventions, released in the 1950s, was the concept of extracting power from the angular motion at the joints of an articulated raft (Enriques, et al., 2016).

10.6 Typical Hydraulic Circuit for Wave Generators

Most newer wave power system designs rely on hydraulic power. The typical circuit for this system is shown in Figures 10.12a and 10.12b (Esposito, 2009). In these figures, imagine a buoy placed on the surface of the ocean allowed to rise or sink with the wave. As in Figure 10.12a, as the two-way piston lowers due to its weight (as well as the fall of the buoy), the water is pushed to a series of check valves. The darker lines denote the direction of the positively pushed water through the piston. Water then goes up to the high-pressure accumulator, where excess energy is stored while the rest of pressurized water goes to the hydraulic motor that drives the generator. Return water (denoted by the red lines) goes back through the low-pressure accumulator, and if the water has more

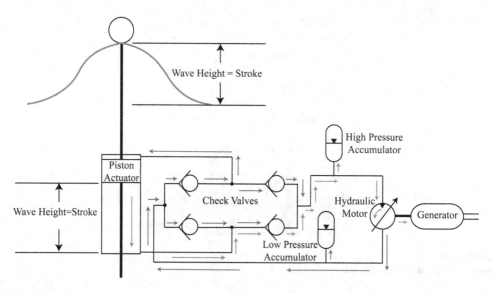

FIGURE 10.12a
The hydraulic circuit for a typical piston-based wave energy converter showing downward stroke of piston cylinder (wave receding).

than enough energy, it can also store excess energy at this point. The low-pressure water will be conveyed to another set of check valves and return some water to the back side of the piston ram. Other low-pressure water contributes to the flow of incoming water from the end cap of the piston.

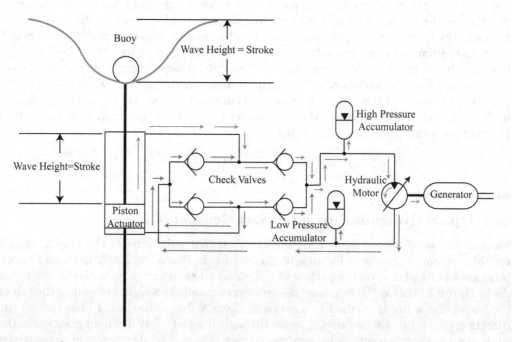

FIGURE 10.12b
The hydraulic circuit for a typical piston-based wave energy converter showing upward stroke of piston cylinder (wave peaking).

As the buoy from the lowest portion of the wave now begins to rise (Figure 10.12b), as shown, water is pushed on the top side of the check valve system while still pushing the water to the high-pressure accumulator and onto the hydraulic motor and generator. Hence, with each movement of the piston, whether up or down, the hydraulic motor moves in just one direction, generating power. Of course, there are minute variations in water pressure as the piston moves up and down. This is where the accumulator becomes crucial. The accumulator balances off the water pressure to make the movement of the hydraulic motor smooth and nearly uniform through the motion of the buoy. The development of these simple hydraulic systems allows for the design of even more complicated units that could generate more power by exceeding the amount of pressure that has not been previously achieved.

Hydraulic motors are rated according to their volumetric displacements (V_D) and are usually defined in units of m³/rev or in³/rev as shown in Equations 10.8 and 10.9:

$$Q_T\left(\frac{m^3}{s}\right) = V_D\left(\frac{m^3}{rev}\right) \times N\left(\frac{rev}{min}\right) \times \frac{1\,min}{60\,s} \tag{10.8}$$

$$Q_T\left(gpm\right) = \frac{V_D\left(\dfrac{in^3}{rev}\right) \times N\left(\dfrac{rev}{min}\right)}{231\dfrac{in^3}{gals}} \tag{10.9}$$

where
Q_t = displacement volume (m³/s or in³/s)
N = revolutions per minute (rpm) of the hydraulic motor

The power generated by the motor is given in Equations 10.10 and 10.11:

$$HP_t\left(Watts\right) = \frac{V_D\left(\dfrac{m^3}{rev}\right) \times p(Pa) \times N\left(\dfrac{rev}{min}\right)}{60} \tag{10.10}$$

$$HP_t\left(Watts\right) = \frac{V_D\left(\dfrac{in^3}{rev}\right) \times p(psi) \times N\left(\dfrac{rev}{min}\right)}{395,000} \tag{10.11}$$

Example 10.9 shows how these hydraulic motor equations are used to estimate the displacement volume for a hydraulic motor and the power produced due to wave action.

Example 10.9: Calculating Displacement Volume of Wave Hydraulic Power System

A wave power–generating device is to be designed such that a generator is to be attached to a hydraulic motor (speed ratio of 9:1). The rpm of this motor is 200. The prime mover must come from a hydraulic motor that would turn this generator at the rate required (usually 1,800 rpm). The power of the hydraulic motor will ultimately come from the power generated from waves as shown in Figure 10.12. Determine the displacement volume of the hydraulic motor if the volumetric displacement is 9.84 × 10⁻⁵ m³/rev [6 in³/rev]. Assume the pressure developed was 681 kPa (681,000 Pa) (or 98.4 psi).

SOLUTION:

   a. The volumetric flow rate of displacement volume is calculated using Equation 10.8:

$$Q_T\left(\frac{m^3}{s}\right) = V_D\left(\frac{m^3}{rev}\right) \times N\left(\frac{rev}{min}\right) \times \frac{1\ min}{60\ s}$$

$$Q_T\left(\frac{m^3}{s}\right) = 9.84 \times 10^{-5}\left(\frac{m^3}{rev}\right) \times 200\left(\frac{rev}{min}\right) \times \frac{1\ min}{60\ s} = 3.28 \times 10^{-4}\ \frac{m^3}{s}$$

$$Q_T\left(\frac{m^3}{s}\right) = 3.28 \times 10^{-4}\ \frac{m^3}{s} \times \frac{1,000\ L}{1\ m^3} = 0.328\ \frac{L}{s}$$

$$Q_T\left(gpm\right) = \frac{V_D\left(\frac{in^3}{rev}\right) \times N\left(\frac{rev}{min}\right)}{231\ \frac{in^3}{gals}}$$

$$Q_T\left(gpm\right) = \frac{6\left(\frac{in^3}{rev}\right) \times 200\left(\frac{rev}{min}\right)}{231\ \frac{in^3}{gals}} = 5.195\ gpm$$

   b. This is the flow that goes through the hydraulic motor.

   c. The theoretical power is calculated using Equation 10.5 or Equation 10.6 since volumetric displacement and pressure are already given:

$$Power\left(kW\right) = \frac{Q_{in}\left(\frac{L}{s}\right) \times p\left(\frac{N}{m^2}\right)}{1,000,000}$$

$$Power\left(kW\right) = \frac{0.328\left(\frac{L}{s}\right) \times 681,000\left(\frac{N}{m^2}\right)}{1,000,000} = 0.223\ kW$$

$$Power\left(hp\right) = \frac{Q_{in}\left(gpm\right) \times p\left(psi\right)}{1,714}$$

$$Power\left(hp\right) = \frac{5.195\left(gpm\right) \times 98.4\left(psi\right)}{1,714} = 0.298\ hp$$

10.7 Approximating Wave Height Using Significant Wave Height, H_s

The wave height to use in many wave energy calculations is not always the mean value or the average value of wave height calculations. Because of variability in wave propagation, statistical analysis is usually performed. The first concept is the use of the significant wave height value, H_s, instead of mean wave height. The mean wave height is simply the average of all wave height measurements. The H_s is approximated by Equation 10.12 (NOAA, 2018):

$$H_s = 4 \times \sqrt{Variance\ of\ Water\ Level} \qquad (10.12)$$

In Equation 10.11, significant wave height is an average measurement of the largest 33% of the waves, measured from trough to crest. H_s is used instead of the mean height due to variability in wave height estimate. According to NOAA, larger waves are more "significant" (or important) than smaller waves. H_s is usually approximated by the Rayleigh distribution (i.e., assumed normally distributed), which is the square root of the water level variance multiplied by 4. The variance must first be calculated from a given set of data, and H_s is used instead of mean wave height. Example 10.10 shows how H_s is estimated with these approximations.

Example 10.10: Estimating Significant Wave Height

The following data were obtained from wave height measurement (in meters) 5, 6, 9, 7, 8, representing the highest one-third of the bigger waves. Estimate the significant wave height H_s. Compare this value with the simple mean wave height.

SOLUTION:

a. The simple mean is calculated as follows:

$$Mean\ Height\ (m) = \frac{(5+6+9+7+8)}{5} = 7\ m\left[2.96\ ft\right]$$

b. The sum of squares (SS) is then calculated as follows:

$$SS = (5-7)^2 + (6-7)^2 + (9-7)^2 + (7-7)^2 + (8-7)^2 = 10$$

c. The degrees of freedom (DF) value is given as follows:

$$DF = (n-1) = 5-1 = 4$$

d. The standard deviation (SD) of the sample is given by the following equation and is calculated as follows:

$$SD = \sqrt{\frac{SS}{(n-1)}} = \sqrt{\frac{10}{(5-1)}} = 1.58$$

e. Hence, the variance, which is simply the square of the standard deviation (SD), will be calculated as follows:

$$(SD)^2 = (1.58)^2 = 2.5$$

f. Thus, the significant wave height is calculated following Equation 10.11 and is shown below:

$$H_s = 4 \times \sqrt{2.5} = 6.32\ meters\left[20.73\ ft\right]$$

g. The value of significant wave height is less than the simple mean of the data. For large volumes of data, one must rely on statistical software or a spreadsheet program, such as Microsoft Excel.

10.8 Beneficial and Non-Beneficial Environmental Impacts of Wave Power

There are various environmental impacts that must be addressed when installing a wave energy converter system. These effects may be classified as beneficial or non-beneficial. These are discussed in this section.

10.8.1 Advantages

Wave power works well with wind energy extractors, and the energy supply is greater in winter, when other renewables may be quite low. Of course, the long-lasting advantage is that this renewable energy does not vary with time of day as much as other renewables, such as solar energy. In the future, wave power may be exchanged as carbon credits, comparing the amount of carbon saved from fossil fuels. While most coastal communities are the first to benefit, the production of electrical power can cross distance barriers. This is a renewable energy that may have the most potential, considering that earth is three-quarters full of salt water.

10.8.2 Disadvantages

The biggest environmental disadvantage is the altering of the coastal areas due to the presence of energy extraction machines and structures. Bhattacharya (1983) calls it "ecological disaster" if not designed properly. The new structures may alter the littoral processes along the coasts. Second would be the erosion and accretion or buildup of sand and other debris due to constant movement and dislocation of nearshore solids. The bottom surface or the ocean floor will also be affected, especially if moorings are widespread. Sediment transport, water quality, and formation of deltas (if placed near the mouth of a river) are also possible disadvantages. Finally, the risks of marine mammals being hit by WECs are considerable. It was also reported that the high-power transport of electrical charges could affect the mammals as well as create underwater noises.

10.9 Year-Round Distribution of Wave Energy

There is a huge difference between the availability of wave power during summer and during winter in some arid countries. Figure 10.13 shows a typical frequency distribution of wave power over a year of observation for a station in India (59°N, 19°W) in the North Atlantic. The Y-axis is the power distribution in units of kW s/m (Mollison, et al., 1976), and the X-axis is the frequency in units per seconds. The yearly average power for this site is around 91 kW/m [37.2 hp/ft]. Note that the distribution is high during winter (dotted line) with peaks close to 600 kW s/m [245 hp/ft-s]. The mean frequency is less than 0.2/s with peaks at less than 0.1/s. The summer season has a very low energy rate with peaks at less than 50 kW s/m [20.4 hp/ft-s]. The curve follows a typical Weibull distribution function previously presented in the wind energy chapter. Analysis for energy generated in a year may be made using that particular statistical distribution.

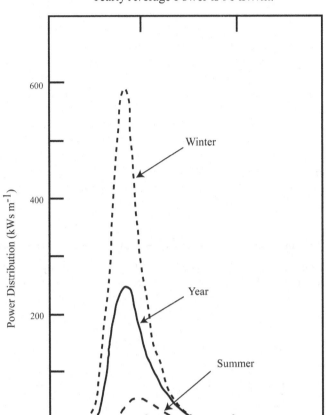

Frequency Distribution of Wave Power for a Year Observation.
Yearly Average Power is 91 kW/m.

FIGURE 10.13
Wave power distribution for a year in India (59°N, 19°W) (Mollison, et al., 1976).

The average for the year has a peak of around 250 kW s/m [102.2 hp/ft-s]. In the design of wave power–generating devices, this frequency distribution must be taken into account since the power requirement is nearly constant for the year, while the resource is not. Proper matching of energy needs and supply from waves must be performed so that deficiency in power is avoided.

10.10 Economic Aspects and Potential Locations

Wave energy converters are said to be a little bit more expensive than nuclear technologies. The current reported capital cost is around $1,000/kW [$746/hp]. The locations with the most potential include western Europe, the north of the United Kingdom, and the Pacific coastline of North and South America, southern Africa, Australia, and New Zealand. The north and south temperate zones are ideal for wave power generation. In these areas, the westerly wind flows are stronger during the wintertime (IRENA, 2014).

Example 10.11 shows the estimated initial capital cost for a facility that generated over $62.5 MW of electrical power from wave energy. Note that in this calculation, the capacity factor and reliability are not yet accounted for. The frequency distribution depicted in Figure 10.13 will show how power output varies through the year. More importantly, since most facilities installed are simply demonstration systems, there is still uncertainty about scaling these systems up.

Because of the vast ocean areas that will be affected, the wave power project must also take into account lost opportunities for people such as fishermen whose livelihoods depend on these areas and trade at industry ports. There was a report that of the 2,700 GW of potential wave power, only less than 18.5% can be captured with the current technology being developed. Hence, more demonstration projects are needed to establish the long-term economic benefits of wave power.

Example 10.11: Estimating Capital Cost of Wave Power Systems per Kilometer

The capital cost for a wave converter is around $1,000/kW [$746/hp]. In an area where the average wave height is 5 meters [16.4 ft] for a period of 6 seconds, how much is the capital cost for every km [mi] of wave distance?

SOLUTION:

a. The power is first calculated:

$$Power\left(\frac{kW}{m}\right) = \left(0.50\frac{kW}{m^3s}\right) \times (5\ m)^2 \times 6\ s = 75\frac{kW}{m}\left[30.65\frac{hp}{ft}\right]$$

b. The power per km is then calculated:

$$Power(kW) = 75\frac{kW}{m} \times \frac{1,000\ m}{km} = 75,000\ kW\left[100,536\ hp\right]$$

c. Hence, the total cost will be as follows:

$$Cost(\$) = 75,000\ kW \times \frac{\$1,000}{kW} = \$75\ M$$

d. Over $75 million is needed for this project.

10.11 Countries with Wave Energy Studies (IRENA, 2014)

The number of countries with wave energy studies is growing at a rapid rate. Australia has built a 19 MW wave power station in Portland, Victoria. China continues to compete with world powers by starting a 100 kW unit and several small units. India has initiated a 150 kW [201 hp] pilot plant in Kerala and Indonesia a 1.1 MW wedge groove wave power plant. Japan has also reported the development of various small units (30, 40, 60, 110 kW) and another 200 kW [268 hp] unit as demonstration systems as well. Norway has built a 350 to 500 kW [469 to 670 hp] unit, and Portugal has initiated an ambitious 2.25 MW unit using the Pelamis wave energy converter technology. Sweden also plans to build 15 kW [20 hp] and 150 kW [200 hp] units, and most projects in the United Kingdom include 3 MW (Scotland, 2007). There is a 20 MW Wave Hub on the north coast of Cornwall, England. In the United States, a 1.5 MW offshore facility near Reedsport, Oregon, is being planned. Some recent projects for each country are summarized in this section, summarized from IRENA (2014).

10.11.1 United Kingdom

There are more than 10 listed wave energy projects in which the United Kingdom (Scotland and England) is involved. One proponent is Albatern WaveNet, based in Scotland. The Albatern group works with their third-iteration devices and deploy these in Scottish fish farms. A six-unit array is also being deployed for full evaluation at Kishorn Port. The company is targeting off-grid markets where diesel generation is presently used. Diesel generation is very popular for the offshore fishing industry. Another project in the United Kingdom was initiated by Anaconda Wave Energy, a converter designed by Checkmate Sea Energy Group. This project is still in its early stage of development. The device is a 200 meter (660 ft) long rubber tube that is anchored underwater. The passing waves will instigate a wave inside the tube that will propagate down its walls and drive a turbine at one end. The United Kingdom's more popular project is the Pelamis wave energy converter, considered the world's first offshore unit.

10.11.2 Australia

One of Australia's main projects is a concept utilizing the oscillating water column technology. Electricity is generated by air flowing through the turbine. This is the third medium-scale demonstration unit installed near Port Kembia, New South Wales, Australia. The unit is grid-connected. One of its projects sank along the port, and newer projects were reinitiated. A full-scale commercial unit that is near the shore is called Green Wave and has a reported capacity of 1 MW, installed in 2013 along Port MacDonnell in South Australia. Another Australian project is the CETO wave power device. The device is being tested off Fremantle, Western Australia. It consists of a single piston pump attached to the sea floor with a float (buoy) connected to the piston. Waves cause the float to rise and fall, generating pressurized water that is piped to an onshore facility to drive a hydraulic generator or to run a reverse-osmosis desalination plant.

10.11.3 Denmark

Denmark has its share of wave energy projects. The first project concept is a surface-following attenuator initiated in 2011. The device consists of two floats connected by a hinge. The units are placed on the ocean surface and follow the waves. The relative mechanical motion of these two devices is translated into electrical power using a power takeoff system. A 1:5 scale prototype was tested near the sea off Frederikshawn in Denmark. Another project is called the wave piston system. This system uses vertical plates to exploit the horizontal movement of waves. Plates are attached in parallel alignment on a single structure as they neutralize each other—inventors call this "force cancellation." Pressurized water is generated and sent to a turbine that is attached to a generator. Modeling studies showed that the design eliminates having to anchor the system on the seabed and also reduces the structure size substantially. The Wave Dragon is another invention from Denmark. This is a type of overtopping device wherein large wind reflectors focus waves up a ramp into an offshore reservoir. The water returns to the ocean by force of gravity, and power is produced via power generators similar to those in hydroelectric power plants. The wave start is also another project using a multi-point absorber. Floats are used that move with waves. This motion is transferred via hydraulics into the rotation of a generator that produces electricity. A half-scale model situated in Hanstholm has been producing electricity and has contributed to the grid since September 2009. Another unit (scale 1:10) was also tested in Nissum Bredning in 2005, although the system was decommissioned in 2011.

10.11.4 United States

The cycloidal wave energy converter is being developed by the Atargis Energy Corporation and uses the principle of a fully submerged wave termination device. The device has a fully submerged 20 meter (66 ft) diameter rotor with two hydrofoils and buoys. Numerical studies showed more than 99% wave power termination capability. It also uses a direct drive generator. Experiments have confirmed the claim of high-power termination capability. The project was initiated in 2006. Another project is the PowerBuoy being developed by Ocean Power Technologies. It also uses a buoy that is placed offshore and uses a hydroelectric turbine to generate electrical power. The project is located in Reedsport, Oregon. It uses a rack and pinion arrangement to convert mechanical motion into a circular motion to run generators. Starting in 2004, Sri International has also been developing and deploying offshore buoys with hydroelectric turbines, using a special polymer called an electroactive polymer artificial muscle.

10.11.5 Belgium

The Flansea Corporation (Flanders Electricity from the Sea) is developing a point absorber buoy and a hydroelectric turbine. A unit has been deployed in the southern North Sea area. The converter works by means of a cable that is being moved constantly by the buoy. This bobbing action generates electricity.

10.11.6 Sweden

The Lysekil Project in Sweden was initiated by Uppsala University and uses a buoy to capture the wave motion. This was deployed offshore using a linear generator in 2002. The action is very simple. The buoy is anchored on the seabed and the buoy on the surface. The up-and-down motion of the buoy translates to a linear action and with a linear generator directly produces the current output. Hundreds of these buoys are needed to generate substantial power.

10.11.7 Ireland

Ireland has at least three known wave energy projects. The first one is an Ocean Energy buoy deployed offshore that uses an air turbine to generate power. The project was initiated in 2006, and in September 2009, the two-year sea trial was completed using a one-quarter scale system. The uniqueness of this system is that it has only one mechanical moving part. Another project was initiated by Sea Power Company, Ltd. This project uses a surface-following attenuator that may be place offshore or nearshore. The project was initiated in 2008, and the intent was to use the system for a reverse-osmosis desalination plant and also for direct production of mechanical drive power. They are still conducting tank studies. Finally, the WaveBob project was also initiated in this country using buoys and direct drive power takeoff. The project was conceptualized in 2009, and they have been conducting ocean trials as well as extensive tank tests. A unique feature is an ocean-going heaving buoy with a submerged tank that captures an additional mass of seawater for added power and tenability and as a safety feature. They call this the tank's "venting" unit.

10.11.8 Israel

Israel has also reported a wave energy converter project using buoys and a hydraulic ram that would be placed near the shore. The design is a breakwater-based wave machine.

10.12 Conclusion

While wave power has been in use since the 18th century, it has not at present gained widespread commercial applications. There are numerous wave farms that are being developed by countries like Portugal, the United Kingdom, Australia, New Zealand, and the United States, but these are largely demonstration systems, even though the systems are on the order of several MW. The reported worldwide potential is over 2 TW and could equal the world's energy consumption (Falnes, 2007).

There are hundreds of companies currently initiating wave-related energy conversion systems, and each company has its own unique and innovative design, the most famous of which is the Pelamis wave energy converter installed in 2004 at the European Marine Energy Center (EMEC) in Orkney, Scotland. This is considered the world's first working Pelamis machine of large scale and the first offshore wave energy conversion device to generate electricity into a national grid anywhere in the world. The main principle makes use of hydraulic power.

One will continue to observe the yearly release of new and innovative designs and a range of power outputs. In the United States, the Pacific Northwest Generating Cooperative, based in Oregon, is funding the construction of a commercial wave power park in Reedsport, Oregon, using buoys. The rise and fall of the waves moves a rack and pinion within a buoy and spins a generator. The electricity is transmitted by a submerged transmission line. The buoys are designed to be installed 1 to 5 miles (8 km) offshore in water 100 to 200 feet (60 m) deep. The peak-rated power output is around 150 kW [201 hp].

Two of the most recent additions to these wave energy demonstration projects include Seabased Industry AB in cooperation with the Swedish Energy Agency and the CCel

Company based in the United Kingdom. CCell uses an oscillating wave surge converter, while Seabased Industry utilizes a buoy. Sweden is building its first wave energy park along the Swedish coast, northwest of Smogen. CCell claims to have invented the world's most efficient wave energy conversion device (IRENA, 2014).

The development of wave energy conversion devices is exciting, with numerous companies tapping into innovative designs and continuing demonstration projects despite the relatively lower price of crude oil. The main reason is the fact that these units can continue to generate power as long as waves (an unlimited resource in some shore areas) are present.

10.13 Problems

10.13.1 Determine the Constant for Wave Power Equation

P10.1   Determine the estimated value of constant such that g and ρ do not appear in the equation (i.e., value of $\rho g^2/64\pi$ or units of kW/m^{3}s) if the density for salt water is 1,027 kg/m^3 instead of 1,000 kg/m^3 (density of pure water).

10.13.2 Basic Power from Wave

P10.2   Determine the power from the wave in units of kW/m if the wave height is 4 meters and the wave period is 8 seconds. Use Equation 10.1 for calculations.

10.13.3 Wave Power in Storms

P10.3   In a major storm, the wave height could be 10 meters for a period of 10 seconds. How much will the total available power be if the wave is 50 meters long?

10.13.4 Total Power from Wave

P10.4   Given a wave height of 5 meters and a wave period of 7 seconds, estimate the power per meter of crest head. Determine the power for every km of wave.

10.13.5 Hydraulic Power Developed from Buoys

P10.5   Determine the pressure generated if the buoy load is 8,909.1 N [2,000 lbs] and the wave height is 6 meters [19.68 ft] with a wave period of 8 seconds. Also estimate the volumetric flow rate generated assuming the diameter of the small piston cylinder is 10 cm [3.94 inch]. Please refer to Figure 10.3 for this arrangement. Finally, estimate the theoretical power generated for this action.

10.13.6 Hydraulic Power

P10.6   Determine the output power or power stored (in hp) for a system assuming the stroke is 3 feet at a rate of one cycle per 5 seconds, matching the wave cycle. The buoy load force is around 500 lbs. Use the dimensions in Figure 10.3 for calculations. Assume the handle's total length is 12 feet and the distance from the handle pivot is 2 feet (the short side distance to the anchor). The pump has a 3-inch-diameter piston, and the load cylinder has a 6-inch-diameter piston. The load will be raised 12 inches. Also determine the number of cycles required to lift this load at this height. Each complete cycle consists of two pump strokes (intake and power).

10.13.7 Hydraulic Jack Power (Metric)

P10.7   Determine the output power or power stored (in kW) for this system assuming the stroke is 6 meters [19.68 ft] at a rate of one cycle per 8 seconds, matching the wave cycle in the previous example. The buoy load force is around 8,909.1 N [2,000 lbs]. Use Figure 10.3 for calculations. Assume the handle's total length is 4 m [13.12 ft] and the distance from the handle pivot is 1 m (the short side distance to the anchor or 3.28 ft). The pump has a 10 cm [3.94 in] diameter piston, and the load cylinder has a 20 cm [7.87 in] diameter piston. The load will be raised 15 m [49.2 ft]. Determine the number of cycles required to lift this load at this height. Each complete cycle consists of two pump strokes (intake and power).

10.13.8 Hydraulic Jack Power (English System)

P10.8   Determine the output power or power stored (in hp) for this system assuming the stroke is 6 meters [19.68 ft] at a rate of one cycle per 8 seconds, matching the wave cycle in the previous example. The buoy load force is around 8,909.1 N [2,000 lbs]. Use Figure 10.3 for calculations. Assume the handle's total length is 4 m [13.12 ft] and the distance from the handle pivot is 1 m (the short side distance to the anchor or 3.28 ft). The pump has a 10 cm [3.94 in] diameter piston, and the load cylinder has a 20 cm [7.87 in] diameter piston. The load will be raised 15 m [49.2 ft]. Determine the number of cycles required to lift this load at this height. Each complete cycle consists of two pump strokes (intake and power).

10.13.9 Piston Power for Surface Attenuator

P10.9   Power Generated in a Piston per Stroke. Determine the power generated in each piston for every complete stroke if the piston diameter is 1 meter and the stroke is 2 meters. The wave motion completes its cycle every 6 seconds (that is, assuming six seconds per stroke). Assume that the pressure was created by a 500-liter volume of water creating about 4,900 N of load onto the piston rod, driving the piston in and out due to the wave movement. The pressure created by this force is approximately 6,239 N/m^2.

10.13.10 Significant Wave Height (H$_s$)

P10.10   The following data were obtained from wave height measurement (in meters): 7, 9, 9, 9, 8, 6, representing the highest one-third of the bigger waves. Estimate the significant wave height H_s. Compare this value with the simple mean wave height.

10.13.11 Capital Cost of Wave Converters

P10.11   The capital cost for a wave converter is around $1,000/kW. In an area where the average wave height is 6 meters for a period of 7 seconds, how much is the capital cost for every km of wave distance?

References

Bhattacharya, S. C. 1983. Lecture Notes in "Renewable Energy Conversion" Class. Asian Institute of Technology, Bangkok, Thailand.

Boyle, G. (Ed). 2004. Renewable Energy: Power for a Sustainable Future. 2nd Edition. Oxford University Press, The Open University, Cambridge, UK.

Damen, M, M. G. se Sousa Prado, F. Gardner and H. Polinder. 2006. Modeling and test results of the Archimedes wave swing. Proceedings of the Institute of Mechanical Engineers, Part A: Journal of Power and Energy 220 (8) 855–856.

Drew, B, A. R. Plummer and M. N. Sahinkaya. 2009. A review of wave energy converter technology. Proceedings of the Institute of Mechanical Engineers, Part A: Journal of Power and Energy. doi:10.1234/09576509JPE782.

Esposito, A. 2009. Fluid Power with Applications. 7th Edition. Pearson, Prentice Hall, Upper Saddle River, NJ.

Falnes, J. 2007. A review of wave-energy extraction. Marine Structures 20 (4): 185–201. doi:10.1016/j.marstruc.2007.09.001.

Gunawaradane, S. D. G. S. P., A. M. P. Uyanwaththa, D. M. A. R. Tennakoon, S. B. Wijekoon and W. M. J. S. Ranasinghe. 2010. Model study on "Pendulor" type wave energy device to utilize ocean wave energy in Sri Lanka. Proceedings of the International Conference on Sustainable Built Environment (ICBE-2010), Kandy, Sri Lanka, December 13–14, 2010.

Henriques, J. C. C., J. C. C. Portillo, L. M. C. Gato, R. F. P. Gomes, D. N. Ferreira and A. F. O. Falcao. 2016. Design of oscillating-water column wave energy converters with an application to self-powered sensor buoys. Energy 112 (2016): 852–867.

IRENA. 2014. Wave Energy Technology Brief. International Renewable Energy Agency (IRENA). Ocean Energy Technology Brief 3. June. Available at: https://www.irena.org/documentdownloads/publications/wave-energy_v4_web.pdf. Accessed June 7, 2019.

Kofoed, J. P. 2006. Vertical distribution of wave overtopping for design of multi-level overtopping based wave energy converters. Proceedings of the 30th International Conference on Coastal Engineering (ICCE 2006), San Diego, CA, September 4–8, 2006.

McGrath, J. 2008. Could Salter's Duck have solved the oil crisis? HowStuffWorks.com. July 14. Available at: https://science.howstuffworks.com/environmental/green-science/salters-duck1.htm. Accessed April 25, 2018.

Mollison, D., O. Buneman and S. Salter. 1976. Wave power availability in the North East Atlantic. Nature 263 (5574): 223–226.

NOAA. 2018. Significant wave height. National Weather Service of the National Oceanic and Atmospheric Administration. Available at: https://www.weather.gov/mfl/waves. Accessed April 25, 2018.

Ocean Energy Council. 2018. Wave energy: what is wave energy? Ocean Energy Council Newsletter. Available at: http://www.oceanenergycouncil.com/ocean-energy/wave-energy/. Accessed April 24, 2018.

Renewable Solar Energy. 2017. What is wave power or wave energy? How wave power works. Renewable Energy Online Newsletter. Available at: http://renewable-solarenergy.com/wave-power-energy.html. Accessed April 24, 2018.

US DOE. 2018. Marine and Hydrokinetic Technology Glossary. OpenEI, MHK Technology Database, National Renewable Energy Laboratory of the U.S. Department of Energy. Available at: https://openei.org/wiki/Marine_and_Hydrokinetic_Technology_Glossary. Accessed April 24, 2018.

11

Ocean Thermal Energy Conversion (OTEC) Systems

Learning Objectives

Upon completion of this chapter, one should be able to:

1. Describe the concept of generating energy and power from the thermal energy of oceans.

2. Enumerate the various types of energy generating systems from ocean thermal energy conversion (OTEC) systems.

3. Enumerate various potential sites for OTEC systems.

4. Describe the various applications of OTEC systems.

5. Describe the various conversion efficiencies in OTEC energy systems.

6. Relate the overall environmental and economic issues concerning OTEC power systems.

11.1 Introduction

The ocean is a vast renewable energy resource. The simplest way to extract energy from the ocean is to exploit the temperature differences between the cold water of deep-ocean regions and the warm water of surface regions. The surface of the ocean is continually being heated by the sun each day and cools at night. In deep-ocean regions, the temperature is almost constant, and there will always be temperature differences between the deeper regions and the heated ocean surface. The energy is extracted by the temperature differential using heat engines that convert the energy into useful (most commonly electrical) power. Because of the vastness of the ocean resource, this thermal energy is projected to be much greater than other energy forms, such as salinity gradient energy, wave energy, and tidal power. However, thermal efficiency is generally low, and the vastness of this ocean resource is a necessity in order to generate the power required for communities. Ocean thermal energy conversion systems (OTEC) are usually used as base load power plants because they can provide continuous power throughout the day without any diurnal interruption. The energy resource estimate for OTEC was reported at 88,000 TWh/year. Even if this energy is utilized, it should not affect the ocean thermal profile (Pelc and Fujita, 2002). The ideal thermodynamic cycle used for energy conversion is the Rankine Cycle. The turbines used are usually the low-pressure types. The theory of OTEC was conceptualized in the 1880s by the French engineer Jacques-Arsene d'Arsonval. However, the first bench-scale demonstration

World's OTEC Resources

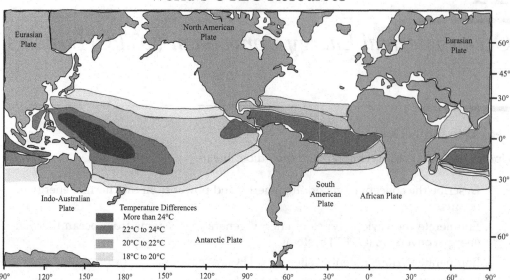

FIGURE 11.1
The world's ocean thermal energy conversion (OTEC) resources.

plant was not constructed until 1926. The plant was initiated by d'Arsonval's student Georges Claude in Matanzas, Cuba, and made operational in 1930 with a power output of 22 kW [29.5 hp] of electricity. A low-pressure turbine was used. Currently, the only operational OTEC plant is located in Japan and is managed by Saga University (IRENA, 2014). However, numerous projects around the world are being installed to capture energy from ocean waters.

Figure 11.1 shows the world's OTEC resources. Note that the higher temperature differential is found in tropical areas near the equator, where the temperature difference is the highest. The locations indicated in dark red have an average temperature difference of more than 24°C [75.2°F]. The temperature differences in the light red portions range from 22°C to 24°C [71.6°F to 75.2°F], while the orange regions have a range from 20°C to 22°C [68°F to 71.6°F]. Regions marked in yellow have temperature differences ranging from 18°C to 20°C [64.4°F to 67°F].

Many demonstration systems have been built one after the other; however, numerous factors have prevented them from expansion and commercialization. The primary hindrance is the popularity of fossil fuels (oil and coal), making the system seem unattractive.

Japan's role in OTEC technology development increased in the 1970s when Japanese engineers built a 100 kW [134 hp] closed-cycle OTEC plant on the island of Nauru. The plant became operational in 1981, producing about 120 kW [161 hp] of electricity, where 75% was used to power the plant and 25% was used to power a nearby school and other communities (Ikegami and Furugen, 2013). Russia and the United States have also had their share of technical development in this field. As seen in the map in Figure 11.1, Hawaii sits as the most ideal place for OTEC development in the United States, initiated by the Natural Energy Laboratory of Hawaii Authority in 1974. This facility is still the leading research facility for OTEC research in the United States.

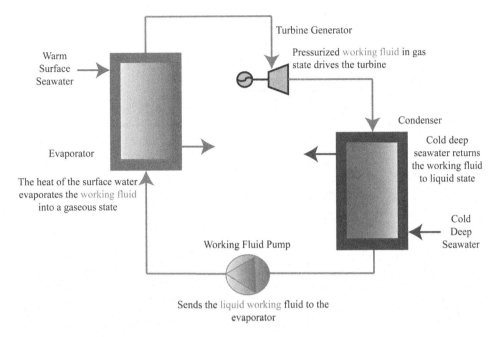

FIGURE 11.2
The basic OTEC thermodynamic cycle (Vega, 2003; Okinawa OTEC, 2013).

11.2 The Basic OTEC System

The basic OTEC system is schematically shown in Figure 11.2. The major component parts include a condenser where cold deep seawater returns the working fluid to a liquid state and this liquid is pumped to the evaporator. The heat of the surface water evaporates the working fluid into its gaseous state. The gas then enters the low-pressure turbine, which drives the generator to produce electrical power. The challenge of this system is the fact that the temperature difference is quite small and seawater will not evaporate readily at low-temperature conditions. As a result, almost always, a working fluid with a low boiling point (e.g., ammonia or a commercial refrigerant) must be used. Note that because of the low-temperature differences available, the overall conversion efficiency is quite low—way below 10%. As mentioned earlier, this requires large volumes of ocean water in order to generate appreciable amounts of power. The energy taken from the warm water will be used to generate power on the turbine side. Work is done to pump the liquid water back to the evaporator while the condenser absorbs the heat to bring the vapor into the liquid phase.

11.3 OTEC Components and Temperature Profiles

The practical temperatures associated with the components of the OTEC system are shown in Figure 11.3. In this schematic, the deep-water temperature may be assumed to be around 5°C [41°F], while the surface water is around 25°C [77°F] for a temperature

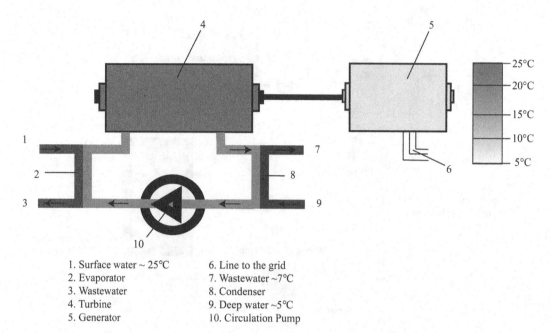

FIGURE 11.3
Component parts of an OTEC system (Vega, 2003).

1. Surface water ~ 25°C
2. Evaporator
3. Wastewater
4. Turbine
5. Generator
6. Line to the grid
7. Wastewater ~7°C
8. Condenser
9. Deep water ~5°C
10. Circulation Pump

difference of at least 20°C [68°F]. Note that the heat energy associated with this profile shows that for the volume of water equivalent to an area of a hectare [2.47 acre], 10 meters [32.8 ft] deep, the energy absorbed by the seawater is still quite substantial and when converted into electrical power could generate a significant amount of power despite the low overall efficiency. This is illustrated in Example 11.1. The simple equation for calculating sensible heat absorbed by the ocean water is shown in Equation 11.1. In this example, it is shown that a hectare of land with a depth of 10 meters [32.8 ft] will absorb energy of at least 7.86 million MJ of energy from a mere 20°C [68°F] temperature difference. If the overall conversion efficiency is assumed to be 5%, the electrical power that can be generated amounts to more than 4.55 MW:

$$Q(kJ) = mC_p\Delta t \qquad (11.1)$$

where
Q = heat absorbed by the ocean (MJ)
m = mass of the body of water (kg)
C_p = specific heat of the body of water, 3.93 kJ/kg°C
Δt = change in temperature, (°C)

Example 11.1: Estimating Heat Absorbed by Ocean Water Bodies

Determine the amount of heat energy absorbed by a hectare of seawater surface water with a depth of 10 meters. Assume the original temperature of 5°C [41°F] and a final temperature of 25°C [77°F]. Assume the specific heat of seawater to be 3.93 kJ/kg°C [0.94 Btu/lb°F]. If the overall conversion efficiency is 5%, estimate the actual power. If the cycle happens in 24 hours, how much power is produced?

SOLUTION:

a. The energy absorbed is simply calculated using the sensible heat energy equation as shown:

$$Q(kJ) = mC_p \Delta t$$

$$Q(MJ) = \frac{10,000\ m^2 \times 10\ m}{1\ m^3} \times \frac{1,000\ kg}{1\ m^3} \times \frac{3.93\ kJ}{kg\,°C} \times 20°C \times \frac{MJ}{1,000\ kJ} = 7,860,000\ MJ$$

Thus, 7,860,000 MJ [7,450 million Btu] of energy is absorbed by this body of water.

b. If the overall conversion efficiency is 5%, the actual energy produced is calculated:

$$Actual\ Energy(MJ) = Theoretical\ Energy \times Conversion\ Efficiency$$

$$Actual\ Energy(MJ) = 7,860,000\ MJ \times 0.05 = 393,000\ MJ\,[372.5\ MBtu]$$

c. The power is calculated from actual energy per unit of time:

$$Actual\ Power(MW) = \frac{Energy(MJ)}{Time(s)} = \frac{393,000\ MJ}{24\ hrs} \times \frac{1\ hr}{3600\ s} = 4.55\ MW$$

The ideal Carnot Cycle efficiency for ocean thermal systems is given in Equation 11.2 (Holman, 1980). In this equation, the ideal efficiency is only a function of temperature. Note that the temperature must be in absolute units (Kelvin). The actual temperature profile found in tropical oceans maintains a reported average temperature of 28°C [82.4°F], whereas beyond 1,000 meters [3,280 ft] of depth (or 1 km) the temperature levels off to around 4.4°C [39.92°F]. Hence, if one uses the ideal Carnot Cycle efficiency at the given temperature difference, the overall efficiency is only about 7.84% (below 10%, as expected):

$$\eta_c\,(\%) = \frac{T_h - T_c}{T_h} \times 100\% \tag{11.2}$$

Example 11.2: Ideal Carnot Cycle Efficiency Calculations

Determine the ideal Carnot Cycle efficiency for an ocean with a warm temperature of 28°C [82.4°F] and a cold temperature of 4.4°C [39.92°F].

SOLUTION:

a. Convert the temperature units into the absolute scale: $T_h = 28 + 273.15 = 301.15$ K, and $T_c = 4.4 + 273.15 = 277.55$ K [499.59°R].

b. Using Equation 11.2 directly, we have the following:

$$\eta_c\,(\%) = \frac{T_h - T_c}{T_h} \times 100\% = \frac{301.15 - 277.55}{301.15} \times 100\% = 7.84\%$$

c. The ideal Carnot Cycle efficiency is only 7.84%.

11.4 Other Applications of OTEC

The main advantage of the OTEC system is the range of applications that arise from harnessing this thermal energy from the ocean. Figure 11.4 shows the various pathways and the benefits resulting from the processes. The main product is usually pure water, if desalination is implemented. The desalination of surface and deep-ocean water will also produce a valuable mineral water resource. If there is a deficiency of potable water on mainland, an OTEC system may provide the solution. The world problem of water scarcity for food production may soon find its solution. The ocean could potentially become a constant source of the water needed on land. On the right-hand portion of the figure, we enumerate several applications:

a. Deep-ocean water ice

b. Extraction of lithium (or in some areas deuterium)

c. Chilling water needs

d. Greenhouse water and energy need

e. Use of water for food-related businesses or projects

f. Aquaculture

On the left-hand side of the figure are further applications. Electricity production is perhaps the most common application, followed by hydrogen production, storage, and fuel cells for producing even more energy. The ocean resource is important not only for power generation but also for the many needs of communities near the shore. Obviously, care should be taken in using massive amounts of ocean water, as marine ecosystems are affected. A large volume of ocean water is required to generate MW range of power, and

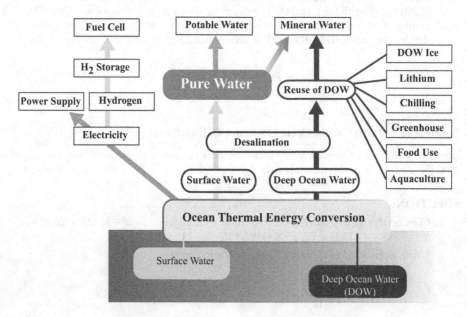

FIGURE 11.4
Applications of OTEC Systems (Vega, 2003).

the magnitude of this water requirement is illustrated in Example 11.3. For a mere MW of power generated, the volume of water needed is several times the volume of water in an Olympic-size swimming pool. In fact, for every MW of electrical power produced, about 10 Olympic-size swimming pools are needed.

Example 11.3: Volume of Water Needed for a MW of Power

Determine the volume of ocean water needed to generate a MW of electrical power, assuming a 20°C [68°F] temperature difference and continuous power production (24 hrs a day and 365 days in a year). Assume an overall conversion efficiency of 5%. Use the simple sensible heat equation for the calculation. Compare this volume of water to the capacity of an Olympic-size swimming pool that measures 50 meters [164 ft] in length, 25 meters [82 ft] in width, and with a depth of 3 meters [9.84 ft]. How many Olympic-size swimming pools are needed to generate a MW of electrical power? Assume the density of seawater to be 1,027 kg/m³ [64 lbs/ft³].

SOLUTION:

a. The efficiency equation is first used to calculate the input energy required using the overall conversion efficiency. If 1 MW of output power is needed, a 5% conversion efficiency shows 20 MW of input power as shown below:

$$Input\ Energy\,(MW) = \frac{Output\ Energy}{Overall\ Efficiency\,(decimal)} = \frac{1\ MW}{0.05} = 20\ MW$$

b. This energy must come from the thermal energy from the ocean, assuming the specific heat of ocean water to be 3.93 kJ/kg°C [0.24 Btu/lb°F] with a 20°C [68°F] difference:

$$Mass\ Flow\left(\frac{kg}{s}\right) = \frac{Power}{C_p \Delta t}$$

$$Mass\ Flow\left(\frac{kg}{s}\right) = \frac{20\ MW}{} \times \frac{kg°C}{3.93\ kJ} \times \frac{1}{20°C} \times \frac{1 \times 10^6\ W}{1\ MW} \times \frac{J}{W-s} \times \frac{kJ}{1,000\ J}$$

$$= 254.5\frac{kg}{s}$$

$$Mass\ Flow\left(\frac{lbs}{min}\right) = 254.5\frac{kg}{s} \times \frac{2.2\ lbs}{1\ kg} \times \frac{60\ s}{min} = 33,588\frac{lbs}{min}$$

c. The volume and weight of seawater that is contained in an Olympic-size swimming pool is calculated below:

$$Volume\,(m^3) = 50\ m \times 25\ m \times 3\ m = 3,750\ m^3\left[132,328\ ft^3\right]$$

$$Weight\,(kg) = 3,750\ m^3 \times \frac{1,027\ kg}{m^3} = 3,851,250\ kg\,[8,472,750\ lbs]$$

d. Hence, if one were to produce 1 MW of electrical power and use 254.5 kg/s [560 lbs/s] of water, one Olympic-size swimming pool would be emptied in 4.2 hours. In terms of how many swimming pools are needed to generate this much power in one day, it would take the volume of about six Olympic-size swimming pools. These values are shown in the calculations below:

$$Time\,(hrs) = 3,851,250\ kg \times \frac{s}{254.5\ kg} \times \frac{1\ hr}{3600\ s} = 4.2\ hrs$$

$$\#\ of\ Pools = \frac{24\ hrs}{day} \times \frac{per\ pool}{4.2\ hrs} = 5.7\ pools \approx 6\ pools$$

One other issue associated with OTEC systems to generate appreciable power is the energy or power needed to move these large volumes of water. For example, moving 254 kg/s [560 lbs/s] of ocean water through pipes would require enormous power, which would be deducted from the output power of the facility. Example 11.4 shows the order of magnitude of this pumping power with this volume of water. This example shows that more than 30% of the output power from the turbine-generator will be used to pump water for every 100 meters [328 ft] of depth. This will reduce the net power output of OTEC facilities. There will be a certain depth where much of the power may be used just to pump ocean water from the ocean bottom to the surface.

Example 11.4: Theoretical Pumping Power

Determine the amount of theoretical power and actual power used to move 254 kg/s [558.8 lbs/s] of water through 100 meters [328 ft] of head. Note that we are moving water from the bottom of the ocean to the top. Assume an overall pumping efficiency of 75%. Use the basic hydro power equation from Chapter 5 for calculations.

SOLUTION:

a. The problem is straightforward using the pumping power equation shown. However, we are calculating the power needed to raise the water against 100 meters of depth without losses:

$$Theoretical\ Power\,(kW) = \frac{QH}{102} = \frac{254\left(\frac{kg}{s}\right) \times 100\ m}{102} = 249\ kW\left[333.8\ hp\right]$$

$$P_t\,(kW) = 254\left(\frac{kg}{s}\right) \times 100\ m \times \frac{Ns^2}{kg-m} \times \frac{9.8\ m}{s^2} \times \frac{W-s}{N-m} \times \frac{kW}{1,000\ W}$$

$$= 249\ kW\left[333.8\ hp\right]$$

b. The actual power to bring this water body close to the surface should be higher. Please remember this is not to generate power but to raise water. The calculation is as follows:

$$Actual\ Power\ to\ Raise\ Water\,(kW) = \frac{249\ kW}{0.75} = 332\ kW\left[445\ hp\right]$$

c. Note that it will take more power to bring the water from the lower elevation to the higher elevation. Thus, the efficiency equation will be used in reverse to find a higher actual power than the lower theoretical power to accomplish the work.

11.5 Uses of OTEC Systems

The primary purpose of an OTEC system is base load electricity production. The western and eastern portions of the United States should benefit from these systems, including the Gulf Coast states of Texas and Louisiana. Note that heat energy stored in vast oceans is a considerable amount that should not be ignored. OTEC works best if the temperature difference between the colder ocean water and warm surface water is at least 20°C [36°F]. These locations around the earth have already been identified. These conditions exist only in the tropical coastal areas (between the Tropics of Capricorn and Cancer). The primary limitation of an OTEC system is the requirement of large-scale conveying of ocean water. This would require a large-diameter intake pipe submerged about a mile or so into the ocean's depth. From the simple example shown earlier, there is potential to generate billions of watts of electrical power.

In the future, as has been hinted at in a previous section, a secondary use of ocean water is the replenishment of water that is used substantially on land. The water-energy nexus will remain the issue for mainland, and the ocean is the last stop for potable water. In the future, one might soon see vast pipelines constructed deep into the ocean, transporting large volumes of water. Beside these water pipelines would be electrical lines that convey electrical power. Tourism could also be developed around such large-scale conversion systems. Habitable communities could sprout near such energy projects, as manpower is still necessary for operation.

A base load plant usually satisfies the minimum requirement of a population, community, or region. For example, a population of 1 million residents in the United States with an average per capita electrical energy of around 12,000 kWh/yr would need a power plant of about 10 MW. The calculation is shown in Example 11.5. The electric power consumption per capita for most counties are as shown in Table 11.1 (OECD/IEA, 2014).

Example 11.5: Base Load Power Calculations

Determine the minimum size of a power plant that would satisfy the requirement of a community for a population of 1 million. The average per capita need of a household was reported to be 12,000 kWh/yr.

SOLUTION:

a. Power is simply the ratio of energy and time. The calculation is a simple conversion from the data given assuming 365 days in a year and 24 hours in a day as shown:

$$Power(MW) = \frac{12,000\ kWh}{hh - year} \times \frac{1\ year}{365\ days} \times \frac{1\ day}{24\ hrs} \times \frac{1,000,000\ hh}{} \times \frac{1\ MW}{1,000\ kW}$$
$$= 1,370\ MW$$

b. Usually, power plants are backed up with additional power in case the requirement peaks more than the base load. For this problem, a 2,000 MW base load plant may be needed. In some developing countries, a base load plant is usually about 5,000 MW.

TABLE 11.1

Electric Power Consumption per Capita for Selected Countries

Country	2006–2010	2011–2015	Country	2006–2010	2011–2015
Albania	2,195	2,118	Iran	2,662	2,762
Algeria	1,122	1,236	Iraq	1,341	1,474
Angola	229	220	Israel	6,930	7,189
Argentina	2,901	2,955	Japan	7,841	7,752
Australia	10,712	10,398	Kenya	157	160
Austria	8,390	8,507	Korea	10,162	10,346
Bahrain	17,093	17,395	Malaysia	4,114	4,345
Bangladesh	258	279	Mexico	2,074	2,012
Belgium	8,021	7,987	Myanmar	151	153
Bolivia	637	663	Norway	23,510	23,658
Brazil	2,394	2,462	Pakistan	457	452
Canada	16,168	15,615	Philippines	651	672
China	3,298	3,475	Qatar	15,800	16,183
Colombia	1,121	1,150	Sweden	14,030	14,290
Costa Rica	1,899	1,957	Switzerland	7,928	7,886
Finland	15,707	15,687	Thailand	2,305	2,465
Germany	7,146	7,270	United Arab Emirates	10,537	10,463
Ghana	321	346	United Kingdom	5,473	5,452
Guatemala	531	529	United States	13,340	12,954
India	698	730	Vietnam	1,099	1,273

Source: OECD/IEA (2014).

11.6 Basic Thermodynamic Cycle: Rankine Cycle

The governing equations for the ideal Rankine cycle are shown in Equations 11.3, 11.4, 11.5, and 11.6. The net work output and the overall efficiency is given in Equations 11.7 and 11.8 (Holman, 1980). The specific heat input Q_e of the working fluid is the difference between the enthalpy at point 3 and point 2, given by Equation 11.3. This heat input comes from solar energy in the form of sensible heat. The associated diagram for thermodynamic points are illustrated in Figures 11.6 and 11.7:

$$Q_e\left(\frac{kJ}{kg}\right) = \left(h_3 - h_2\right) \tag{11.3}$$

The specific heat loss to the surroundings of the condenser is the difference between the enthalpy at point 4 and at point 1, given by Equation 11.4:

$$Q_c\left(\frac{kJ}{kg}\right) = \left(h_4 - h_1\right) \tag{11.4}$$

The specific turbine work is the difference between the enthalpy at point 3 and at point 4, given by Equation 11.5:

$$W_t \left(\frac{kJ}{kg} \right) = (h_3 - h_4) \tag{11.5}$$

The pump work (Equation 11.6) is the difference between the pressures at point 2 and at point 1 and the specific volume. The specific volume is in m^3/kg and the pressure in kPa (kN/m^2). Note that kN-m is also kJ, and hence the units are also in kJ/kg, and no conversion constant is needed:

$$W_p \left(\frac{kJ}{kg} \right) = \int vdp = v \times (P_2 - P_1) \tag{11.6}$$

The net work output is the difference between the turbine work and the pump work as shown in Equation 11.7:

$$W_{net} \left(\frac{kJ}{kg} \right) = (W_t - W_p) \tag{11.7}$$

The overall efficiency is shown in Equation 11.8:

$$Cycle \; Efficiency \, (\%) = \frac{(W_t - W_p)}{Q_e} \times 100\% \tag{11.8}$$

A typical temperature profile in an OTEC system is shown in Figure 11.5. The cold intake water may be as low as 9°C [48.2°F], and the output may be a couple of degrees lower. The warm intake water may be at 25°C [77°F] and as it goes through the evaporator may also be a couple of degrees lower. As the water goes through the turbine, there will be a big drop in temperature, and the output temperature at this stage may be more than 10°C lower.

The ideal thermodynamic cycle is that shown in Figure 11.6. This figure is a temperature-entropy diagram showing processes that occur in each component of the OTEC system. The associated energy transfers occur at the condenser, evaporator, and pump to move the ocean water, and the work is delivered by the turbine. There are two associated temperatures provided—the warm water (T_h) and the cold water (T_c). The solid lines indicate the saturated ocean water on the left and the saturated steam on the right. When ocean water is evaporated in a flash vessel, the property of the liquid is in the slightly superheated vapor region at the upper right. The liquid coming out of the turbine is a mixture of liquid and steam at point 4. The properties of the ocean water and whichever working fluid is used must be taken from the saturated liquid and gas tables as well as the vapor region. The actual thermodynamic chart shown in Figure 11.7 shows the slight change during actual conditions. The pump work is slightly different, while most other points approach the state points shown.

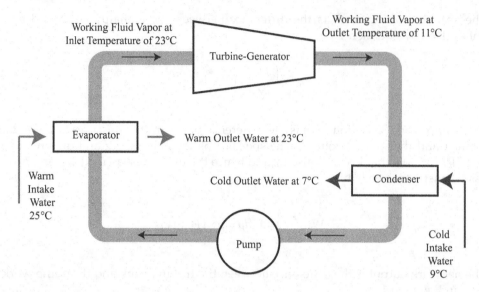

FIGURE 11.5
Basic data for an ideal OTEC system (Vega, 2003).

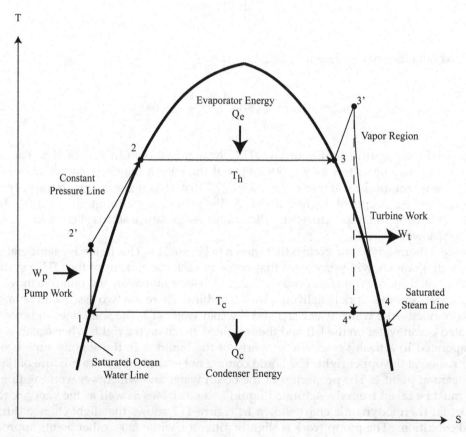

FIGURE 11.6
Ideal TS diagram for an OTEC system.

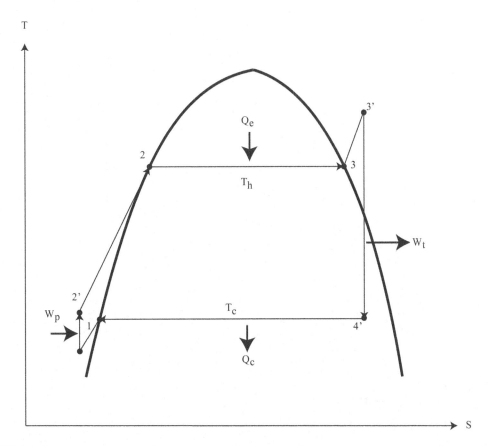

FIGURE 11.7
Heat and work energy of the ideal thermodynamic cycle.

11.7 OTEC Power Generation Systems

There are two general pathways for OTEC systems—closed cycle and open cycle. The basic thermodynamics of the systems are very similar, and the main difference would be the type of working fluid used.

11.7.1 Closed Cycle

In closed-cycle power generation, the main fluid used is a low-boiling-point chemical (e.g., ammonia and most refrigerants), to be converted into vapor in order to rotate the turbine to generate electricity. Warm surface seawater is pumped through a heat exchanger, where the low-boiling-point fluid is vaporized. The expanding vapor turns the turbine blades (attached to a generator) to generate electricity. Figure 11.8 shows this system. There will be no contact between salt water and the working fluid. The major heat transfer will occur in the heat exchangers. As seen from Figure 11.8, the warm intake water on the left side of the diagram will be passed through evaporators where the refrigerant will be allowed to evaporate. Warm outlet water will be conveyed to the back end of the evaporator. The

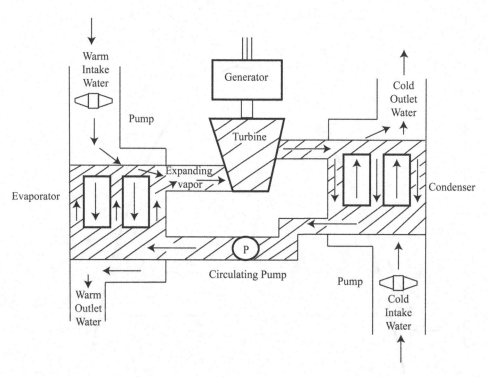

FIGURE 11.8
Schematic of an operational closed OTEC system.

common heat exchangers used are the counter flow types. The expanding vapor drives the turbine, and the turbine is attached to a generator for power production.

The working fluid is then condensed. The condensation is aided by the cold water being pumped from the bottom of the ocean. The absorption of energy converts the working fluid back to the liquid state. A circulating pump returns the working fluid to the evaporator, and the whole cycle is repeated. The main component parts are quite simple, and newer companies have a wide range of refrigerants to use. Example 11.6 shows how heat and work parameters are estimated, including how efficiency is calculated. Data for this example were taken from thermodynamic charts for the working fluid (either water, ammonia, or similar refrigerant). Actual thermodynamic cycle calculations are also shown in Example 11.7 relating real temperatures that may be encountered on actual OTEC systems. The ideal thermodynamic cycle efficiencies range from 7% to 8%, while actual OTEC system efficiencies range from 3% to slightly over 4%. Thus, the original examples showing 5% overall conversion efficiencies are valid assumptions.

Example 11.6: OTEC Closed Cycle Examples

An example of the use of ammonia as working fluid is shown in Figure 11.9. In this example, the properties of ammonia are found in the thermodynamic table (e.g., Holman, 1980). The ideal Rankine Cycle is used for calculations with operating temperatures between (T_1) 5°C [41°F] and (T_3) = 26.9°C [80.42°F]. Determine the following: (a) Q_e, (b) Q_c, (c) pump work, and (d) turbine work.

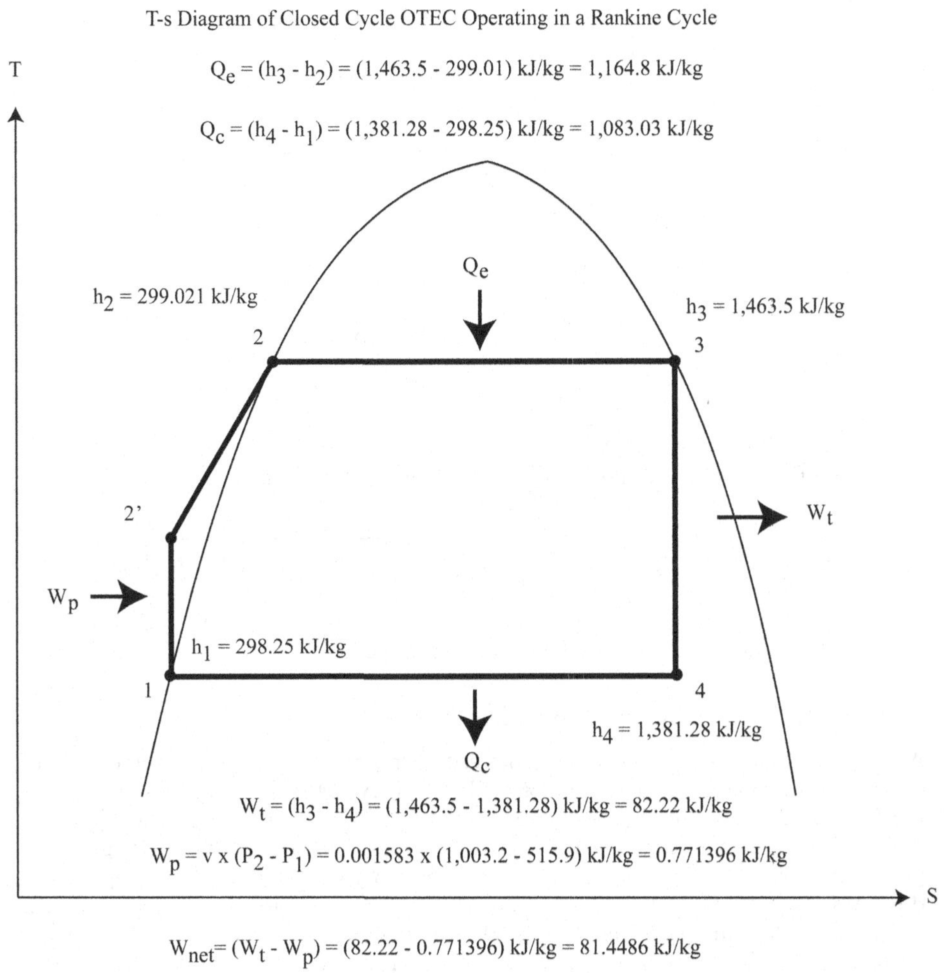

T-s Diagram of Closed Cycle OTEC Operating in a Rankine Cycle

$Q_e = (h_3 - h_2) = (1,463.5 - 299.01)$ kJ/kg = 1,164.8 kJ/kg

$Q_c = (h_4 - h_1) = (1,381.28 - 298.25)$ kJ/kg = 1,083.03 kJ/kg

$h_2 = 299.021$ kJ/kg

$h_3 = 1,463.5$ kJ/kg

$h_1 = 298.25$ kJ/kg

$h_4 = 1,381.28$ kJ/kg

$W_t = (h_3 - h_4) = (1,463.5 - 1,381.28)$ kJ/kg = 82.22 kJ/kg

$W_p = v \times (P_2 - P_1) = 0.001583 \times (1,003.2 - 515.9)$ kJ/kg = 0.771396 kJ/kg

$W_{net} = (W_t - W_p) = (82.22 - 0.771396)$ kJ/kg = 81.4486 kJ/kg

FIGURE 11.9
Ideal thermodynamic cycle for OTEC systems.

SOLUTION:

The values for the enthalpy and entropy data are also shown in the figure, including all associated values of heat input and heat output as well as work output and work input. These property values are also summarized below:

   a. $h_1 = 298.25$ kJ/kg [871.31 Btu/lb]
   b. $h_2 = 299.01$ kJ/kg [128.83 Btu/lb]
   c. $h_3 = 1,463.5$ kJ/kg [630.55 Btu/lb]
   d. $h_4 = 1,381.28$ kJ/kg [595.12 Btu/lb]
   e. $v = 0.001583$ m³/kg [0.0254 ft³/lb]
   f. $P_2 = 1,003.2$ kPa (kN/m²) [145.6 psi]
   g. $P_1 = 515.9$ kPa (kN/m²) [74.88 psi]
   h. Heat input to the evaporator, $Q_e = (h_3 - h_2) = (1,463.5 - 299.01) = 1,164.8$ kJ/kg [501.85 Btu/lb]

i. Heat absorbed by the condenser = $Q_c = (h_4 - h_1) = (1,381.28 - 298.25) = 1,083.03$ kJ/kg [466.62 Btu/lb]

j. Pump work = $v(dP) = v(P_2 - P_1) = 0.001583 (1,003.2 - 515.9) = 0.7714$ kJ/kg [0.3324 Btu/lb]

k. Turbine work = $(h_3 - h_4) = (1,463.5 - 1,381.28) = 82.22$ kJ/kg [35.42 Btu/lb]

11.7.1.1 Efficiency Calculations

From the given data in the figure, the overall cycle efficiency is calculated:

$$\frac{\left(W_t - W_p\right)}{Q_e} \times 100\% = \frac{(82.22 - 0.771396)}{1,164.8} \times 100\% = 6.99\%$$

The ideal cycle efficiency is only about 7%.
The Carnot efficiency is calculated below and comes to about 8%:

$$\frac{(T_h - T_c)}{T_h} \times 100\% = \frac{((29.6 + 273.15) - (5 + 273.15))}{29.6 + 273.15} \times 100\% = 8.03\%$$

Note that the above calculations are based on ideal conditions.

Example 11.7: Actual Thermodynamic Cycle Examples

Actual cycles have numerous losses from each component part of the system. In this example, the efficiency calculations are made for every 1,000 kg/s [2,200 lbs/s] of working fluid flow. While the deep-sea temperature is assumed to be at 5°C [41°F], when the liquid reaches the condenser, there may be a loss in temperature of perhaps a 2°C [35.6°F] difference. The example will be recalculated with this initial loss in temperature during transport to the upper ocean layer. On the surface region, while the originally assumed temperature is around 26.9°C [80.42°F], there may also be temperature losses in the boiler, and this can be assumed to cause another couple of degrees of difference. The temperature after the boiler losses can be assumed to be 24.9°C [76.82°F]. The vapor reaching the turbine may also be reduced by a couple of degrees to an assumed value of 22.9°C [73.22°F]. On the pump side, the liquid ammonia is assumed to have a temperature of 10°C [50°F]. The new values for working fluid properties are then shown in Figure 11.10. In this example, the overall ideal cycle efficiency, and Carnot efficiency are recalculated. The property values for all points are also summarized below:

a. $h_1 = 227.8$ kJ/kg [98.15 Btu/lb]

b. $h_3 = 1,463.9$ kJ/kg [630.72 Btu/lb]

c. $v = 0.0016$ m³/kg [0.0257 ft³/lb]

d. $P_2 = 913.4$ kPa (kN/m²) [132.6 psi]

e. $P_1 = 615.2$ kPa (kN/m²) [89.29 psi]

f. $S_3 = 5.072$ kJ/kg [2.185 Btu/lb]

g. $S_{2f} = 0.881$ kJ/kg [0.380 Btu/lb]

h. $S_{fg} = 4.3266$ kJ/kg [1.8641 Btu/lb]

i. $h_{fg} = 1,225.1$ kJ/kg [527.83 Btu/lb]

j. Pump work = $v(dP) = v(P_2 - P_1) = 0.001583 (1,003.2 - 515.9) = 0.7714$ kJ/kg [0.3324 Btu/lb]

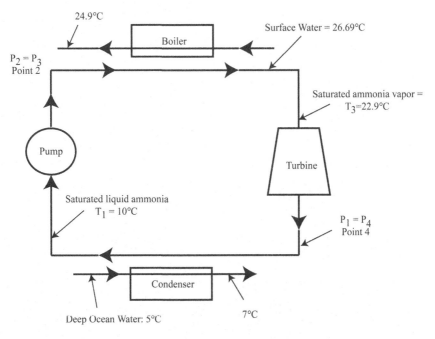

24.9°C

Boiler

Surface Water = 26.69°C

$P_2 = P_3$
Point 2

Saturated ammonia vapor =
T_3=22.9°C

Pump

Turbine

Saturated liquid ammonia
T_1 = 10°C

$P_1 = P_4$
Point 4

Condenser

Deep Ocean Water: 5°C

7°C

FIGURE 11.10
A closed-cycle OTEC system using liquid with low boiling point.

SOLUTION:

The actual cycle calculations are as follows:

a. At point 3, the temperature of the working fluid is reduced to 22.9°C [73.22°F] due to heat exchange losses from the condensed ammonia coming to the warm surface. It cannot be the same temperature as the surface water temperature. The properties at this point will be

h_3 = 1,463.9 kJ/kg [630.55 Btu/lb]
s_3 = 5.072 kJ/kg [2.1853 Btu/lb]
$P_3 = P_2$ = 913.4 kPa [132.6 psi]

b. The enthalpy, entropy, and pressures 1 and 4 at point 1 are found from saturated ammonia vapor steam tables.

c. At point 1 in the schematic, the temperature is 10°C [50°F], and the ammonia is in a saturated liquid state.

d. Values for enthalpy, entropy, and specific volume of the ammonia can be obtained from the saturated ammonia tables. From the saturated ammonia tables we have the following:

h_1 = 227.8 kJ/kg [98.15 Btu/lb]
s_1 = 0.881 kJ/kg [0.3796 Btu/lb]
$P_1 = P_4$ = 615.2 kPa [89.3 psi]
v = 0.0016 m³/kg [0.02566 ft³/lb]

e. To solve the enthalpy at point 2, we need to use the equation for quality (x_2) as follows:

$$s_3 = s_{2,f} + x_2\left(s_{fg}\right)$$
$$5.072 = 0.881 + x_2\left(4.3266\right)$$
$$x_2 = 0.9686 = 96.86\%$$

f. Hence, h_4 is calculated as follows:

$$h_4 = h_{1,f} + x_2 \left(h_{fg} \right)$$

$$h_4 = 227.8 + 0.9686 \times (1,225.1) = 1,414.43 \frac{kJ}{kg} \left[609.41 \frac{Btu}{lb} \right]$$

g. The power output of the turbine is calculated next:

$$W_t = \dot{m} \left(h_3 - h_4 \right)$$

$$W_t = 1,000 \frac{kg}{s} (1,463.9 - 1,414.43) \frac{kJ}{kg} = 49,470 \ kW \left[66,314 \ hp \right]$$

h. The pump power input and total pump power input will be calculated along with enthalpy at point 4:

$$W_p = -v \left(P_2 - P_1 \right)$$

$$W_p = -0.0016 \times (913.4 - 615.2) = -0.47712 \ kW \left[0.64 \ hp \right]$$

$$w_p = \left(h_1 - h_2 \right)$$

$$-0.47712 = 227.8 - h_2$$

$$h_2 = 228.277 \frac{kJ}{kg} \left[98.353 \frac{Btu}{lb} \right]$$

i. Hence, h_2 is equal to 228.277 kJ/kg [98.353 Btu/lb], and the total pump input is then calculated:

$$W_{pump-total} = -\dot{m} \left(W_p \right)$$

$$W_{pump-total} = -1,000 \frac{kg}{s} \left(0.47712 \frac{kJ}{kg} \right) = -477.12 \ kW$$

$$W_{pump-total} = -477.12 \ kW \times \frac{hp}{0.746 \ kW} = 639.6 \ hp$$

j. The heat supplied and specific heat to the heat exchanger may now be calculated:

$$q_h = \left(h_3 - h_2 \right)$$

$$q_h = \left(h_3 - h_2 \right) = 1,463.9 - 228,277 = 1,235.62 \frac{kJ}{kg}$$

$$q_h = 1,235.62 \frac{kJ}{kg} \times \frac{1 \ kg}{2.2 \ lbs} \times \frac{1,000 \ J}{1 \ kJ} \times \frac{Btu}{1,055 \ J} = 532.4 \frac{Btu}{lb}$$

$$Q_h = \dot{m} \left(q_h \right) = 1,000 \frac{kg}{s} \times 1,235.62 \frac{kJ}{kg} = 1,235,620 \ kW$$

$$Q_h = 1,235,620 \ kW \times \frac{hp}{0.746 \ kW} = 1,656,327 \ hp$$

k. The actual cycle efficiency may now be calculated as follows:

$$\eta_{cycle}\left(\%\right) = \frac{w_{net}}{Q_h} \times 100\% = \frac{W_t - W_p}{Q_h} = \frac{\left(49,470 - 477.12\right)}{1,235,620} \times 100\% = 3.97\%$$

l. The Carnot efficiency is calculated:

$$\eta_{Carnot}\left(\%\right) = \frac{\left(T_h - T_c\right)}{T_h} \times 100\% = \frac{\left(\left(22.9 + 273.15\right) - \left(10 + 273.15\right)\right)}{22.9 + 273.15} \times 100\% = 4.35\%$$

Note that the actual efficiency is quite low, within 3% to 4%. To achieve significant output power, vast volumes of ocean water must be used. This will require larger pipe diameters for conveying the required working fluid as well as greater pumping power to circulate the liquid.

11.7.2 Open Cycle

The open cycle, on the other hand, uses the tropical oceans' warm water to make electricity as it is placed in a low-pressure container where it would boil. This drives a low-pressure turbine that drives a generator to produce electricity. The key for this system is the use of low-pressure turbines. The schematic for this system is shown in Figure 11.11. The main components are similar to that of the closed-cycle system except for the working fluid used. Warm intake water is fed into the evaporator. This component is kept at low pressure to evaporate salt water and convert it into expanding vapor. The expanding vapor is received in the low-pressure turbine, where the turbine drives the generator for power

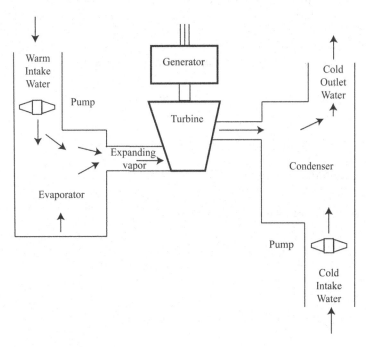

FIGURE 11.11
The OTEC open-cycle system.

production. The vapor is condensed with the aid of cold water from deep in the ocean. Note that there is no recycling of working fluid as in the closed cycle. As a result, pump energy is conserved, and the efficiency is also conserved.

Since the temperature difference is quite low, some flash evaporation of the warm seawater is enhanced by reducing the pressure below the saturation value. The pressure may be from 1% to 3% of the atmospheric values. Leakage must be prevented at all costs. Flash evaporation is therefore a unique feature of open-cycle systems. This event requires complicated heat and mass transfer processes and is beyond the scope of analysis for this chapter.

The overall conversion efficiencies of open-cycle systems are similar to those of the closed cycle. The main difference is the volume of water to be used. To achieve the same energy output, more water is required in the open system than in the closed system.

11.7.3 Hybrid Systems

Hybrid OTEC systems combine the features of the open and closed systems. The main advantage of this system is its slightly increased overall efficiency. The system uses a low-boiling-point working fluid as well as ocean water to generate power. A more common design involves bringing warm water in a flash vessel to generate vapor or steam. The vapor is then used in a heat exchanger to vaporize a low-boiling-point working fluid (such as ammonia or refrigerant) to generate power. Refrigerant-grade anhydrous ammonia is usually at least 99.95% pure. In this state, its boiling point will be around −33.3°C [−28°F], and it can be inferred that any cold ocean water will be sufficient to evaporate pure ammonia. At the freezing temperature of water close to 0°C [32°F], the density of pure ammonia is around 0.6833 kg/L [5.69 lbs/gal]. The vapor pressure at −17.8°C [0°F] is only around 110 kPa [16 psi] and is close to atmospheric pressure.

The input energy required to change the state of a liquid into gaseous form at constant temperature is called the latent heat of vaporization. The reported value for ammonia is 1,369 kJ/kg [589 Btu/lb]. The value for water is around 2,257 kJ/kg [970.4 Btu/lb]. The heat required to evaporate a fluid can be calculated using Equation 11.9. Example 11.8 shows the heat required to evaporate 10 kg [22 lbs] of water and ammonia:

$$q(kJ) = mh_e \qquad\qquad (11.9)$$

where
  q = heat of evaporation (kJ, Btu)
  m = mass of liquid (kg, lb)
  h_e = heat of evaporation (kJ/kg or Btu/lb)

Example 11.8: Heat of Evaporation Calculations

Calculate the heat required to evaporate 100 kg [220 lbs] of water and 100 kg [220 lbs] of liquid ammonia. The heat of evaporation for water and ammonia are 2,257 and 1,369 kJ/kg [972.4 and 589.8 Btu/lb], respectively. Comment on which liquid is easier to vaporize.

SOLUTION:

  a. Using Equation 11.8 directly, the values for water are as follows:

$$q(kJ) = mh_e = 100 \; kg \times 2,257 \frac{kJ}{kg} = 225.7 \; MJ \, [213,933.65 \; Btu]$$

b. The heat of vaporation for ammonia will be as follows:

$$q(kJ) = mh_e = 100 \ kg \times 1,369 \frac{kJ}{kg} = 136,900 \ kJ \ [129,763 \ Btu]$$

c. It will take less energy to evaporate ammonia than water, and hybrid systems have this slight advantage when using the energy from water to vaporize ammonia.

11.8 Projects Under Way for OTEC Systems (IRENA, 2014)

There are several OTEC projects that are currently under way around the globe. Leading the OTEC systems is the Makai Ocean Engineering Project, the world's largest operational system located in Hawaii. The facility was inaugurated in August 2015 at the Natural Energy Laboratory of Hawaii Authority (NELHA). The other projects are discussed in this section.

11.8.1 Natural Energy Laboratory of Hawaii Authority (NELHA)

The Hawaii Ocean Science and Technology (HOST) Park was tasked by the U.S. government to develop and diversify resources and facilities for energy- and ocean-related research, education, and commercial activities. The current project is a 100 kW [134 hp] OTEC facility whose output may be enough to power about 120 Hawaii homes annually. A larger 1 MW is in the planning stage. The design is based on a true closed-cycle OTEC system, and this will ultimately be connected to the grid. The project will also provide large amounts of cold and warm water for other secondary uses (e.g., marine algae and shrimp production). The research and development of the plant was funded by the Office of Naval Research (ONR) through the Hawaii Natural Energy Institute (HNEI), and the infrastructure was funded by the Naval Facilities Engineering Command (VANFC). The accrued electricity revenue from this power plant could sustain further research and development of OTEC technologies (Vega, 2003). Example 11.9 shows the calculation for satisfying the power needs of around 120 households.

Example 11.9: Number of Households Serviced by an OTEC System

Determine the number of households that will be served by a 100 kW [134 hp] power plant. Assume that each household will use around 7,500 kWh of energy per year.

SOLUTION:

a. This is a straightforward energy and power calculation, as power is the energy per unit of time.

$$Number \ of \ Households = 100 \ kW \times \frac{HH - yr}{7,500 \ kWh} \times \frac{365 \ days}{yr} \times \frac{24 \ hr}{day} = 117 \ households$$

b. This power plant can serve close to around 120 households.

11.8.2 OTEC Projects in Japan

Japan has an OTEC demonstration project located in Okinawa, Japan. The project started in 2013 with the installation of a 100 kW [134 hp] OTEC facility in Kumejima (Kume Island, Okinawa, Japan). The established surface water temperature was 25.2°C [77.36°F], and the deep-sea water temperature was 8.9°C [48.02°F]. The surface temperature was measured within 15 meters [49.2 ft] from the surface, while the cold water was measured at an around 612 meters [2,007.36 ft] of depth. The surface temperature ranged from a low just above 20°C [68°F] (January to February) to a high just below 30°C [86°F] (July to August). With these stable temperatures, the power output is likewise expected to be stable, and prediction of potential power generation in the future should be more reliable.

It was reported that the potential power generation in Japan is 5,952 MW within 30 km [18.65 mi] of the shore (IRENA, 2014). With no offshore distance limitation, the potential was reported to be 173,569 MW. In Okinawa alone, the potential was reported to be 2,797 MW within a 30 km [18.645 mi] distance and 79,992 MW without offshore distance limitations. Currently, Okinawa's power generation is around 2,000 MW, and an OTEC facility in this area should handle the city's energy needs. The Institute of Ocean Energy at Saga University was instrumental in designing the OTEC system and providing research studies. The IHI Corporation (Tokyo, Japan) provided the actual unit used for the demonstration project (Okinawa, OTEC, 2013).

11.8.3 OTEC Facility in India

The National Institute of Ocean Technology (NIOT) in India succeeded in constructing a 1 MW pilot OTEC plant in 2001 (Avery and Wu, 1994) with the cooperation of Saga University in Japan. The design was based on a floating unit with a mooring arrangement to bring output power to the mainland. The design uses the closed cycle with ammonia as the refrigerant. Makai Ocean Engineering (Hawaii) provided the conceptual designs. This Indian facility is the largest conceptualized system, similar to that intended for the Hawaii facility. The materials used were quite expensive with most heat exchangers made out of titanium. The evaporator area was reported to be 3,720 m² [40,021 ft²] and the condenser 3,410 m² [36,686 ft²]. The turbine system consists of four-stage axial flow turbines with a design rpm of 3,000. The cold-water pipe tubing was made of high-density polyethylene (HDPE) with an inner diameter of 0.90 meters [2.95 ft] and a length of around 1,100 meters [335.4 ft]. The unit was constructed, deployed, and operated for the seashore of Hawaii in the U.S. Mini-OTEC program. Note that even though the size is the largest amongst other systems, this project is still considered to be in the demonstration stage. Full-scale systems are expected to be in the range of 20 to 50 MW.

11.8.4 Other OTEC Projects Around the World

At the International OTEC Symposium held in September 2013, France presented its vision for an OTEC facility to be installed in Martinique (Brochard, 2013). The size of the facility was originally planned to be within 10 MW electrical output but was eventually scaled down to 5.7 MW. A self-funded feasibility study was initiated in 2008 followed by numerous other studies. The final funding was expected to come from the NER 300 European funding program. This facility was expected to cost around

$183 million. The project was called NAUTILUS and was jointly supported by the French Minister, Akuo Energy (France), DCNS Group (A Naval Defense Company based in France), and Entrepose Group (VINCI Construction, France), originally a steel tube manufacturer.

In South Korea, a team of experts from the Korea Research Institute of Ships and Ocean Engineering, affiliated with the Korea Institute of Ocean Science and Technology (KIOST), had successfully finished the design of a 20 kW [26.81 hp] OTEC pilot plant for public demonstration. The 20 kW [26.81 hp] system was a scaled-down version of the 1 MW design developed by a Korean OTEC team that consisted of 13 national institutes and universities and presented in 2012. A year later, parts were manufactured and fabricated, and public demonstrations were held after a series of multiple runs. The final goal of the project was to design a 1 MW system by 2015 (OTEC News, 2014).

Perhaps the most ambitious planned OTEC system is one envisioned in China. A 10 MW OTEC system is being planned off the coast of Hainan Island in southern China. The project is being jointly developed by the Beijing-based resort developer Reignwood Group and the U.S.-based defense and aerospace company Lockheed Martin. A memorandum of agreement was signed between the two companies in April 2013. The construction of the offshore power plant was expected to start in 2014 and is scheduled for completion in 2017. If this pushes through, this would be the world's largest OTEC facility. The plant was configured to be a closed-cycle OTEC system with ammonia as the working fluid. The turbine system of the plant will be placed above the water surface, with warm water passing through the heat exchanger and boiling the working fluid to create vapor. The vapor then passes through the underwater heat exchanger to be condensed back into liquid ammonia. Cold water is pumped from 800 to 1,000 meters below the ocean surface. The system will operate with the cyclical process and cool the working fluid in a closed-loop system. The electricity generated would be supplied to the nearby resort on Hainan Island, which will be built by the Reignwood Group. The total electricity needs of the resort community will be met by the OTEC plant. This resort would be marketed as a low-carbon real-estate development. The project developers hope that this project would pave the way for an increasing number of commercial deployments ranging between 10 and 100 MW of capacity off the coast of southern China. The prototype could provide data for further technological developments to increase efficiency and lower capital costs for worldwide adoption (Quick, 2013).

11.9 Technical Limitations and Cost (IRENA, 2014)

The ideal energy conversion efficiency for a temperature difference of at least 20°C [68°F] was shown to be only between 7% and 8%. Actual efficiencies may only be between 3% and 4% and are clear technical limitations of OTEC systems. It was also shown that pumping power for the ocean water from various depths would require enormous amounts of liquid. For example, approximately 4 m^3/s [63,348 gpm] of warm seawater and 2 m^3/s [31,674 gpm] of cold water (ratio of 2:1) with a nominal temperature difference of 20°C [68°F] are required per MW of exploitable or net electricity (Vega, 1995). Hence, building large-capacity OTEC plants will require large conveying systems as well as heat exchangers.

Another design limitation is the required size of pipes and tubing. For example, a 100 MW OTEC plant would use 400 m³/s [6.3 million gpm] of 26°C [78.8°F] water flowing through a pipe with a 16 meter [52.48 ft] inside diameter extending to a depth of 20 meters [65.6 ft] or more. About 200 m³/s [3.15 million gpm] of 4°C [39.2°F] water flowing through an 11 meter [36.08 ft] diameter pipe extending to a depth of 100 meters [328 ft] and a 20 meter [65.6 ft] diameter pipe is required for the mixed water return. This pipe discharge depth will be around 60 m [196.8 ft] or even deeper. As large volumes of ocean water are needed, corresponding large volumes of refrigerant are likewise necessary for the closed-loop system. While pressure will not be a huge factor, leaks within the system should not be discounted. The system would require very sturdy material for the heat exchangers. Several materials, such as titanium, have been mentioned. The biggest barrier to adoption is the initial capital cost and the massive construction of the facilities.

The cost for OTEC system may be approximated by the graph shown in Figure 11.12. The relationship representing the cost function in units of $/kW is given in Equation 11.11.

The empirical relationship for the capital cost of OTEC system is given in Equation 11.10 (modified from Vega, 2003). Example 10 shows how this equation may be used to estimate initial capital cost for a given size of OTEC power plant:

$$Cost\left(\frac{\$}{kW}\right) = 373 \times ln\left(Size, MW_{net}\right) + 21,687 \qquad (11.10)$$

where
    $Cost$ = $/kW
    $Size$ = MW_{net}

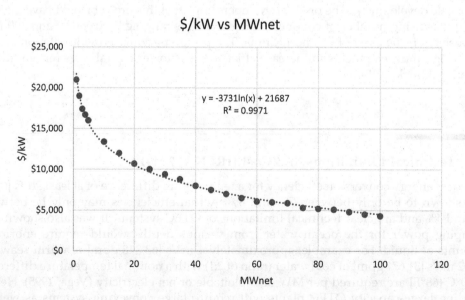

$/kW vs MWnet

y = -3731ln(x) + 21687
R² = 0.9971

FIGURE 11.12
Relationship between initial capital cost ($/kW) and power output of OTEC systems (modified from Vega, 2003).

Example 11.10: Capital Cost Calculations for OTEC System

Determine the initial capital cost for a 10 MW OTEC power plant using the empirical Equation 11.11.

SOLUTION:

a. The direct use of the empirical equation yields the following results:

$$Cost\left(\frac{\$}{kW}\right) = 373 \times ln(10) + 21,687 = \$22,545/kW$$

$$Cost\left(\frac{\$}{hp}\right) = \frac{\$22,545}{kW} \times \frac{0.746\ kW}{hp} = \$15,818.6/hp$$

b. For this facility, the initial capital cost is around $335 million, as shown below:

$$Cost(\$) = \frac{\$22,545}{kW} \times 10\ MW \times \frac{1,000\ kW}{MW} = \$335.5\ M$$

A study by the U.S. DOE (Gritton, et al., 1980) pegged the total system capital cost to be around $3,400/kW [$2,536.4/hp]. If this is the long-term capital cost estimate, the 10 MW power plant would cost roughly $3.4 million/MW, and the 10 MW facility will cost $34 million. Notably, there is a 10-fold difference between the cost used in the empirical equation and the sweeping cost provided in the DOE study. As such, there is still a lot of uncertainty in the cost estimates for OTEC systems. Until numerous demonstration plants are established, the capital cost will remain variable. Note that this initial estimate from the DOE study was done in the 1980s, while the empirical equation was used to report in 2003.

11.10 Conclusion

OTEC systems have perhaps the largest renewable energy potential for countries within the Tropics of Cancer and Capricorn. Careful site selection is key to keeping the environmental effects of OTEC minimal. It has been made clear that OTEC requires substantial capital investment at the beginning of the project, although long-term levelized cost should be manageable. There are only a few hundred land-based sites in the tropics where deep-ocean water is close enough to shore to make OTEC plants feasible. However, several tens of MW may be generated and may provide the base load power for many communities near the shore.

There are other advantages in the utilization of OTEC systems, the foremost of which is aquaculture. In conjunction with OTEC plants, cultivation of marine organisms, such as microalgae, may become popular for additional fuel, valuable chemicals, and food uses. One advantage of open or hybrid OTEC plants is the production of fresh water from salt water to alleviate water problems on the mainland as well as increase mineral water production. All these water deficiencies on the land may be solved by simply generating power and using this power to desalinate ocean water. There are therefore numerous possibilities for OTEC systems.

Perhaps one vision for OTEC systems is the creation of ocean communities where these power plants are located to make power plant communities sustainable and independent from land resources. The appeal of tourism can also be taken advantage of. The prospect of generating power, food, and fresh water by communities in the middle of oceans is exciting.

11.11 Problems

11.11.1 Heat Capacity of the Ocean

P11.1 Determine the amount of energy absorbed by a hectare of seawater surface water with a depth of 5 meters. Assume the original temperature of 5°C and the final temperature of 25°C. Assume the specific heat of sea water to be 3.93 kJ/kg°C. If the overall conversion efficiency is 3%, estimate the actual power. If the cycle happens in 24 hours, how much power is produced?

11.11.2 Ideal Carnot Cycle Efficiency

P11.2 Determine the ideal Carnot Cycle efficiency for an ocean with a warm temperature of 25°C and a cold temperature of 5°C.

11.11.3 Volume of Water Needed for a 100 kW of Power

P11.3 Determine the volume of ocean water needed to generate a 100 kW of electrical power assuming a 15°C temperature difference and power produced continuously for 24 hours a day and 365 days in a year. Assume an overall conversion efficiency of 5%. Use the simple sensible heat equation for the calculation. Compare this volume of water to the capacity of an Olympic-size swimming pool that measures 50 meters in length, 25 meters in width, with a depth of 3 meters. How many Olympic-sized swimming pools are needed to generate a 100 kW of electrical power in a day? Assume the density of seawater to be 1,027 kg/m³.

11.11.4 Calculating Water Pumping Power

P11.4 Determine the amount of theoretical power and actual power used to move 50 kg/s of water through 100 meters of head. Assume an overall pumping efficiency of 75%. Use the basic hydro power equation from Chapter 5 for calculations.

11.11.5 Base Load Power Calculations

P11.5 Determine the minimum size of a power plant that would satisfy the requirement of a community with a population of 12 million. The average per capita use of a household was reported to be 670 kWh/yr.

11.11.6 OTEC Closed Cycle Calculations

P11.6 The following data are provided for an OTEC system operating at temperature between (T_1) 4°C and (T_3) 28°C using Refrigerant 717 as working fluid. The properties of the refrigerant at these temperatures are found on its thermodynamic table. Calculate the overall cycle efficiency as well as the Carnot efficiency. The values for the properties are summarized below:

a. $h_1 = 199.7$ kJ/kg
b. $h_2 = 313.4$ kJ/kg
c. $h_3 = 1,467.8$ kJ/kg
d. $h_4 = 1,385$ kJ/kg
e. $v = 0.001173$ m³/kg
f. $P_2 = 1,099$ kPa (kN/m²)
g. $P_1 = 497.5$ kPa (kN/m²)
h. Heat input to the evaporator, $Q_e = (h_3 - h_2) = (1,467.8 - 313.4) = 1,154.4$ kJ/kg
i. Heat absorbed by the condenser $= Q_c = (h_4 - h_1) = (1,385 - 199.7) = 1,185.3$ kJ/kg
j. Pump work $= v(dP) = v(P_2 - P_1) = 0.001173 (1,099 - 497.5) = 0.7056$ kJ/kg
k. Turbine work $= (h_3 - h_4) = (1,467.8 - 1,385) = 82.8$ kJ/kg

11.11.7 Actual OTEC Cycle Examples

P11.7 Actual Rankine cycles have numerous losses from each component part of the system. In this problem, the efficiency calculations will be made for every 1,000 kg/s of working fluid flow. The low-temperature region will be at 7.5°C and the high-temperature region will be assumed to be at 25°C. Recalculate the OTEC system efficiency as well as the Carnot efficiency for these conditions. The values for working fluid properties are listed below. Use the diagram in Figure 11.6 and Figure 11.7 for the points.

a. $h_1 = 263$ kJ/kg
b. $h_3 = 1,463.7$ kJ/kg
c. $v = 0.001592$ m³/kg
d. $P_2 = 958.3$ kPa (kN/m²)
e. $P_1 = 565.6$ kPa (kN/m²)
f. $S_3 = 5.049$ kJ/kg
g. $S_{2f} = 0.799$ kJ/kg
h. $S_{fg} = 4.984$ kJ/kg
i. $h_{fg} = 1,353.9$ kJ/kg

11.11.8 Heat of Evaporation Calculations

P11.8 Calculate the heat required to evaporate 100 kg of ethylene glycol and 100 kg of Refrigerant R-11. The heat of evaporation for ethylene glycol and R-11 are 800 and 232 kJ/kg, respectively.

11.11.9 Estimating the Number of Households Served by OTEC

P11.9 Determine the number of households that will be served by a 1 MW OTEC power plant. Assume that each household will use around 12,000 kWh of energy per year.

11.11.10 Estimating the Initial Capital Cost of OTEC

P11.10 Determine the initial capital cost for a 100 kW OTEC power plant using empirical Equation 11.11. Compare this cost with the DOE's estimate of $3,400/kW.

References

Avery, W. H. and C. Wu. 1994. Renewable Energy from the Ocean: A Guide to OTEC. Oxford University Press, New York.

Borgnakke, C. and R. E. Sonntag. 2009. Fundamentals of Thermodynamics. 7th Edition. John Wiley & Sons, Hoboken, NJ.

Brochard, E. 2013. DCNS roadmap for OTEC. Proceedings of the First International OTEC Symposium held on September 9–11, 2013, Honolulu, Hawaii. Side Event of the Asia Clean Energy Summit and Co-hosted by the National oceanic and Atmospheric Administration (NOAA) Office of Ocean and Coastal Resource Management (OCRM).

Engels, W. and F. Zabihian. 2014. Principle and preliminary calculation of ocean thermal energy conversion. ASEE 2014 Zone I Conference, April 3–5, 2014, University of Bridgeport, Bridgeport, CT.

Gritton, E. C. R. Y., R. Y. Pei, A. Aroesty, M. M. Balaban, C. Gazley, R. W. Hess and H. Krase. 1980. A quantitative evaluation of closed-cycle ocean thermal energy conversion (OTEC) technology in central station applications. USDOE Report R-2595-DOE prepared for the U.S. Department of Energy by the Rand Corporation, Santa Monica, CA. p. 126.

Holman, J. P. 1980. Thermodynamics. 3rd Edition. McGraw-Hill, New York.

Ikegami, Y. and K. Furugen. 2013. 100 kW CC-OTEC plant and deep ocean water applications in Kumejima, Okinawa, Japan. Proceedings of the International OTEC Symposium and the Islands and Isolated Communities Congress on Energy, Security, Business and Sustainability, held September 9–11, 2013, at the Hawaii Convention Center, Honolulu, HI.

IRENA. 2014. Ocean Thermal Energy Conversion Technology Brief. International Renewable Energy Agency (IRENA). Ocean Energy Technology Brief 1. June. Available at: http://www.irena.org/DocumentDownloads/Publications/Ocean_Thermal_Energy_V4_web.pdf. Accessed April 25, 2018.

OECD/IEA. 2014. Electric Power Consumption (kWh per Capita). U.S. Energy Information Agency, U.S. Department of Energy, Washington, DC. Available at: http://www.iea.org/stats/index.asp.

Okinawa OTEC. 2013. Okinawa OTEC renewable energy for the future. Technology Brief. Okinawa Prefecture. Available at: http://otecokinawa.com/en/OTEC/index.html. Accessed April 25, 2018.

OTEC News. 2014. 20 kW OTEC pilot plant public demonstration in South Korea. OTEC News: Your In-Depth Source of Information for Ocean Thermal Energy Conversion. January 9. Available at: http://www.otecnews.org/2014/01/20kw-otec-pilot-plant-public-demonstration-south-korea/. Accessed April 25, 2018.

Pelc, R. and R. M. Fujita. 2002. Renewable energy from the ocean. Marine Policy 26: 471–479.

Quick, D. 2013. World's largest OTEC power plant planned for China. New Atlas Newsletter. April 18. Available at: https://newatlas.com/otec-plant-lockheed-martin-reignwood-china/27164/. Accessed April 25, 2018.

Vega, L. A. 2003. Ocean thermal energy conversion primer. Marine Technology Society Journal 6 (4): 23–35.

12

Human and Animal Power, and Piezoelectrics

Learning Objectives

Upon completion of this chapter, one should be able to:

1. Describe the concept of generating energy and power from humans and animals.
2. Estimate the animal and human power available.
3. Enumerate the various scenarios where human and animal power may be of significant use.
4. Describe the concept of piezoelectric energy generators.
5. Describe the various conversion efficiencies in piezoelectric systems.
6. Relate the overall environmental and economic issues concerning human and animal power and piezoelectric systems.

12.1 Introduction

Almost all geographical regions began their agriculture and industry using human and animal power sources. Some authors (Fuller and Aye, 2012) call human and animal power the "forgotten renewables." It is true that there is still a significant use of human and animal power in many developing countries. Animal power makes up a big portion of agricultural power needs on farms and in fields. After biomass, human power and animal power are the most popular renewable energy resources in developing countries. There are numerous international forums that provide connections, help, and support for those who use draft animals in their daily lives. One of them is the Draft Animal Power Network (DAPNet), a U.S.-based registered 501(c)(3) non-profit organization (https://www.draftanimalpower.org/). They hold annual gatherings for members, distribute newsletters, and host regional educational events. The goal of this group is to provide year-round educational and networking opportunities for all interested members as well as highlight the ongoing efforts of people throughout the world who are educating, mentoring, and building community around animal power and renewable land use. There is a considerable number of people who have given up their tractors and have replaced them with horses and mules. They maintain livelihood projects based solely on animal power. The term "draft animal power" refers to applications where animals are used to pull loads. However, the energy and power derived from animals is very limited. On a given day, mature draft animals may sustain work at a speed of around 3.6 km/hr (2 miles per hour, mph) and carrying a load from 150 to 250 lbs [68.2 to 113.6 kg]. In fact, the definition of "horsepower" was originated by James Watts and derived from an actual horse's power (Encyclopedia Britannica, 2018). He observed that a horse could carry a load of 150 lbs (68.2 kg) walking at a speed of 2.5 mph (40.2 kilometers per hour, kph). This became the constant used to define a horsepower as shown in Example 12.1.

Example 12.1: Calculating Power Generated from a Work Horse

Determine the amount of power generated from a horse carrying a load of 150 lbs [68.2 kg] and walking at a rate of 2.5 mph [40.2 kph].

SOLUTION:

a. The power is the product of force and speed:

$$Power\left(\frac{ft-lb}{min}\right) = Force \times Speed = 150 \ lbs \times \frac{2.5 \ mi}{hr} \times \frac{5,280 \ ft}{mile} \times \frac{1 \ hr}{60 \ min} = 33,000 \ \frac{ft-lb}{min}$$

b. The value of 33,000 ft-lb/min is the conversion constant for a horsepower [746 watts].
c. Thus, 1 horsepower is equal to 33,000 ft-lb/min. This constant is used at present.
d. In the metric system, 1 horsepower is equivalent to 746 watts.

The available power from human beings will be a small fraction of what a draft animal can pull; a mature human being can only generate useful continuous work of about one-tenth of a draft animal. In this case, that measures to 0.1 hp or 74.6 watts (Irfan, 2012). The primary contribution of human beings is of course their intellect, and thus they are the ubiquitous element with the use of animal power. The comparative power of various draft animals is shown in Table 12.1. This table shows the sustainable power of individual animals in good working condition and their corresponding energy output per day in units of MJ. The horse still has the highest energy output in a given day. In terms of power, Table 12.2 shows the range of power available from these draft animals, including the potential of humans to generate continuous power in a day. On average, a mature human being can only sustain about 60 watts of power continuously (Mother Earth News, 2018) and 78 W or 0.1 hp in Table 12.1. The calculation is shown in Example 12.2. This calculation shows that indeed, human power is one-tenth of animal power, or close to 0.1 hp (around 61 watts in the example), a very small amount of power.

TABLE 12.1

Sustainable Power of Various Work Animals Using Maximum Reported Weights and Constant Work Speed (Modified from Lovett, 2007)

Animal	Maximum Weight (kg) [lbs]	Pull-Weight Ratio	Typical Pull (kg$_f$)	Power Output (Watts) [hp]	Working Hours/Day (hrs)	Energy Output/Day (MJ) [Btu]
Horse	1,000 [2,200]	0.13	130	1,274 [1.7]	10	46 [43,601]
Ox	1,100 [2,420]	0.11	121	1,186 [1.6]	6	26 [24,645]
Buffalo	990 [2,178]	0.12	118.8	1,164 [1.56]	5	21 [19,905]
Camel	600 [1,320]	0.13	78	764 [1.02]	6	17 [16,114]
Mule	450 [990]	0.13	58.5	573 [0.77]	6	12 [11,374]
Donkey	480 [1,056]	0.13	62.4	612	4	9 [8,531]

TABLE 12.2

Sustainable Power of Draft Animals and Humans at Similar Work Speed (Modified from Lovett, 2007)

Animal	Force Exerted (kg)[lbs]	Power (Watts)[HP]	Working Hours per Day	Energy Output per Day (MJ) [Btu]
Horse	55 [121]	539 [0.72]	10	19 [18,009]
Ox	55 [121]	539 [0.72]	6	12 [11,374]
Mule	27 [59]	265 [0.35]	6	6 [5,687]
Donkey	14 [31]	137 [0.18]	6	3 [2,844]
Man	8 [17.6]	78 [0.10]	8	2.2 [2,085]

Example 12.2: Calculating the Amount of Power Generated by Humans

Determine the amount of power generated by a human being carrying a load of 18 lbs [8.2 kg] and walking at a rate of 2.5 feet/s [45.7 m/min].

SOLUTION:

a. The power is the product of force and speed as follows:

$$Power(hp) = Force \times Speed$$

$$Power(hp) = 18 \ lbs \times \frac{2.5 \ ft}{s} \times \frac{hp - min}{33,000 \ ft - lbs} \times \frac{60 \ s}{min} = 0.082 \ hp$$

b. The value 0.082 hp is equivalent to 61 watts as shown below:

$$Power(Watts) = 0.082 \ hp \times \frac{746 \ Watts}{hp} = 61 \ Watts$$

As seen from Example 12.2, the simple equation to determine the horsepower if the load is in lbs and speed in feet/s is shown in Equation 12.1. The constant of proportionality is 550, which is the ratio of 33,000 ft-lbs/hp-min and 60 s/min. The conversion from horsepower (hp) to the metric units of watts is shown in Equation 12.2:

$$Power(hp) = \frac{Force(lbs) \times Speed(ft/s)}{550} \tag{12.1}$$

$$Power(Watts) = hp \times \frac{746 \ Watts}{hp} \tag{12.2}$$

One emerging miniscule sustainable power source, although not related to human and animal power, is piezoelectric power. Some materials may generate electric potential due to a response in temperature or if they are deformed. There are many naturally occurring crystals (quartz, topaz, etc.) and synthetics (langasite and gallium orthophosphate) that exhibit piezoelectricity, and this will be discussed in this chapter as well. In fact, dry bone also exhibits some piezoelectric properties. This is an area that may be of importance in the future in generating small amounts of power. Numerous research projects envision placing such devices on the soles of shoes for humans to be able to generate the power they need for their portable electronic devices as well as using beams in certain buildings that are deformed by regular cycles and as a result are able to generate lighting power during off hours. The types of applications could be endless.

Perhaps the most significant applications of piezoelectricity are in industrial and manufacturing businesses, followed by applications in the automotive industry. Additional concepts can be extended into the use of some medical instruments and in the telecommunication industry. Many of these applications will be discussed in the last part of this chapter.

12.2 Animal Power

Draft animals are defined as large, domesticated four-footed animals that are used for power in agriculture. The primary draft animals currently being used in agriculture are as follows:

   a. Oxen
   b. Water buffalo
   c. Horses
   d. Mules
   e. Camels
   f. Donkeys

The advantages and disadvantages of the use of animals as power sources on farms are discussed in this section.

Advantages of Animals as a Source of Power

   a. These animals are a great source of reserve power for emergencies and temporary loads.
   b. They are in various sizes within a narrow range.
   c. They can be adapted to practically all draft work in the farm.
   d. They can operate in dry or wet soils and can be submerged in muddy soils.
   e. The fuel (food) they use is readily available on most farms.
   f. They can be reproduced at the farm level.
   g. They provide a relatively cheap type of power in areas where there are surpluses of both grain and roughage.

Disadvantages of Animal as Source of Power

   a. They require feed and care when not working.
   b. They can work on heavy loads only for a limited period of time. They require frequent resting periods.
   c. In hot weather, they cannot work efficiently.
   d. The working speed of animals is limited and relatively low.
   e. For most stationary work, they are not very efficient.
   f. They require relatively large spaces for shelter and feed storage.

g. These animals are difficult to manage when used in large units.

h. They utilize productive land and pasture that could be diverted toward other purposes.

12.2.1 Draft Animal Performance Compared with Mechanical Tractors

In developing countries, draft animals are advantageous compared to tractors due to the absence of high costs for fuel and the fact that animals can be reproduced at the farmland themselves. Most farmers in the third world simply cannot afford the luxury of tractors. The performance efficiencies of draft animals have been established as quite low compared with tractors. The field efficiency of most tractors with implements could be as low as 50% for sugar beet harvesters (ASABE Standards D497.6, 2009) to as high as 90% for most tillage implements. However, for draft animals, the reported performance efficiencies vary from 9% to 10% for bovines and 10% to 12% for the horse family. A rule of thumb in calculating available draft from work animals is to take approximately 10% of the animal's body weight (Lovett, 2007).

One clear difference between maintaining a tractor and a work animal is the fact that work animals are living creatures that must be tended to properly in order to provide the best power source in the farm. The performance of a draft animal depends on many characteristics listed below.

Characteristics of Draft Animals That Affect Performance

a. Breed

b. Weight

c. Sex

d. Age

e. Health

f. Training

g. Quality of feed

12.2.2 Draft Horsepower Capability of Various Animals

The maximum tractive requirement for plowing is done at a rate of one-seventh to one-tenth of the animal weight. The performance is reduced by an average of between 10% and 20% below the optimum standard if the animals work in groups or teams. Unlike mechanical tractors that are additive, work animals in large numbers in fact have reduced effectiveness. Table 12.3 shows the pull, speed, and power developed from work animals. These data may be used for making estimates of field capacities and production processes from work animals. Note that these data are for individual animals. When working in groups, the value called "harnessing factor" is reduced. That is, one cannot simply multiply the power of each animal by 5 if there are 5 animals in a group. Campbell (1990) has developed a table for harnessing factor, and the fitted equation is shown in Equation 12.3, where the variable X refers to number of animals in a group and an expanded table is given in Table 12.4. Harnessing factors are used when animals are employed in a team. The harnessing factor should be developed for each animal in question. Examples 12.3 and 12.4 show how the data from Tables 12.3 and 12.4 may be used:

$$Harnessing\ Factor = -0.0679X^2 + 1.0293X + 0.06 \tag{12.3}$$

TABLE 12.3

Drawbar Pull and Power Developed from Various Work Animals (Modified from Campbell, 1990)

Animal	Average Weight, kg [lbs]	Approximate Pull, kg [lbs]	Average Speed, m/s [ft/min]	Power Developed	
				Watts	hp
Light horse	550 [1,210]	70 [154]	1 [196.8]	686	0.92
Bullock	700 [1,540]	70 [154]	0.7 [137.8]	480	0.64
Buffalo	650 [1,430]	65 [143]	0.85 [167.3]	541	0.72
Mule	425 [935]	55 [121]	0.94 [185]	507	0.68
Camel	475 [1,045]	45 [99]	1.11 [218.4]	490	0.66
Cow	500 [1,100]	55 [121]	0.69 [135.8]	372	0.50
Donkey	250 [550]	35 [77]	0.69 [135.8]	237	0.32

Refer to Figure 12.1 for the harnessing factor relationship derived from Cambell's (1990) reported table.

Example 12.3: Estimating the Amount of Power Developed by a Work Horse

Calculate the amount of power that can be developed by a 600 kg [1,320 lb] horse pulling an 80 kg [176 lb] load and traveling at an average speed of 0.9 m/s [2 mph]. Provide your answers in kg-m/s, watts, and horsepower.

SOLUTION:

a. Equation 12.1 may be used but with different units and a different constant:

$$Power\left(\frac{kg-m}{s}\right) = \frac{Force(kg) \times Speed(m/s)}{constant}$$

$$Power\left(\frac{kg-m}{s}\right) = \frac{80\ kg \times 0.9\ m}{sec} = 72\frac{kg-m}{s}$$

b. The conversion of the above units into horsepower may be done as follows:

$$Power(Watts) = 72\frac{kg-m}{s} \times \frac{N-s^2}{kg-m} \times \frac{9.8\ m}{s^2} \times \frac{W-s}{N-m} = 705.6\ Watts$$

TABLE 12.4

Reworked and Expanded Harnessing Factors for Work Animals (Modified from Campbell, 1990)

Number of Animals	Revised Harnessing Factor
1	1.02
2	1.85
3	2.54
4	3.09
5	3.51
6	3.79
7	3.94
8	3.95

Harnessing Factor Equation
(Developed from Campbell, 1990)

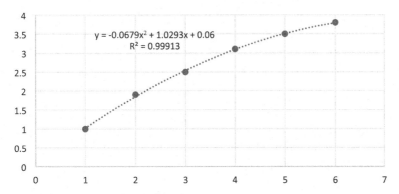

$y = -0.0679x^2 + 1.0293x + 0.06$
$R^2 = 0.99913$

FIGURE 12.1
Harnessing factor relationship derived from Cambell's (1990) reported table.

c. The conversion to horsepower is the reverse of the previous example:

$$Power(hp) = 705.6 \ Watts \times \frac{hp}{746 \ Watts} = 0.95 \ hp$$

Example 12.4: Calculating the Draft Provided by an Ox

If an ox can provide a draft of 60 kg [132 lbs], how much draft can be provided by 6 oxen together? Use the harnessing factor for work animals in Table 12.4 for your calculations. Report your units in Newton. If the animals' speed is 3 m/s [6.7 mph], determine the power developed from the group.

SOLUTION:

a. The solution is straightforward:

$$Total \ Draft(kg) = \frac{Draft}{ox} \times No. \ of \ oxen \times Harnessing \ Factor$$

$$Total \ Draft(kg) = \frac{60 \ kg}{ox} \times 6 \ oxens \times \frac{3.79}{6.0} = 227.4 \ kg \ [500.3 \ lbs]$$

b. Note that the harnessing factor value in the table means that this number is out of the total population, that is, 3.79 out of 6, 3.51 out of 5, and so on.
c. The power generated from this group is calculated as follows:

$$Power(Watts) = 227.4 \ kg \times \frac{3m}{s} \times \frac{Ns^2}{kg - m} \times \frac{9.8m}{s^2} \times \frac{W - s}{N - m} = 6,685.6 \ Watts \ [9 \ hp]$$

12.2.3 Unique Perspectives of Animal Power

Perhaps other hidden perspectives related to the use of animal power are the energy-related products, co-products, and by-products of animals for energy use. For example, animals are also a source of energy in the form of milk. Each mature cow averaged about 70 lbs [31.8 kg] of milk each day, or about 8 gallons [30.3 L] per day (Dairy Moos, 2017). The energy content of whole milk is around 150 kilocalories (kcal), and 1 kcal is equivalent to 4,184 Joules [3.97 Btu] of energy. The United States is one of the leading dairy-producing countries in the world, and American cows are among the most productive cows in the world. The average cow in the United States produces about 21,000 lbs [9,545.5 kg] of milk per year, or nearly 2,500 gallons [661 L] a year. A record-producing cow in Wisconsin produced 72,000 lbs [32,727.3 kg] of milk in a year, or about 8,000 gallons of milk [30,280 L] (Dairy Moos, 2017). The energy produced by an average cow each year from the milk is roughly close to 24,000 MJ [227.5 million Btu]. The cow also generates manure, and Example 12.5 shows that an additional 9,627 MJ [9.125 MBtu] is likewise produced. This manure may be used to produce electrical power via thermal conversion as discussed in an earlier chapter on biomass conversion.

Example 12.5: Calculating the Energy Produced by a Single Cow

Determine the energy in MJ [million Btu] produced by a cow in the form of milk. Assume the energy content of milk is 150 kcal [594 Btu] per 250 g [0.55 lb] and 21,000 lbs [9,545.5 kg] was produced in a year. If the same cow produces an average of 5 lbs [2.27 kg] of dry manure each day, or 1,825 lbs [829.5 kg] in a year, with a heating value of 5,000 Btu/lb [11.61 MJ/kg], how much energy in MJ is likewise produced? Which cow product produced more energy each year?

SOLUTION:

   a. The milk energy is simply a conversion problem:

$$Energy(MJ) = 21,000 \ lbs \times \frac{150 \ kcal}{0.250 \ g} \times \frac{4,184 \ J}{kcal} \times \frac{MJ}{1 \times 10^6} \times \frac{kg}{2.2 \ lbs} = 23,963 \ MJ \ [22.7 \ MBtu]$$

   b. The manure energy is also estimated the same way:

$$Energy(MJ) = 1,825 \ lbs \times \frac{5,000 \ Btu}{lb} \times \frac{1,055}{Btu} \times \frac{MJ}{1 \times 10^6} = 9,627 \ MJ \ [9.1 \ MBtu]$$

   c. More energy is produced by a cow in the form of milk than from the manure generated.

12.3 Human Power

Say a human being absorbs a total power of 373 watts (0.5 hp) from the food he eats. Of this amount, about one-tenth hp, or 74.5 watts, is available for useful work (Campbell, 1990). The remaining energy ingested by humans is expended for the usual bodily functions. This information was based on an adequately fed 35-year-old male European laborer working an 8-hour day and 48-hour week. A man aged 20 can generate approximately 15% more useful energy than this average European male, while a 60-year-old male can achieve 20% less. Campbell (1990) has also established an equation relating the power available from a human being, and this is shown in Equation 12.4 (based on a test that lasted from

4 minutes to 8 hours of human work in the field). This is an empirical equation and hence would vary to other human body types. Example 12.6 shows how this formula is used:

$$Power(hp) = 0.35 - 0.092 \times log(t) \tag{12.4}$$

where
  t = time (min)

Example 12.6: Calculating the Output Power Generated by Humans

A person is asked to do work for 4 hours. Estimate the power that can be generated by this person in units of horsepower (hp) and watts.

SOLUTION:

  a. Equation 12.4 is simply used:

$$Power(hp) = 0.35 - 0.092 \times log\left(4\ hrs \times \frac{60\ min}{hr}\right) = 0.131\ hp$$

  b. The equivalent power in watts is calculated:

$$Power(Watts) = 0.131\ hp \times \frac{746\ Watts}{hp} = 98\ Watts$$

  c. Only a small amount of power is generated by this person.

12.3.1 Advantages of Humans for Energy Use

What humans possess over any other power source is their intelligence—conscience and the ability to make judgments—as well as the physical deftness and dexterity in controlling other mechanical devices on the farm. A list is enumerated below:

  a. Humans can operate engines, tractors, and self-propelled machines and control larger machinery.
  b. When using draft animals, humans provide all manual operation of the plows and harrows that are pulled by draft animals.
  c. Humans are able to perform small-scale planting and transplanting of crops.
  d. Humans can perform manual spraying and weeding.
  e. Humans can harvest using sickles.
  f. Humans can clean and maintain fields.
  g. They can operate threshing and shelling machines.
  h. Humans are able to operate mechanical dryers and dry crops traditionally through sun exposure.
  i. Humans can load and unload crops and products manually.
  j. Humans can perform simple milling and storage operations.
  k. Humans can execute small-scale pumping.

12.3.2 Disadvantages of Humans for Energy Use

The main disadvantage of humans is the very low amount of physical power they can offer. Other disadvantages include the following:

a. The power provided by humans diminishes with time.

b. Harsh environmental conditions are not suited for humans.

c. The limited amount of power and strength limits the range of activities they can do on the farm.

d. Higher human body weight does not translate into increased power capability.

e. Physical strain affects their performance.

f. They require energy in the form of food, where the conversion efficiency is 25% at best.

A human's lower body is an important source of power. Campbell (1990) has developed an equation relating the range of power human legs can provide. This is given in Equation 12.5. The range of power is between 0.27 and 0.53 hp (201 and 395 watts). This equation is similar to Equation 12.4 in its use and indicates that the efficient use of legs after long periods of time greatly diminishes:

$$Power(hp) = 0.53 - 0.13 \times log(t) \tag{12.5}$$

where
t = time (min)

> **Example 12.7: Estimating the Output Leg Power Generated by Humans**
>
> A person is asked to do work for 8 hours using his legs. Estimate the power that can be generated by this person in units of horsepower (hp) and watts.
>
> **SOLUTION:**
>
> a. Equation 12.4 is simply used:
>
> $$Power(hp) = 0.53 - 0.13 \times log\left(8\ hrs \times \frac{60\ min}{hr}\right) = 0.181\ hp$$
>
> b. The equivalent power in watts is calculated:
>
> $$Power(hp) = 0.181\ hp \times \frac{746\ Watts}{hp} = 135\ Watts$$
>
> c. As expected, a small amount of power generated from a human's legs is higher than that generated using human arms/hands. If one were to compare this with Equation 12.4, the power is 0.22 hp or 164 Watts for 4 hours of legwork, compared with 98 Watts for mainly arm work.

12.3.3 Human Factors in Energy and Power: The Ergonomic Factors

While the power available from humans is quite low, their ability to control, make judgments, and operate large machinery is an inestimable advantage (Gustafson, et al., 2017).

The age of robotics has not yet infiltrated agriculture in significant amounts, and machinery design takes into consideration the human ability factor and incorporates this into the systems' basic engineering. The systematic consideration of human factors, including human-machine interfaces, should be a significant part of the machinery design process and should continue throughout the final design stages. When human factors are considered thoughtfully in the design, there is a noticeable difference in the prevention of personal injury and dangerous or harmful consequences.

Risk assessments must be made in the design of machines to be operated by humans. Some criteria for evaluation considering human factors in the design of machines are the following:

a. Individual risk perception.

b. Correct estimation of human capabilities.

c. Estimating the consequences of familiar hazards.

d. Recognizing subtle changes to familiar tasks.

e. Control information must be clearly communicated and understood.

f. Design configurations should match human operator expectations.

g. Human response time to some operations.

The ergonomic factor is a human-related factor that characterizes a machine in terms of ease of convenience of operation (Berolo, et al., 2011). The machine must be designed for ease of operation based on its technical design, and human factors must always be considered. Some of these factors affecting the human as operator are as follows:

a. Physical condition of the operator: the machine must take into account the weight, age, and strength of the operator.

b. Heartbeat ratio: the stress imparted onto humans is a strong function of this factor. There are ideal heartbeat ratios before and after the operation of a machine.

c. Body movement: the body movement of humans must be considered when designing machines. The operation of a complicated machine must take into consideration the possible required body movements, utilizing a number of muscle groups, such as those in the arms, shoulders, back, and legs, without bringing exaggerated or overbearing movements to a particular set of muscles.

When humans are the main source of power on a farm, the various agricultural operations will certainly require several man-hour requirements. Tables 12.5a and 12.5b show the various man-hour requirements per hectare for various agricultural field operations. These tables were revised from ranges of data reported by Campbell (1990).

Preparing land would usually need the highest amount of man-hours. Clearing the field takes up the most—slash and burn takes between 240 and 360 man-hours/hectare [97.2 and 145.7 man-hours/acre]. Additionally, individual activities such as spading soil for 25 cm [9.84 in] depth will expend about 500 man-hours/hectare [202.4 man-hours/acre] of work (Campbell, 1990). Humans could cover a hectare of land in several days. The productivity is really quite low compared with the output with the aid of mechanical farm equipment, such as tractors or even single-cylinder engine-driven machines. Clearly, when humans control such equipment, farm operation productivity increases even with a minimal power requirement.

TABLE 12.5a

Reworked Man-Hour Requirements per Hectare [Acres] of Various Farm Activities

Operation and Rates of Work	Average Man-Hours/Hectare	Average Man-Hours/Acre
Manual tillage		
Slash and burn	300	121
Tillage with hoe	200	81
Hoeing, flooded soil	150	61
Spade, 25 cm depth	500	20
Manual planting		
Broadcasting	3.3	1.34
Using a dibble stick	160	65
Seeding in pre-marked rows and covering by foot	80	32
Push- or pull-type planter in dry soils	20	8
Manual weeding and pest control		
Hand weeding transplanted rice	220	89
Hand weeding rice in broadcast field	1150	466
Hand weeding rice in dibbled field	380	154
Hand weeding rice in drilled field	321	130
Rotary push-type weeder in rice	90	36

TABLE 12.5b

Man-Hour Requirements per Hectare [Acre] of Various Farm Activities (Modified and Updated from Campbell, 1990)

Operation and Rates of Work	Man-Hours/Hectare	Man-Hours/Acre
Knapsack sprayer	14	5.7
Manual harvesting, threshing, and processing		
Harvesting rice with sickle or knife	70	28
Reaping with a scythe	34	13.7
Bundling rice into sheaves	24	9.7
Hauling sheaves to thresher	39	15.8
Threshing rice with hand sticks	182	73.7
Threshing rice with flail	25	10
Threshing rice on a bamboo ladder	45	18

Humans are on the brink of designing unmanned machinery capable of covering thousands of hectares, and thus, while man's pure power reserve is low, his ability to design robotics and artificial intelligence–controlled machines can raise productivity exponentially. In the end, it is the human intellect that will provide progress to agriculture and related industries.

12.4 Piezoelectrics

Certain solid materials accumulate electric charge. Crystals, ceramics, and even biological material such as bones have this characteristic. DNA and various proteins have electric charges, especially in response to mechanical pressure. The electrical charges are quite

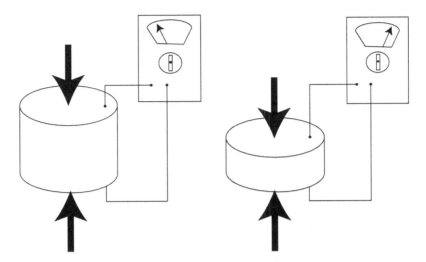

FIGURE 12.2
Deformation of a piezoelectric material and the generation of electric voltage or power.

small but when used for applications where electrical power requirements are also quite low should become a useful source of energy. "Piezo" is a Greek root word that means "pressure." Piezoelectricity is in fact considered an ancient source of electric charge. It was discovered in 1880 by French physicists Jacques and Pierre Curie (Manbachi and Cobbold, 2011).

The two most common applications of piezoelectricity have been in electric cigarette lighters or charcoal igniters. Pressing a lighter button causes a spring-loaded hammer to scratch a piezoelectric crystal. This action generates high-voltage electric current that flows across a small spark-plug gap, creating a spark. If there is combustible gas present within the vicinity of the spark, combustion takes place. Numerous household gas burners have been ignited this way. There are numerous materials, both natural and synthetic, that can be used for such production of small amounts of energy and power. A list is presented in the next section.

A visual presentation of piezoelectric power is shown in Figure 12.2. The oscillation of a material creates pressure and generates electrical voltage as shown.

Naturally Occurring Crystals with Piezoelectric Behavior (Beijing Ultrasonics, 2012).

a. Quartz

b. Berlinite ($AlPO_4$), a rare phosphate mineral that is structurally identical to quartz

c. Sucrose (table sugar)

d. Rochele salt

e. Topaz

f. Tourmaline-group minerals

g. Lead titanate ($PbTiO_3$) (occurs in nature as mineral macedonite)

Other Naturally Occurring Materials (Lee, et al., 2012)

a. Bone

b. Tendon

c. Silk

d. Some wood

e. Enamel

f. Dentin

g. DNA

h. Viral proteins, including those from bacteriophage

Synthetic Crystals and Langasite ($La_3Ga_5SiO_{14}$), a Quartz Analog Crystal

a. Gallium orthophosphate ($GaPO_4$), another quartz analog crystal

b. Lithium niobate ($LiNbO_3$)

c. Lithium tantalite ($LiTaO_3$)

Synthetic Ceramics (Beijing Ultrasonic, 2012)

a. Barium titanate ($BaTiO_3$) (the first piezoelectric ceramic discovered)

b. Lead zirconate titanate ($Pb[Zr_xTi_{1-x}]O_3$) (more commonly known as PZT and the most common piezoelectric ceramic in use at present)

c. Potassium niobate ($KNbO_3$)

d. Sodium tungstate (Na_2WO_3)

e. $Ba_2NaNb_5O_5$

f. $Pb_2KNb_5O_{15}$

g. Zinc oxide (ZnO)

12.4.1 Applications of Piezoelectricity

The most ubiquitous applications of piezoelectricity at present are as short-term high-voltage power sources in industrial and manufacturing businesses and in the automotive industry. Numerous other piezoelectric-related devices are also utilized as sensors and actuators. It is worth noting that the standards of time are still based on quartz clocks due to their accurate frequency or oscillation. Finally, piezoelectrics are now being used in conjunction with other renewable energy sources, such as solar photovoltaic cells, to improve their efficiency of conversion.

12.4.2 High-Voltage Power Sources

Substances like quartz are direct sources of short potential energy in the form of electricity. Voltage in the thousands may be generated. In a project called "Energy Harvesting" by the Defense Advanced Research Projects Agency (DARPA) of the U.S. Department of Defense (DOD), piezoelectric generators embedded in soldiers' boots generated 1 to 2 watts of power from their continuous walking (Lebrun, 2013). A goal was to also generate enough electrical power to operate smaller battlefield equipment, such as radios, low-power microprocessors for GPS devices, and other military gadgets. The project was abandoned due to the impracticality and discomfort from the additional energy expended by the person wearing the shoes.

Some dance floors are also being converted into energy-generating devices (Gupta, et al., 2016). The energy generated by the movement of dancers on the floor is being used to power LED lights installed on the floor, even leaving excess power to some small electronic devices. The same article also reported that a researcher from the Israel Institute of Technology in Haifa has installed a piezoelectric generator system 100 meters [328 ft] beneath a four-lane highway. The vibrations caused by normal traffic on the road system were said to recover an estimated 400 kWh of energy. This application could be extended to airport runways, racetracks, and other asphalt surfaces that deform when in use. Note that there is a limitation on the power that is generated from piezoelectricity. While the voltage produced may be quite high, the amperage is rather low. Thus, sensitive wires and capacitors must be used to limit losses of power. In most instances, expensive materials such as platinum are used to improve electrical conductivity, and this may render the whole device too expensive.

Example 12.8 shows some calculations on power output from a small piezoelectric device.

The voltage or power from a piezoelectric device and stored in a capacitor is calculated using Equations 12.6 and 12.7:

$$Energy\,(Joules) = \frac{1}{2}CV^2 \qquad (12.6)$$

where

$Energy$ = energy stored in the capacitor (Joules)

C = capacitance (farads, or if microfarads, μF, is used, divide by 1×10^6)

V = voltage measured across the capacitor (volts)

The energy from several cycles of deformation is given by Equation 12.6:

$$Energy(Joules)_{cycle} = \frac{1}{2}C \times \left(V_1^2 - V_o^2\right) \qquad (12.7)$$

Example 12.8: Calculating the Energy Produced from a Piezoelectric Element

Determine the energy produced from numerous cycles on a piezoelectric element. The measured voltage across the capacitor after a series of cycles was found to be 8.5 volts, and the capacitor has a rating of 220 microfarad and 35 V. Provide your answer in Joule.

SOLUTION:

a. Equation 12.6 is simply used:

$$Energy(Joules) = \frac{1}{2}CV^2 = \frac{1}{2} \times \frac{220\ farads}{1 \times 10^6} \times (8.5\ V)^2 = 0.00795\ Joule$$

b. A small amount of energy is stored in a capacitor after a series of cycles.

Example 12.9: Calculating the Energy Produced for Each Tap of Piezoelectrics

Determine the energy produced for each single tap on a piezoelectric element. The initial voltage was found to be 2.0 volts, and after each tap, the final voltage was found to be 2.05 volts. The capacitor has a rating of 220 microfarad and 35 V. Provide your answer in Joule.

SOLUTION:

a. Equation 12.7 is simply used:

$$Energy\left(\frac{Joule}{cycle}\right) = \frac{1}{2}CV^2 = \frac{1}{2} \times \frac{220\ farad}{1 \times 10^6} \times \left[(2.05\ V)^2 - (2\ V)^2\right] = 0.000022\frac{Joule}{cycle}$$

b. A small amount of energy is stored from each deformation cycle of a piezoelectric device, a very small fraction of a Joule.

The above examples show a very small amount of energy being stored in a capacitor. For example, a typical cell phone battery will have about 18,000 Joules [17.1 Btu] of energy. If we replace this battery with a capacitor from a piezoelectric device and determine how many cycles are needed to store this energy, the answer would be close to a billion cycles. That is, take the ratio of 18,000 Joules and 0.000022, and you would get a little over 810 million cycles. If you do this cycle five times a second, it would still take several years to fully charge this cell phone. This is obviously unfeasible for such an application. Hence, research is currently under way to improve this energy storage using efficient materials as well as applications where cycles are more pronounced each second. If one were to make some simple assumptions such as those shown in Example 12.10, your cell phone could be fully charged in a matter of days instead of years.

Example 12.10: Estimating the Time to Charge a Cell Phone via Piezoelectrics

Determine how long it would take to charge a cell phone battery with a fully charged load requirement of 18,000 Joules [17.1 Btu]. The voltage across the capacitor was improved from 2 volts to 12 volts with a single cycle. Assuming this is done continuously in a day, in how many days will one achieve full charge? The capacitor has a rating of 220 microfarad and 35 V. One can generate 5 cycles in a second on this device.

SOLUTION:

a. Equation 12.6 is simply used:

$$Energy\left(\frac{Joule}{cycle}\right) = \frac{1}{2}CV^2 = \frac{1}{2} \times \frac{220\ farad}{1 \times 10^6} \times \left[(12\ V)^2 - (2\ V)^2\right] = 0.0154\frac{Joule}{cycle}$$

b. To generate 18,000 joules of energy, the number of cycles is calculated:

$$Cycles = 18,000\ Joules \times \frac{cycle}{0.0154\ Joules} = 1,168,831\ cycles$$

c. The number of days required to get this fully charged is calculated as follows:

$$Days = 1,168,831\ cycles \times \frac{sec}{5\ cycles} \times \frac{hr}{3600s} \times \frac{day}{24\ hrs} = 2.7\ days$$

d. With this improvement, it would only take about 2.7 days to fully charge this cell phone. Of course, the given voltage increase is quite a remarkable feat, and a 2.7-day charging period is still highly impractical.

Another application of piezoelectricity is in a piezoelectric transformer. This component's mechanism is different from commonly used conventional electromagnetic induction since it uses acoustic coupling instead. The most commonly used material is PZT (lead zirconate titanate) bar (APC International, 2016). An input voltage is applied across the PZT bar, creating alternating stress and causing the bar to vibrate. The vibration frequency usually ranges from 100 kHz to 1 MHz. A higher output voltage is then generated across another section of the bar by the piezoelectric effect. Step ratios could be as high as 1000:1. These devices can be used in direct and alternative current inverters to drive cold cathode fluorescent lamps. Piezoelectric transformers are very compact but could be a source of high voltage.

12.4.3 Use of Piezoelectric Devices as Sensors

One may consider piezoelectric devices as alternatives to strain gauges. In many instances, strain gauges are technically piezoelectric devices. Common strain gauges transform the deformation of materials into forces as piezoelectric devices do. Piezoelectric devices can measure tensile and compressive forces in the same manner that strain gauges do. However, in piezoelectric applications, pressure changes are converted into energy and power by virtue of the sound wave energy produced. Many ultrasonic devices are related to this piezoelectric principle.

Detection of pressure variations due to sound is another common purpose of piezoelectric devices. Examples include the use of piezoelectric material in many microphones as well as for picking up sounds from acoustic-electric guitars. These devices are called contact microphones. The sound waves are brought about by vibrations of guitar strings creating changes in voltage and are amplified as sound waves for people to hear. In the medical industry, there are numerous uses of piezoelectric devices, such as in sono-probes and for detection of muscle movements and responses. If one needs a sensitive microbalance, piezoelectric devices are also used as sensors.

Finally, the automotive industry has started to use many piezoelectric devices as sensors for detecting engine behaviors such as the occurrence of "knocks." The devices are called "knock sensors" or detonation sensors. These sensors pick up a certain range of frequencies associated with the unique "pinging" sound when excess fuel is combusted in the internal combustion engine chamber. In some fuel injection systems, piezoelectric sensors measure the engine manifold pressure response after each combustion episode. This sensing system is referred to as manifold absolute pressure (MAP) profile. This pressure response is then converted to determine engine load at each combustion cycle related to the fuel injection timing. These events happen in a very small fraction of time, usually in milliseconds, and piezoelectric devices are very useful for this application (APC International, Ltd, 2015).

12.4.4 Piezoelectric Devices as Tiny Actuators

In the field of nanotechnology, piezoelectric devices are making great strides. Very high electric fields respond to precise changes in these devices, which provide accuracy in the micron range. Piezoelectric devices are being used as precision tools for positioning objects with impressive accuracy. In fact, most common stepper motors are slowly being replaced with piezoelectric motors due to the preferred level of accuracy. The applications are limited to extremely small distances and movements. Piezoelectric components are also being utilized in laser mirror alignments. On the same principle, piezoelectric devices are useful for positioning concentrated solar panels or solar collectors. They have the ability to move heavy solar devices and move them precisely over small distances or arcs for effective solar capture. In many sophisticated medical devices such as CT (computerized tomography) scanners or MRI (magnetic resonance imaging) scanners, piezoelectric elements are used because the electromagnetic signals (or noises) of other electric motors affect the output signals of these devices. There is a list of numerous other actuators that are being replaced by piezoelectric devices, ranging from loudspeakers, inkjet printers, atomic force microscopes or scanning tunneling microscopes, diesel engines fuel injectors, X-ray shutters, and crystal earpieces for low-power radios to high-intensity focused ultrasound for patients' bodies (Yoichi, 2006).

12.4.5 Piezoelectric Motors

There are various types of motors that are slowly being replaced by piezoelectric device counterparts. These are enumerated below:

a. Traveling wave motors used in auto-focus cameras

b. Inchworm motors for simple precise linear motors

c. Rectangular four-quadrant motors with high power density (2.5 W/cm^3) and speed ranging from 10 to 800 mm/s

d. Stepper piezomotors

The principle behind designing piezoelectric motors is the opposite of the piezoelectric effect discussed above. In piezoelectric motors, an electric field is applied to the piezoelectric material (instead of the material producing energy). One surface is usually fixed, causing the other one to move and turn like an electric motor. The piezoelectric signal is emitted by the sine wave of the alternating current (McMahon and Vorndran, 2010).

12.4.6 Potential Future Applications of Piezoelectricity

A potential emerging use of piezoelectric materials is in antennas, which can in turn generate electromagnetic radiation (Sinha and Gehan, 2015). On a larger-scale application, piezoelectric floors have been commercially developed, and harvesting kinetic energy from walking pedestrians has been investigated. In Tokyo (and Shibuya train stations), these piezoelectric devices have been installed to generate electricity from foot traffic used to run the automatic ticket gates and electronic display systems (Cafiso, et al., 2013). In London, these devices have been installed in many dance floors to power basic lighting and sound systems (Arjun, et al., 2011). In all of these applications, the power generation efficiencies are generally low (Li and Strezov, 2014) due to the low frequencies provided by foot traffic. Simulation studies have been made, and in one educational building application in Australia, some 1.1 MWh/yr of energy may be generated. The number may seem high, but this value represents about 0.5% of the annual energy needs of the building. We still categorize piezoelectric power under low-level energy applications.

12.5 Conclusions

In this chapter, we have discussed three important alternative energy sources—human, animal, and piezoelectric power. Animals and humans are fundamental sources, especially in developing countries. While their power output and energy is minimal, some countries cannot simply afford to invest in expensive machinery for agriculture and industry, nor is it ideal for them to automate certain farm tasks. We have seen that the most an animal could generate is a little over 1 horsepower, or 746 watts, and humans just a fraction of this value (about 10%). The collective use of animals as farm power is also not directly or proportionally related to the number of animals. As the number of collective animals are used, the harnessing factor decreases.

The major advantage of human beings is their intellect. A human can control and operate farm machinery more than 5,000 times its own power output. Humans can design unmanned machinery. However, as humans operate through the day, their power output declines. Their ability to maintain their optimal power output and efficiency decreases. Their judgment is also affected, and humans must be very careful when operating machinery. Expectedly, human arms have less power (though more dexterity) than human legs, and one will see numerous farm machinery systems operated with leg strength. Research on the use of human power focuses on reducing drudgery from a full day of work. Accidents have been found to happen when the human operator is fatigued during the latter part of the workday.

Another important power source discussed in this chapter is piezoelectricity. While the voltage output of these devices can be quite high, the amperage is low, and the cycle of operations must be maximized accordingly. In the coming years, multiple applications for piezoelectric devices will emerge. Some devices will be used for their high-voltage power, some as sensors for precision applications, and some as tiny actuators and motors. Some applications are sourced from continuous vibrations, which could provide a sustainable supply of small power. This chapter has shown that with further innovation, one may be able to derive power from vibrations made by walking or dancing to supply human electronic devices such as cell phones and laptops. These tiny devices may also be useful in increasing the efficiencies of solar photovoltaic cells.

12.6 Problems

12.6.1 Power from Animals

P12.1 Determine the amount of power (in units of ft-lbs/min) generated from an ox carrying a load of 120 lbs [54.5 kg] and walking at a rate of 2.5 feet/s [1.7 mph]. Convert this unit into horsepower and watts.

12.6.2 Power from Humans

P12.2 Determine the amount of power generated by a human being carrying a load of 25 lbs [11.4 kg] and walking at a rate of 2 feet/s [1.36 mph].

12.6.3 Various Units of Power

P12.3 Calculate the amount of power that can be generated by a 600 kg [1,320 lb] dairy cow pulling a load of 60 kg [132 lbs] and traveling at an average speed of 0.7 m/s [1.565 mph]. Provide your answers in kg-m/s, watts, and horsepower.

12.6.4 Power from Groups of Animals

P12.4 If a small horse can provide a draft of 80 kg [176 lbs], how much can be provided by 5 small horses together? Use the harnessing factor for work animals in Table 12.4 for your calculations. Report your units in kg and in Newtons. If the animals' speed is 2 m/s [4.5 mph], determine the power generated by the group.

12.6.5 Energy Output of a Cow in the Form of Milk

P12.5 Determine the energy in MJ produced by a cow from Wisconsin in the form of milk. Assume the energy content of milk is 150 kcal [594 Btu] per 250 g [0.55 kg] and 72,000 lbs [32,727.3 kg] was produced in a year. If the same cow produces an average of 7 lbs [3.2 kg] of dry manure each day, or 2,555 lbs [1,161.4 kg] in a year, with a heating value of 6,000 Btu/lb [13.93 MJ/kg], how much energy in MJ is likewise produced? How much total energy is produced by this cow in MJ?

12.6.6 Power of Humans over Longer Periods of Time

P12.6 A person is asked to do work for 8 hours. Estimate the power that can be generated by this person in units of horsepower (hp) and watts. Compare this value with the results of Example 12.6.

12.6.7 Power from Arms and Legs of Humans

P12.7 Compare a person's power output if he works for 10 hours a day using his arms or his legs (i.e., use Equations 12.3 and 12.4). Comment on the power output in the results of calculations. Provide results in hp and watts.

12.6.8 Basic Piezoelectric Power from Numerous Repeated Cycles

P12.8 Determine the energy produced for numerous cycles by a piezoelectric element. The measured voltage across the capacitor after a series of cycles was found to be 10 volts, and the capacitor has a rating of 220 microfarad and 35 V capacitor. Provide your answer in Joule.

12.6.9 Piezoelectric Power from Single Tap

P12.9 Determine the energy produced for each single tap on a piezoelectric element. The initial voltage was found to be 2.0 volts, and after each tap, the final voltage was found to be 3.5 volts. The capacitor has a rating of 220 microfarad and 35 V. Provide your answer in Joule.

12.6.10 Charging a Cell Phone with Piezoelectric Power

P12.10   Determine how long it would take to charge a cell phone battery with a fully charged load requirement of 15,000 Joules [14.22 Btu]. The voltage across the capacitor was improved from 2 to 6 volts with a single cycle. Assuming this is done continuously in a day, in how many days can one achieve full charge? The capacitor has a rating of 220 microfarad and 35 V. One can generate 3 cycles a second on this device.

References

APC International Ltd. 2015. Top uses of piezoelectricity in everyday applications. APC International Ltd. Knowledge Center, Piezo Applications. Available at: https://www.americanpiezo.com/blog/top-uses-of-piezoelectricity-in-everyday-applications/. Accessed April 26, 2018.

APC International Ltd. 2016. What is PZT? APC International Ltd. Knowledge Center, Piezo Theory. Available at: https://www.americanpiezo.com/piezo-theory/pzt.html. Accessed April 26, 2018.

Arjun, A., A. Sampath, S. Thiyagarajan and V. Arvind. 2012. A novel approach to recycle energy using piezoelectric crystals. International Journal of Environmental Science and Development 2: 488–492. doi:10.7763/IJESD.2012.V2.175.

ASAE Standards D497.7. 2015. Agricultural Machinery Management Data. ASAE D497.7 MAR2011 (R2015). American Society of Agricultural and Biological Engineers, St. Joseph, MI.

Beijing Ultrasonic. 2012. Principles and applications of piezoceramics. Available at: https://www.bjultrasonic.com/ultrasonic-technical-info/piezoelectric-ceramic-technical/. Accessed April 26, 2018.

Berolo, S., R. P. Wells and B. C. Amick III. 2012. Muscoskeletal symptoms among mobile hand-held device users and their relationship to device use. A preliminary study in a Canadian university. Applied Ergonomics 42: 371–378.

Cafiso, S., M. Cuomo, A. Di Graziano and C. Vecchio. 2013. Experimental analysis for piezoelectric transducers applications into roads pavements. Advanced Materials Research 684: 253–257. doi:10.4028/www.scientific.net/AMR.684.253.

Campbell, J. K. 1990. Dibble Sticks, Donkeys and Diesels: Machines in Crop Production. International Rice Research Institute, Manila, Philippines.

Dairy Moos. 2017. How much milk do cows give? Dairy Moos Dailo Blog: The Blog of a Third-Generation CA Dairy Farmer. August 4, 2013. Available at: http://www.dairymoos.com/how-much-milk-do-cows-give/. Accessed April 26, 2018.

Encyclopedia Britannica. 2018. Horsepower. Available at: https://www.britannica.com/science/horsepower. Accessed April 26, 2018.

Fuller, R. J. and L. Aye. 2012. Human and animal power—The forgotten renewables. Renewable Energy Volume 48: 326–332.

Gupta, A., M. Imran, R. Agarwal, R. Yadav, P. Jangir, and R. Poonia. 2016. Energy harvesting through dance floor using piezoelectric device. International Journal of Engineering and Management Research. 6 (Issue 2): 36–39. Vandana Publications.

Gustafson, E., S. Thomee, A. Grimby-Ekman and M. Hagberg. 2017. Texting on mobile phones and musculoskeletal disorders in young adults: A five-year cohort study. Applied Ergonomics 58: 208–214.

Irfan, U. 2012. Scientists harness human power for electricity. Scientific American Newsletter. June 26. Available at: https://www.scientificamerican.com/article/scientists-harness-human-power-electricity/. Accessed April 26, 2018.

Jensen, H. 1986. Calculations for Piezoelectric Ultrasonic Transducers. Riso National Laboratory, Roskilde, Denmark.

Lebrun, K. 2013. DARPA: Energy harvesting. Nanotechnology Online Newsletter. May 28. Available at: https://revolution-green.com/darpa-energy-harvesting-using-rectenna-arrays-update/. Accessed April 26, 2018.

Lee, B. Y., J. Zhang, C. Zueger, W-J. Chung, S. Y. Yoo, E. Wang, J. Meyer, R. Ramesh and S-W. Lee. 2012. Virus-based piezoelectric energy generation. Nature Nanotechnology. May 13. Available at: https://www.nature.com/articles/nnano.2012.69.pdf. Accessed April 26, 2018. doi:10.1038/NNANO.2012.69.

Li, X. and V. Strezov. 2014. Modelling piezoelectric energy harvesting potential in an educational building. Energy Conversion and Management. 85: 435–442. doi:10.1016/j.enconman.2014.05.096.

Lovett, T. 2007. Draft animal power. Animal Power Online Newsletter. June. Available at: http://worldwideflood.com/ark/technology/animal_power.htm. Accessed April 26, 2018.

Manbachi, A. and R. S. C. Cobbold. 2012. Development and application of piezoelectric materials for ultrasound generation and detection. Ultrasound 19 (4): 187–196. doi:10.1258/ult.2012.011027.

McMahon, J. and S. Vorndran. 2010. Piezo motors and actuators: Medical device performance. Motors and Motion Control Newsletter. August 6. MD+DI Medical Device and Diagnostic Industry Newsletter. Available at: https://www.mddionline.com/piezo-motors-and-actuators-medical-device-performance. Accessed April 26, 2018.

Mother Earth News. 2018. Human-powered machines resources list: Pedal to the metal. Mother Earth News: The Original Guide to Living Wisely. Available at: https://www.motherearthnews.com/renewable-energy/other-renewables/human-powered-machine-zl0z1211zrob. Accessed April 26, 2018.

Sinha, D. and G. Amaratunga. 2015. Electromagnetic radiation under explicit symmetry breaking. Physical Review Letters 114 (14): 147701. doi:10.1103/physrevlett.114.147701.

Yoichi, M. 2006. Applications of piezoelectric actuator. Piezoelectric Devices. NEC Technical Journal 1 (5): 82–86.

13

Cold Fusion and Gravitational Energy

Learning Objectives

Upon completion of this chapter, one should be able to:

1. Describe the concept of generating energy and power from cold fusion and gravitational field energy technologies.
2. Estimate available energy and power from cold fusion and related energy systems.
3. Enumerate important researchers and studies on cold fusion and gravitational field energy.
4. Describe the various conversion efficiencies in cold fusion and gravitational energy systems.
5. Relate the overall issues concerning cold fusion and gravitation as well as barriers against its adoption.

13.1 Introduction

Cold fusion and gravitational energy are two sources worth considering with an open mind despite their controversial standing in the scientific community. These two technologies appear to go against the laws of physics and thermodynamics and seem opposed to what we learn in basic science courses. Solar energy results from extremely hot fusion of two nuclei of deuterium, an isotope of hydrogen. This fusion event and energy release occurs every instant on the surface of the sun and is called solar radiation. Even if this event happens at a relatively "colder" state, the release of energy still occurs. Remember that the effective temperature of the surface of the sun is about over 5,000K [9,000°R] and the fusion reaction may happen very easily. If we replicate this event on earth, we call it cold fusion due to the lower relative temperatures in which fusion is achieved. The earth's ocean is abundant with deuterium. The overall percentage of deuterium may be quite low (0.0156% or 0.0312% on a mass basis), but with the vast ocean resources, one could find an area where deuterium accumulates in abundance (Horibe and Ogura, 1968). Harold Clayton Urey (a chemist at Columbia University) is credited with discovering deuterium. He earned a Nobel Peace Prize in 1934 when his group produced samples of "heavy water" in which deuterium had been highly concentrated (Khan, 2009). It would be promising if deuterium could undergo fusion reaction at cold states and thereby produce an endless supply of energy.

It was in 1989 when two scientists from the University of Utah, Martin Fleischmann and Stanley Pons, reported that their apparatus had produced anomalous heat (called "excess heat") of magnitude they asserted would defy explanation except in terms of nuclear fusion processes conducted at relatively low temperature (Fleischmann, et al., 1989). The scientific community was abuzz following this claim, and numerous groups of researchers tried to replicate their device, to no success. Despite the initial excitement and after months of rigorous vetting, the scientific community considered the results erroneous. Still, numerous scientists continue the pursuit of this angle, and it is worth discussing for the purpose of inquiry and advancement of science.

Another possible resource of note is gravitational field energy. The impetus for further research into this topic was a book written by German physician Hans A. Nieper in the early 1980s. The book compiled the proceedings of an energy meeting held in Hannover, Germany, in November 1980 where scientists from all around the world gathered to present work related to gravitational field energy, also called tachyon field energy (Nieper, 1983). The book details discussion of an innovative automobile designed by Nikola Tesla. Tesla reported to have manufactured an "energy receiver"—an antenna that replaced the engine of a Pierce Arrow automobile as an energy source. The motor achieved a maximum of 1,800 rpm, and Tesla claimed that the unit would generate enough power to illuminate an entire house. Tesla tested the car for a week, reaching top speeds of 90 miles per hour [144.8 kph], with performance data similar to that of a gasoline-powered automobile. Purportedly, there were no gases exiting the exhaust pipe, and Tesla attributed this to the lack of an engine (Nieper, 1983). Tesla knew that this invention would contradict the precepts of that time and avoided discussions surrounding the technology. Unfortunately, the production of this vehicle model was stopped due to economic pressures of the time. Today, there are a handful of scientists working in this field, and it is again worth discussing with the hope that future scientists will apply the correct science behind these events and prove some outstanding theories.

13.2 The Cold Fusion Theory

The cold fusion theory at this point is of course still a theory, and numerous research groups use different designations for their research related to this topic, perhaps using different key words. The concept is similar—that is, to conduct nuclear reactions at or near room temperature. These types of reactions are "low-energy nuclear reactions" (LENR) or "condensed matter nuclear science" (CMNS) (Biberian, 2007).

Figure 13.1 shows a schematic of the original Pons and Fleischmann experiment. The silver coating shown in the diagram prevents radiation losses. In the system shown, deuterium and oxygen gas are lost during electrolysis because it is an open system. The cathode is made of palladium and the anode of platinum. They place the whole setup in a constant temperature bath. The system includes a calorimeter that measures energy generated by the reactions in the form of heat. The heat energy equation was presented in the previous chapter and is reutilized here in Equation 13.1. The medium in this experiment is heavy water (apparently containing deuterium) electrolytes with 0.1 M LiOD:

$$Q(kJ) = mC_p\Delta t \qquad (13.1)$$

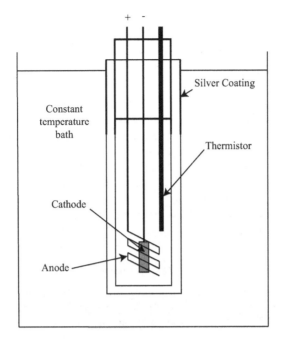

FIGURE 13.1
Schematic diagram of the original experiment by Pons and Fleischmann (Fleischmann, et al., 1989).

where

Q = energy absorbed by the bath (kJ)

m = mass of the fluid in which the setup is immersed (kg)

C_p = specific heat of liquid (kJ/kg°C)

Δt = temperature difference (°C)

In the setup, Equation 13.2 shows the input power calculation using voltage and current:

$$Power(W) = E \times I \tag{13.2}$$

where

Power = input power to the electrolysis setup (watts)

E = measured electrical voltage (volts)

I = measured electrical current (amps)

The product of the power and duration of the experiment is the total energy given to the setup using the input voltage. Equation 13.3 shows this calculation for energy. Note that one must use the consistent units to estimate efficiencies, either in power or in energy units:

$$Energy\,(kJ) = 3.6 \times Power\,(Watts) \times Time\,(hrs) \tag{13.3}$$

where

Energy = energy input to the electrolysis setup (kJ)

Power = power supplied to the setup (kW)

Time = duration of experiment (hours)

Example 13.1 shows how to calculate the energy supplied to the setup and compare it with the energy absorbed by the medium. One would use the equations presented above to conduct energy balancing. In the example, there is more energy in the output than the input—a clear violation of the laws of conservation of energy. Data from the example were taken directly (but not all) from the paper of Pons and Fleischmann (Fleischmann, et al., 1989). Note that with very limited data, one would not be able to evaluate where the excess energy is coming from. In this case, there are more questions generated than scientific answers. Are other system masses being converted into additional energy? Was the input energy constant, or did it vary within the time of the experiment?

Example 13.1: Determining Energy Balance in Electrolysis

Determine the difference between the energy provided to the electrolysis setup similar to that shown in Figure 13.1 from the energy generated by the medium using Equations 11.1 to 11.3. The measured data are as follows:

Voltage = 5.201 volts
Amperage = 0.256 amps
Duration of experiment = 30 minutes
Rise in temperature = 0.42°C
Amount of medium = 1.5 kg [3.3 lbs]
Specific heat of medium = 4.2 kJ/kg°C [1.00315 Btu/lb°F]

SOLUTION:

a. From the input voltage and amperage data, the power supplied to the setup using Equation 13.2 is found:

$$Power(W) = E \times I = 5.201 \ Volts \times 0.256 \ Amperes = 1.33 \ Watts$$

b. The energy used during the experiment is calculated using Equation 11.3 as shown:

$$Energy(kJ) = 3.6 \times 1.33 \ Watts \times 0.50 \ hrs = 2.394 \ kJ$$

c. The energy absorbed is calculated using Equation 13.1 as shown:

$$Q(kJ) = mC_p\Delta t = 1.5 \ kg \times 4.2 \frac{kJ}{kg°C} \times 0.42°C = 2.646 \ kJ$$

d. The excess heat is the difference between the heat absorbed by the medium and the heat input to the setup as follows:

$$Energy \ Gain = Energy \ Generated - Energy \ Supplied$$
$$Energy \ gain = 2.646 \ kJ - 2.394 \ kJ = 0.252 \ kJ$$

e. Clearly, there is excess in energy absorbed by the medium. This represents about 9.5% excess energy. Hence, the value is quite debatable considering other factors beyond the control of the scientist.

In Example 13.1, there is about 10% excess energy from this experiment. The scientist may call this an anomaly, though there may be other unaccounted factors during the experiments. Such artifacts may cause the rise in overall temperature. The type of setup is an open calorimetric experiment, and as such, other external factors come into play. After more than 25 years of work by numerous curious scientists, several proofs have continued to demonstrate excess energy produced by using different types of cathodes and anodes. At present, no pilot or commercial-scale systems are available in the market. There have not been any recent notable breakthroughs. This topic is still in the theoretical stage until complete energy and mass balances are established and losses are more precisely accounted for. The science behind calorimetry and measuring equipment has advanced significantly in the past 30 years. There ought to be retrials done with the advanced measuring instruments of today. Unfortunately, because of the negative publicity surrounding it, no further funding has been devoted to cold fusion research. Fleischmann gave up his citizenship and moved back to France, where he was joined by his American partner Stanley Pons.

13.3 Calorimetry

The errors associated with the increase in temperature in vessels wherein fusion occurs are studied in calorimetry (the process of measuring the heat of chemical reactions or observing physical changes including heat capacity). There are now more sophisticated differential scanning calorimeters that do a good job of measuring the increase in energy due to fusion reactions. The most common device for this phenomenon is the bomb calorimeter. Bomb calorimeters are used to report on heating values of solid biomass materials or liquid fuels. Manufacturers of calorimeters base their measurements on heat that may be generated (exothermic processes), consumed (endothermic processes), or simply dissipated by the sample (Parr, 2016). Temperature measurements are the most common. However, with much advancement in electronics and control systems, many state-of-the-art calorimeters are now designed for conditions that would have been highly impractical several decades ago. Newer cold fusion experiments must take advantage of such technological advancements in order to prove minimal instrumentation errors. For example, Fleischmann and Pons would have had more crude measurements of temperature, and their data would have unsurprisingly been prone to errors. In such experiments, there should be prior calculations to establish the heat capacity of the experimental vessels, ensuring minimal errors in data taking. Heat capacity is defined as the amount of heat required to raise the temperature of the entire calorimeter by 1 Kelvin and is usually determined experimentally before or after the actual measurements of the heat of reaction (Cengel and Boles, 1989). The temperature differences are small; hence, extremely sensitive thermal measuring instruments must be used. The correct protocol for the characterization of a calorimeter is illustrated in Examples 13.2 to 13.4.

Example 13.2: Estimating Heat Capacity of a Calorimeter

A 100 watt heater was used to heat a bomb calorimeter. The temperature rise was reported to be 30K in 5 minutes. Calculate the heat capacity of the calorimeter in units of J/K.

SOLUTION:

a. The energy utilized is first calculated:

$$Energy(Joule) = Power(W) \times Time(min) \times \frac{60\ s}{min}$$

$$Energy(Joule) = 100\ Watts \times 5\ min \times \frac{60\ s}{min} \times \frac{J}{W-s} = 30,000\ J$$

b. The heat capacity is simply the energy used divided by the rise in temperature:

$$Heat\ Capacity\left(\frac{J}{K}\right) = \frac{30,000\ J}{30\ K} = 1,000\ \frac{J}{K}$$

c. If water is used as a medium to absorb the heat, the following calculations are also correct. Note that the specific heat of water is 4.184 kJ/kg-K and one mole of water weighs 18 g, so that the heat capacity of water is also equivalent to 75.312 J/K mol as shown below:

$$Water\ Heat\ Capacity\left(\frac{J}{Kmol}\right) = \frac{4.184\ kJ}{kg \times K} \times \frac{18\ g}{mol} \times \frac{1\ kg}{1,000\ g} \times \frac{1000\ J}{kJ} = 75.312\ \frac{J}{Kmol}$$

d. One can actually determine the number of moles of water used in the calorimeter as shown below:

$$\#\ Moles\ Water\,(mol) = \frac{1,000\ J}{K} \times \frac{Kmol}{75.312\ J} = 13.3\ mol\ water$$

Example 13.3: Heat Released by a Combustible Compound in a Calorimeter

A calorimeter with a heat capacity of 1,000 J/K was used to measure the heat of combustion from half a gram of sugar with a molecular formula of $C_{12}H_{22}O_{11}$. The temperature increase was found to be 8.25K. Calculate the heat released in units of Joules and the amount of heat released by half a gram of sugar in units of kJ/mol.

SOLUTION:

a. The heat released in Joules is calculated with the equation:

$$Heat\ Released(J) = 8.25\ K \times \frac{1,000\ J}{K} = 8,250\ J$$

b. The amount of heat released by half a gram of sugar is calculated as follows:

$$Heat\ Released\left(\frac{kJ}{kg}\right) = \frac{8,250\ J}{0.5\ g} = 16,500\ \frac{J}{g} \times \frac{kJ}{1,000\ J} = 16.5\ \frac{kJ}{g}$$

c. The molecular weight of sugar is found to be $(12 \times 12 + 22 \times 1 + 11 \times 16 = 342 \text{ g/mol})$ 342 g/mol. The heat released by half a gram of sugar in units of kJ/mol is calculated as follows:

$$Heat\ Released\left(\frac{kJ}{kg}\right) = \frac{16.5\ kJ}{g} \times \frac{342\ g}{mol} = 5,643\,\frac{kJ}{mol}$$

Example 13.4: Estimating the Enthalpy of Combustion of Sugars

The heat released by a mole of sugar with a chemical formula $C_{12}H_{22}O_{11}$ was reported to be 5,643 kJ/mol. Determine the enthalpy of combustion per mole of the sugar in air (i.e., 21% O_2 and 79% N_2).

SOLUTION:

a. The balanced chemical reaction is shown below:

$$C_{12}H_{22}O_{11} + 12 \times \left[(O_2 + 3.76N_2)\right] \rightarrow 12CO_2 + 11H_2 + 12 \times 3.76N_2$$

b. Since the total number of moles of gases in the product (i.e., 12) is the same in the reactant, the change in enthalpy is equal to the internal energy generated, which is equal to 5,643 kJ/mol.

c. When a person ingests this sugar, the body will be able to use this much energy and convert it into useful work.

In all of the above examples, complete energy and mass balances must be established and the system 'closed' on where the input energy is coming from and output energy produced from the chemical reactions. Bomb calorimeters are known as devices that measure the heating value of solid and liquid chemicals.

13.4 Cold Fusion by Other Names

In general, cold fusion describes the form of energy generated when hydrogen gas interacts with any kind of metal, such as palladium or nickel and a few others, and this specialty falls under the field of condensed matter nuclear science (CMNS) (Biberian, 2007). In recent years, cold fusion research trials have also been called low-energy nuclear reactions (LENR) (Nagel, et al., 2005). This new terminology used to refer to cold fusion research helps veer focus away from the negative connotations the term "cold fusion" carried from the late 1980s. In some research publications, cold fusion is also called anomalous heat effect (or AHE) (Takahashi, 2016). This is likely what the pioneering researchers Fleischmann and Pons reported in their early findings. That is, excess heat can easily be turned into low-grade power or used for low-energy required processes.

Today there are numerous companies involved with research on new energy technologies, and some are offshoots of the cold fusion fever in the 1980s. For example, the company Brilliant Light Power (PLP, Cranbury, NJ), founded by Randell L. Mills, reported to have discovered a new energy source called "hydrinos." These hydrinos are electrons in a hydrogen atom that can drop below the lowest energy state known as the ground state, which many other researchers have believed does not exist (Ritter, 2016). This new concept was incompatible with mainstream quantum mechanics, and many people simply could not accept it. Some researchers argue that the data were flawed due to outdated equipment that did not have the capability to detect the lowest energy level.

Scientists continue to conduct studies related to this unusual anomaly in the generation of heat when two atoms are fused together. There are books written on this topic as well, mainly in the area of nuclear research. These researchers continue to believe that this energy source could one day change our perception of renewable energy for the growing world population. In the next section, we will discuss a few other key players in the scientific field who may be crucial to the development of cold fusion as an alternative renewable energy source.

13.5 Key Figures in Fusion Energy Research

While Martin Fleischmann and Stanley Pons arguably initiated research into cold fusion in 1989, numerous other researchers around the world followed and continued to conduct further work in this area. Many are still making notable attempts which are circulated in scientific communities around the internet. In 1992, Fleischmann and Pons moved to France to continue their work at the laboratory of Technova Corporation, which is a subsidiary of Toyota (Petit, 2009). In 1995, Fleischmann retired and returned to his native England. He has continued to co-author numerous other papers on the topic. In March 2006, a new company called D2Fusion, Inc., based in Silicon Valley, California, announced in a press conference that Fleischmann, then 79 years old, would be acting as their senior scientific adviser (Park, 2006). However, Fleischmann died of natural causes at his home in Tisbury, Wiltshire, England, on August 3, 2012. Dr. Pons remained in France even after the Toyota-related company closed down in 1998. Research has since shifted from fusion toward neutron-induced reactions and nuclear research.

13.5.1 Randell L. Mills, Brilliant Light Power, New Jersey

Howard J. Wilk, a synthetic organic chemist, long-term unemployed due to downsizing of the pharmaceutical industry and living in Philadelphia, has been tracking the progress of the New Jersey–based company called Brilliant Power (BLP). BLP is developing new energy technologies that are closely related to cold fusion methods. In 1991, BLP's founder, Randell L. Mills, announced at a press conference in Lancaster, Pennsylvania, that his company devised a theory in which the electron in hydrogen could transition from its normal ground energy state to previously unknown lower and more stable states, liberating an abundant supply of energy in the process. It is not fusion, but it is a type of nuclear reaction. Wilk studied Mills's theory, and from this the concept of "hydrino" emerged (Mills, 2016). Cold fusion reactions were rebranded as low-energy nuclear reactions (or LENR), and Wilk became invested in this new energy source. Mills predicted that they would have a commercial version of their technology on the market in late 2017 (Casual Chemist, 2017). However, there have been no further developments since, and we must wait to see if this technology will revolutionize sustainable energy.

13.5.2 Michael McKubre, Energy Research Center, SRI International

Michael McKubre, PhD, a native of New Zealand, is an electrochemist who is another key figure in the development of cold fusion energy. He was the director of the Energy Research Center at SRI International in 1998 (Menlo Park, California). He conducted research in cold fusion from 1989 to 2002 at SRI International. In 2004, he and his colleagues at SRI asked the U.S. Department of Energy (USDOE) to provide a review on the recent achievements of his group in the area of cold fusion. The 2004 review concluded similar inconsistent or unvalidated results reported in 1989 despite further sophistication in the use of advanced calorimeters (USDOE, 2004). McKubre began to call his results low-energy nuclear reactions. The reviewers identified a number of basic science research areas that could be helpful in resolving some controversial issues in the field. These are enumerated below:

a. Improvement in material science aspects of deuterated metals using modern characterization techniques
b. Study of particles reportedly emitted from deuterated foils using state-of-the-art pieces of apparatus and methods

The USDOE solicited the help of 18 individual reviewers, and they came up with the same conclusions—recommending material science improvement and suggesting additional equipment for future research directions. We can hope that the USDOE will begin to provide seed funding to these new types of related technologies.

13.5.3 David J. Nagel, George Washington University

David J. Nagel is an electrical and computer engineering professor at George Washington State University. He was a research manager in naval research. He prefers to call the LENR phenomenon "lattice-enabled nuclear reactions," primarily because the reactions take place within the crystal lattice of an electrode (Nagel, 2015). He compared this technology to that which has been used in southern France—the International Thermonuclear Experimental Reactor (ITER). ITER actually means "the way" in Latin, and this significance will be discussed later. Nagel believes that the LENR technology will continue to grow internationally. The biggest hurdles continue to be inconsistent results and lack of funding. For the LENR technology, many research reports suggest that numerous thresholds must be reached and a minimum amount of deuterium or hydrogen for the reaction to proceed (Nagel and Swanson, 2015). The electrode materials may need to be prepared with specific crystallography and surface morphology to be able to react. This is a common problem with heterogeneous catalysts used in the refining industry and the production of transport fuels. Funding then becomes a big issue with inconsistencies in technology and research.

The commercialization of this type of technology can also be problematic, primarily with the development of advanced prototypes. Researchers like Nagel are still struggling with the lack of support from funding agencies. There have been limited LENR-based companies formed in recent years.

13.5.4 Rossi's E-Cat

A company based in Miami, Florida, has attempted to commercialize LENR technologies. In 2011, engineer Andrea Rossi of Leonardo Corporation in Miami and his colleagues announced at a press conference in Bologna, Italy, that they had built a tabletop reactor called the Energy Catalyzer (E-Cat). E-Cat was said to produce excess energy via a nickel-catalyzed process. They held demonstrations of the technology for potential investors. E-Cat apparently features a self-sustaining process in which electrical power input initiates the fusion of hydrogen and lithium from a powdery mixture of nickel, lithium, and lithium aluminum hydride to form a beryllium isotope. The short-lived beryllium decays into two α-particles with excess energy given off as heat. In their reports, some of the nickel is claimed to turn into copper. The company claimed that no waste is produced during the process and no radiation is detected outside of the apparatus. However, some authors have challenged Rossi's invention as fraudulent (Krivit, 2016).

Rossi had a contract with the U.S. Army for the delivery of heat-generating devices, but apparently the delivered units did not perform as expected. In 2012, Rossi announced the completion of a 1 MW system that could be used to heat or power large buildings. The company anticipated in earlier reports that by 2013, the company would be producing 1 million 10 kW [13.4 hp] household units about the size of a laptop computer (Ritter, 2016). However, these units never materialized.

In 2014, Rossi worked with a company called Industrial Heat, licensed his technology, and formed a private company called Cherokee. Unfortunately, things went sour, and the plan to fund $100 million to build a 1 MW system did not materialize. Today, both parties are in the process of suing each other in court due to apparent violations in either party's agreement. At present, there is no word about whether the 1 MW system will ever be built (Ritter, 2016).

13.5.5 International Thermonuclear Experimental Reactor

The International Thermonuclear Experimental Reactor (ITER) is perhaps one of the most ambitious energy projects in the world today (Clercq, 2018). Thirty-five nations have collaborated to build the world's largest tokamak, a magnetic fusion device that has been designed to prove the feasibility of fusion as a large-scale and carbon-free source of energy, based on the same principle that powers our sun and stars. The United States is also part of this 35-nation collaboration. The term "tokamak" is a Russian acronym that stands for "toroidal chamber with magnetic coils." The experimental machine designed to harness the energy of fusion is located in the south of France. Inside the tokamak is a doughnut-shaped vacuum chamber where fusion reactions occur. Hydrogen is the fuel gas. The reactor goes under extreme pressure and temperature, and the gaseous hydrogen fuel becomes a plasma. Under these conditions, hydrogen atoms can be brought to fuse and yield energy. The energy produced through the fusion of atoms is absorbed as heat in the walls of the vessel. Outside of the vessel is nothing more than a conventional power plant where heat is used to produce steam and eventually electricity by way of turbines and generators. These fusion reactions then closely simulate what happens at the sun's surface and core to generate the solar energy that we receive each day. ITER claims to be the world's largest tokamak. It is twice the size of the largest machine currently in operation, with 10 times the plasma chamber volume (ITER, 2016).

The key membership countries of ITER include the European Union, China, India, Japan, Korea, Russia, and the United States. ITER's "First Plasma" is scheduled to be powered on in December 2025 (Clery, 2015). The deuterium-tritium operation is scheduled to begin in 2035 (ITER, 2016). These milestones could change the way the world generates power. The original ITER system relies on hydrogen as a fuel. From where to generate hydrogen sustainably is a topic in itself that requires more research.

13.6 The Gravitational Power Potential

Gravitational field energy is another important source—in theory, similar to cold fusion, in which more energy is generated than applied—in apparent violation of our knowledge of conservation of energy. It is interesting to note that at around the same time as Tesla performed his experiments, highly educated American physicist Dr. Henry T. Moray was able to draw off continuous power of up to 70 kW from a box the size of a wine crate because of this gravitational field energy. Likewise, another American physicist, Bruce De Palma, built a prototype called N-machine to take advantage of this unique energy source (Nieper, 1983). The Faraday disc is the basic operating principle behind the N-machine. This device was also called the unipolar generator. If one rotates a ring magnet very quickly, one can derive an electrical current from the external periphery of the magnet. In addition, the system changes its gravitational properties. It could gravitate in an optimized form. We have known this phenomenon for many years with respect to the Faraday disc. However, a new concept is the claim that the energy set free by such a rotating magnet can become greater than that which is required to drive them. This is where skeptics do not agree, as it is in violation of our laws of conservation of mass and energy.

N-machines produce very low voltage and very high current (Nieper, 1983). The cross section of the output cable must be large enough to lower the resistance as much as possible. The method of tapping the current from the periphery of the magnet is a considerable problem. Usually this is done by brushes, as normal electric motors would have. However, because of high current, a very durable material must be used. Today, the best material is a very sturdy graphite. Because of additional heat being generated, this graphite must be submerged in good conducting material, and the best would be the use of mercury bath technology. Today, nobody would want to use mercury because of its toxicity. Perhaps as new materials are developed, one can overcome these issues. Moreover, at very high revolutions, the magnets may easily burst. Other than these material limitations, the technology should have great potential in the future (De Palma, 1990). Likewise, in the example, there is more energy produced than initially input to the device. Because of this, there may be missing data, which could account for the excess energy. Example 13.5 shows actual experimental data for the N-machine, showing the input and output power for the unit. The unit has an electric motor on the front end that drives an electromagnet. The electromagnet generates electrical power. Sophisticated instrumentation measures the input and output power from the electric motor and the generator, respectively. The example provided shows actual reported data. The example calculations show that the N-generator has very low drag (De Palma, 1990). The output power is close to input power, a surprising violation of practical friction loss in most conventional power generation systems.

Example 13.5: Calculating Energy Input and Output in an N-Machine

The drive motor for the N-machine is a standard 40 hp [29.84 kW], 440 VAC, 60 Hz, three-phase unit. The motor speed is 3,485 rpm with maximum draw rated at 50.5 amperes at full output. A double drive belt connects the motor to the N-machine. The motor pulley diameter is 9⅞ inches, and the generator pulley diameter is 5¾ inches. The step up ratio is then 1.7174 times the motor speed to rotate the generator at the design speed of 6,000 rpm.

After attaining the operation speed, the drive motor current was measured to be 15 amps at 440 VAC with a power correction of 1.7174 (ratio of diameters) and a power factor of 70% (i.e., 0.70). This is the input energy to the N-machine generator. The output was as follows:

Machine speed	= 6,000 rpm
Drive motor current at no load	= 15 amperes
Drive motor current increase when N-machine is loaded	= ½ ampere
Voltage output of N-generator when loaded	= 1.05 VDC
Current output of the N-generator	= 7,200 amperes
Voltage output of the N-machine at no load	= 1.5 VDC
Current output at no load	= 23,000 amps

Determine the input energy and power output of the N-machine when loaded and at no load.

SOLUTION:

a. Calculate the input energy in watts and in hp.
The input energy is shown below:

$$P_{in} = 15 \ Amps \times 440 \ V \times 0.70 \times 1.73 = 7,992.6 \ W \ [10.71 \ hp]$$

b. Calculate the power output of the N-generator when loaded using the measured voltage and current. Compare this with the measured input power.
When loaded, the power output is as follows:

$$P_{out} = 7,200 \ Amps \times 1.05 \ V = 7,560 \ W \ [10.13 \ hp]$$

Note that the power output is nearly the input value.

c. Calculate the power output of the N-generator at no load using the measured voltage and current. Compare this with the measured input power.
When not loaded, the power output is as follows:

$$P_{in} = 23,000 \ Amps \times 1.5 \ V = 34,500 \ W \ [46.2 \ hp]$$

Note that the power output exceeds power input, a curious violation of our basic energy conservation law.

Nikola Tesla became famous not just because of his automobile and the gravitational field energy source of his automobile's power. He became well known because of his contributions to modern alternating current (AC) electricity supply systems. In 1888, Westinghouse licensed Tesla's AC induction motor and related polyphaser AC. This invention became the cornerstone of the Westinghouse market. This source of income allowed him to develop numerous other inventions and gadgets. He knew very well that some of his inventions would be controversial and did not share much information with engineers and scientists. He seemed to do this type of work out of the impulse to tinker and create (Nieper, 1983).

In 1960, the General Conference on Weights and Measures named the SI unit of magnetic flux density after Tesla to honor him. The Tesla (symbol T) is a unit of magnetic induction or magnetic flux density in SI units. One Tesla is equal to 1 Weber per square meter. Weber units apply to the amount of flux that produce an electromotive force of 1 volt. Hence, 1 Tesla represents 1 kilogram per second per square second per ampere (kg/s²A). In practice, Tesla is a large unit and used primarily for large industrial electromagnetics. For practical purposes, a smaller unit of flux density is used. One would then use the unit gauss (G)—there are 10,000 G in 1 Tesla (1T = 10^4 G) (Encyclopedia Britannica, 2015). We use Equation 13.4 to relate Tesla to other units in the SI system.

$$T = \frac{kg}{A \times s^2} = \frac{V \times s}{m^2} = \frac{N}{A \times m} = \frac{Wb}{m^2} = \frac{kg}{C \times s} = \frac{N \times s}{C \times m} \tag{13.4}$$

where
  A = ampere
  C = coulomb
  kg = kilogram
  m = meter
  N = Newton
  s = second
  T = Tesla
  V = volt
  Wb = Weber

13.7 Tachyon Field Energy

Another hypothetical source is tachyon field energy or simply tachyon. Tachyon is defined as a quantum field with an imaginary mass. Tachyon particles are particles that supposedly move faster than light—an unrealized hypothetical even at this time. Gerald Feinberg, a Columbia University physicist, coined the word "tachyon" (Lee, 1993). Feinberg predicted the existence of the muon neutrino, an elementary subatomic particle with no net electric charge. Some parapsychologists advocated his concept of tachyon to explain precognition or psychokinesis. However, there is yet no scientific evidence that they truly exist. These concepts are simply pseudoscientific. In his book, Nieper relates the tachyon field to the Feinberg field and connects it to the concept of gravitational energy discussed earlier. He posted the simple question—why did the object dropped by Galileo at the Leaning Tower of Pisa have an acceleration due to gravity of 9.81 m/s² [32.2 ft/s²]? This is known and accepted today as gravitational acceleration, but Nieper believed that this way of measurement was confusing and was convinced that this gravitational field acceleration is not really an "attraction" phenomenon. Rather, it is generated by a push- or pressure-based mechanism that would have to be analyzed in more detail. He later mentioned that the capability of a body to be accelerated by gravity was a property not of the body itself but of a relationship between bodies that has yet to be discovered (Nieper, 1983).

  The classical law of universal gravitation or gravitational acceleration was established by Isaac Newton. The law states that gravitational acceleration is proportional to the mass and inversely proportional to the square of the distance as shown in Equation 13.5. In another way, it states that a particle attracts every other particle in the universe with a force that is directly proportional to the product of their masses and inversely proportional to the square of the distance between their centers. Example 13.6 shows how Equation 13.5 is used (with proper units), a problem typically discussed in high school physics. Example 13.7 shows that this gravitational force is also equal to that when Equation 13.6 is used and with acceleration due to gravity equal to 9.81 m/s² [32.2 ft/s²]:

$$F = G \times \frac{m_1 \times m_2}{r^2}$$
(13.5)

where
    F = force between the masses (N)
    G = gravitational constant (6.674×10^{-11} N × (m/kg²))
    m_1 = mass of first particle (kg)
    m_2 = mass of second particle (kg)
    d = distance between their centers (m)

$$F = mg$$
(13.6)

where
    F = force (N)
    g = acceleration due to gravity (9.81 m/s²)

Example 13.6: Calculating the Gravitational Force Between the Earth and a Person

Determine the gravitational force between the earth and a person at sea level. Assume the mass of earth is 5.98×10^{24} kg [1.32×10^{25} lbs] and that of the person is 75 kg [165 lbs]. The distance from the mean sea level and the earth center is assumed to be 6.38×10^6 meters [2.093×10^7 ft]. Compare this force to when the person is at the top of Mount Everest, which is 8,848 meters [29,021.4 ft] from mean sea level. Will the gravitational force increase or decrease with distance?

SOLUTION:
    a. The solution is simply to substitute values:

$$F = \frac{\left[\left(6.674 \times 10^{-11} Nm^2 / kg \right) \times \left(5.98 \times 10^{24} kg \right) \times (75\ kg) \right]}{\left(6.38 \times 10^6 m \right)^2} = 735\ N$$

    b. If the distance between the masses is increased, the new gravitational force is shown below:

$$F = \frac{\left[\left(6.674 \times 10^{-11} Nm^2 / kg \right) \times \left(5.98 \times 10^{24} kg \right) \times (75\ kg) \right]}{\left(6.388848 \times 10^6 m \right)^2} = 733\ N$$

    c. Clearly, the gravitational force decreases with increased distance between masses.

Example 13.7: Calculating the Gravitational Force Exerted by a Person on Earth

Determine the gravitational force exerted by a 75 kg [165 lbs] person on earth. This value is also equivalent to the weight of the person in Newtons.

SOLUTION:

a. Equation 13.6 is used and values are substituted as follows:

$$F = 75 \ kg \times 9.81 \frac{m}{s^2} = 735 \ N$$

b. One will realize that the value of acceleration due to gravity, g, is simply the ratio of (g × mass of earth/square of distance to earth's center). Upon careful analysis, this means that the acceleration due to gravity will change as the person is farther from the earth's mean sea level.

Scientists who believed in tachyon energy do not agree with the concept of "gravitational attraction" or "Earth's attraction" and say it does not exist. What exists, they argue, is a universal space filled with energy-rich charges or gravitational fields or called "gravity gravitons" as described by Feinberg (Nieper, 1983). Perhaps this concept was used by Tesla in the development of his vehicle, which did not have an engine and only had an electric motor that supposedly took advantage of this energy-rich gravitational field. This concept is very similar to the electromagnetic induction concept by Faraday. But no scientist has connected these two concepts.

Another aspect of tachyon field is the theory behind the source of geothermal energy. The theory is that the interior heating of planets, such as that of the earth, is generated by the conversion of gravitational radiation in the form of heat. These scientists believe that the phenomenon is due to the increased density of the Feinberg field toward the sun. Solar energy is supposed to be the main source of this energy. They believe that geothermal energy is nothing more than trapped gravitational energy that is constantly renewed from the outside (Nieper, 1983).

Again, we approach these new theories with caution. At present, we have to rely on solid scientific bases before we consider the endless supply of renewable energy from space. It is our hope that new research funding is made available to evaluate these theories, either to prove or to disprove the claims. An ultimate goal of humankind is an endless source of cheap and renewable energy. Gravity field energy may be an interesting avenue to conduct research only if a large majority of scientists believe that this area of research is a worthy undertaking and funding agencies like the USDOE support the claim. There should be a move to further investigate previous scientific laws, such as Faraday's Law, and consider other unique possibilities that we cannot comprehend at present.

13.8 Len's Law and Faraday's Law

Faraday's Law relates the change in magnetic flux through an electric coil loop to the magnitude of the electromotive force induced in the loop. Equation 13.7 shows Faraday's Law (Lucas, 2016). The negative sign simply indicates the direction of the current induced due to the application of conservation of energy. Faraday's Law determines the magnitude of the voltage produced, but Len's Law denotes the direction in which the current will flow. Len's Law states that the direction is always opposing the change in the produced flux (Encyclopedia Britannica, 2018). Example 13.8 shows how this equation is used to relate to voltage and current:

$$Electromotive\ Force(EMF) = -N \times \frac{\Delta\Phi}{\Delta t} \qquad (13.7)$$

where

N = number of turns of the wire

Φ = magnetic flux through a single loop

t = time

Example 13.8: Calculating Voltage and Current Induced in Electrical Coils

Determine the voltage and the current induced in a group of conducting coils with an area of 2 m² and a constant magnetic field of 5 Teslas perpendicular to the surface of the ring. Assume that there is only one turn or loop of the wire. Over the next 5 seconds, the magnetic field has increased to 10 Teslas. The conducting coil has a resistance of 2 ohms. Determine (a) the change on flux over this period, (b) the voltage, and (c) the current induced.

SOLUTION:

a. The initial and final magnetic flux is calculated as the product of the magnetic field and the area as shown:

$$\Phi_i = 5\ Teslas \times 4m^2 = 20T - m^2$$
$$\Phi_i = 10\ Teslas \times 4m^2 = 40T - m^2$$

b. We calculate the voltage induced using the Faraday-Len's Law:

$$Voltage(V) = -N \times \frac{\Delta\Phi}{\Delta t} = -1 \times \frac{(40-20)Tm^2}{5\ s} = -4\frac{Tm^2}{s} = -4V$$

c. Since the voltage and resistance are known, we may calculate the current as shown:

$$Current(Amps) = -\frac{4V}{2\ Ohms} = 2\ Amps$$

13.9 Other Scientists Investigating Gravitational Field Energy and Other Renewables

The study of gravitational field energy began with Tesla. However, there are numerous other scientists who made contributions to the theory. In this section, we will enumerate a few noted scientists and their related theories.

13.9.1 Dr. T. Henry Moray, American Physicist

In the early 1900s, Dr. T. Henry Moray of Salt Lake City developed an energy-generating device called Radiant Energy, which he claimed derived power from the cosmos to the earth. The Moray device captures energy that is supposedly available in our space (Bearden, 1988). His son, Dr. John E. Moray (2011), described his father's invention in his book, stating that it was able to produce 50,000 watts [67 hp] of AC power for several hours using a 300 foot [91.5 m] antenna. Unfortunately, this technology was not released due to patent rejections (and supposedly without ample investigation). The technology was considered a primitive version of the transistor, using "germanium triode." Moray's son published *The Sea of Energy in Which the Earth Floats* (Moray, 2011), a book similar to what Moray wrote during his research days. Moray was also inspired by the work of Tesla.

Moray studied abroad and underwent examinations for his doctorate in electrical engineering from the University of Uppsala in Sweden from 1912 to 1914. He returned home after World War I and began to develop the device in the 1920s and 1930s. He refused to divulge the specs of his detector tube. He claimed to have developed a transistor-type valve in 1925, way ahead of the discovery of the transistor. Inside his device was a small rounded pellet, a mixture of triboluminescent zinc, a semiconductor device, and a radioactive material that was not named. He filed a patent application on July 13, 1931, but was never issued one. Bell Laboratories did not introduce the transistor device the company is now known for until many years after this patent application (Bearden 1988; Moray, 2011). The U.S. Patent Office refused to grant Moray his patent because his device used a cold cathode in the tubes (not heated, as the examiner thought was common knowledge at that time) and because he failed to identify the source of energy (Bearden, 1988). Despite this obstacle, Moray demonstrated his radiant energy device to electrical engineering professors, congressmen, dignitaries, and numerous visitors of his lab. He even brought the unit outside of his lab to prove that he did not use an unknown signal from his facility to generate the output power. He even asked other people to dismantle the unit and put it back together to prove there were no tricks involved. In all of the tests, he was successful in demonstrating that the device could produce energy output without any appreciable energy input. Even with all these demonstrations, no commercial unit was ever made or sold, and none have yet explained the extraordinary electromagnetic theory behind the device (Moray, 2011).

Unfortunately, in 1939, Dr. Moray's device was destroyed by interested groups who wanted full disclosure of construction details. Dr. John Moray, his son, has since established and operated a research institute in Salt Lake City to continue his father's quest for this alternative energy source. Dr. T. Henry Moray died in May 1974 without commercializing his technology. The Patent Office still would not grant him the patent, even though his son keeps the patent application current (Bearden, 1988).

13.9.2 Professor Shinichi Seike, Director, Gravity Research Laboratory, Japan

Professor Seike introduced the concept of ultra-relativity in his book titled *The Principles of Ultra Relativity* (Seike, 1983). Seike is from Uwajima City, Ehime Prefecture, Japan. He based his design off an energy source that is clearly gravitational. Researchers like Professor Seike believed that there exists a relationship between gravity and electricity production, even though Faraday had not found any (Assis, 1992). Seike designed a device to prove his theory using electrical coils of peculiar winding, also called Klein coils, with some transistor coils (Seike, 1983). As Seike's device produces power or energy, the weight of the coils was reported to gradually decrease while electric current flows through the coils. Professor Seike's theory is also related to the Lorentz force, a very familiar term in physics, particularly electromagnetism, or the combination of electric and magnetic fields, but perhaps with a deeper spin or more complicated twist.

Seike's device introduced the concept of nuclear electrical resonance (NER) as similarly applied to our basic nuclear magnetic resonance (NMR) equipment for substance examination for health and chemistry. NMR uses the charges in spatial electron spins due to the application of magnetic field. The material being examined is placed in a very high frequency magnetic field, and a given chemical or its molecules would produce a unique absorption response. The uniqueness of Seike's device is that it utilizes both the polar and the axial spin, where the polar spin, Seike claimed, is directly related to the gravitational field. His device is virtually a rotating electrical AC field superimposed on a direct current magnetic field. He then claims that there is a significant increase in the "negative gravitational energies" occurring at certain resonance frequencies. What this means is that the energy from the earth's gravitational field "enters" the system of the secondary artificial field created by this anti-gravity motor. The negative G-energies cause a weakening of the earth's gravitational field, thereby cancelling it altogether. Additional depolarization then causes the device to be repulsed by the larger gravitational body, which in this case is the earth.

Perhaps the reason Seike's device has not been manufactured or commercialized is the fact that the conditions he described for his device to work occur only at very high electrical voltages with corresponding ultra-high alternating current frequencies. Seike proposes the use of ferromagnetic materials for his device. Examples include ferrite and ferromagnetic components such as barium-strontium-titanate (BST). This material is now a common nano-particle used in electronics, ceramics, and batteries (American Elements, 2017; CAS#12430-73-8 with Linear Formula BaO_4SrTi and EC No. 235-659-5). In the design, three spherical condensers are alternately charged and discharged by three magnetic coils. The resulting output is the conversion of gravitational field energy into mechanical and electrical energy. Seike calculated the power output of around 3×10^9 kW [4.02×10^9 hp] for an anti-gravity engine using approximately one tonne each of ferrite and BST for the design (Seike, 1983). At present, no commercial system is yet in place.

13.9.3 Bruce De Palma's N-Machine

Dr. Bruce De Palma graduated from the Massachusetts Institute of Technology in 1954. He devoted most of his career to conducting experiments with a rotating machine that he claimed produced more output power than the input power, again opposed to the Law of Conservation of Energy. He based his design on Faraday's disc, the device invented by the famous physicist Michael Faraday. He called this the N-machine, or Homopolar Generator, which is nothing more than a rotating magnetized gyroscope. He fabricated a 100 kW [134 hp] generator and claimed this unit could produce five times more power than it consumed (De Palma, 1990).

In 1978, a large N-machine was fabricated and tested in Santa Barbara, California. This machine was independently tested by Dr. Robert Kincheloe, Professor Emeritus of Electrical Engineering at Stanford University, following a request by the USDOE. Dr. Kincheloe presented his report in 1986 to the Society for Scientific Exploration held in San Francisco, California, on June 21, 1986, with the following summary: "De Palma may have been right in that there is indeed a situation here whereby energy is being obtained from a previously unknown and unexplained source. There is a conclusion that most scientists and engineers would reject out of hand as being a violation of accepted laws of physics and if true has incredible implications" (Nieper, 1983).

A senior fellow and noted physicist at the Institute for Advanced Studies in Austin, Texas, Dr. Harold Puthoff, also commented the following: "it isn't clear where the reported excess energy is coming from—whether out of the electromagnetic field or as a result of some anomaly associated with rotating bodies in terms of inertia. The De Palma machine needs to be replicated on a broad scale to see if it actually works" (Nieper, 1983). Dr. Puthoff believes that a new, non-polluting energy source may be achieved by tapping the force of random fluctuations of jostling particles within a vacuum. This theory is now called zero point energy (Puthoff, et al., 2002). Zero point energy is the general term applied to theories that attempt to explain the concept of tapping into the abundant power available directly from the vacuum of space itself. At present, the scientific community remains skeptical of this concept or has dismissed this concept as a hoax.

De Palma's career in the United States was full of controversy, and in 1994 he sought out a more favorable audience in Australia and New Zealand, where he spent the last years of his life as a permanent resident demonstrating his invention. During the last five years of his life, he would build three prototypes and one fabricated in New Zealand that was claimed to produce 1.25 watts of energy for every 1 watt of power input. However, during tests in Auckland, the electrical output was scattered, and most of the output energy was lost as heat. Where the claimed excess energy would have come from is still a mystery. De Palma died in October 1997 in New Zealand without seeing his invention come into commercialization. However, he was the inspiration for a noted Indian nuclear engineer, Paramahamsa Tewari, who after De Palma's death continued this type of work (Tewari, 2007).

13.9.4 Paramahamsa Tewari of India and His Space Power Generator

Professor Tewari is a top nuclear engineer and scientist in India. He was a former executive director (nuclear projects) of the Nuclear Power Corporation of India (NPCIL). He is the author of a unique theory in physics, the Space Vortex Theory (SVT). His theory produced an invention called the Tewari Reactionless Generator (T-RLG), based on his longtime collaboration with Dr. Bruce De Palma and the N-machine device. The device is also called the Space Power Generator (SPG). This device claims to produce between 200% and 300% of input power (Tewari, 2007).

The SPG consists of a soft-iron core around which electromagnet coils are wound. Both the coils and the iron core are rotated together such that no relative motion between the magnetic field and the soft iron core conductor exists. A vortex of space is apparently formed in the rotating iron, and due to this event, electrical charge is formed, interacts with the orbital electrons of the iron atoms, and sets them free. The free electrons interact with the magnetic field creating positive and negative polarities. This is where power output is at a high current and a few volts are drawn (Christyn, 2015). The unit is very similar to the N-machine of De Palma. The unit requires initial input energy before it can sustain itself. The possible inputs are conventional gasoline or diesel engines, batteries, or any electrical power source (Tewari, 1984).

After years of work, we have not seen widespread commercial application, and the work remains a theory. Just like Bruce De Palma, Professor Tewari passed away on November 27, 2017, in his hometown region in India close to Varanasi without seeing his invention adopted by the free market. He was 80 years old (Tewari, 2017).

We could continue to name and discuss numerous others scientists working on the principle of free energy and watch for further developments on the internet. Listed below are some notable inventions and theories behind them. Again, these theories must be taken with a degree of skepticism:

a. John Bedini and his Kromrey Converter, the Free Energy Converter
b. The Tom Bearden Website and his Search for Free Energy (http://cheniere.org/)
c. The Tesla Switch
d. The Swiss ML Converters—A Masterpiece of Craftsmanship and Electronics Engineering
e. Tom Valone and Zero-Point Energy from the Quantum Vacuum

As we discuss persons who are currently investigating free energy sources around the world, we may note that they all have one thing in common. They believe that it is possible to produce more energy than the initial input energy through proper design of conversion devices, either using gravitational field energy or fusion energy from combining atoms. The scientific community has not fully backed the goals of these researchers primarily because their theories are based upon the opposite to the universal belief that energy can never be created or destroyed. Our current laws of physics and energy state that we cannot create something out nothing. Until we see widespread, highly vetted use of such technologies, we can only continue to hope for an unlimited source of cheap and free energy.

Many of the inventions above require an understanding of motor efficiencies and power as a function of load. In evaluating the circuits, one must be aware of the complexities of electric motors. Table 13.1 shows the different power equations for various types of electric motors. Examples 13.9 and 13.10 show how these equations are used. The usual claim of many gravitational field conversion devices is that the output power well exceeds the input power or that the battery that runs the motor has more energy than what the motor has provided. Complications arise when evaluating motors with or without the load. The efficiency of an electric motor varies significantly with the presence of load. Most electric motors are designed to run at 50% to 100% of its rated load (USDOE, 2014). For AC motors, one would use the nomenclature called power factor (PF). PF is defined as the ratio of real power versus apparent power in the circuit. The reason that power is reduced in alternating current circuits is that the voltage and current waveforms are not in phase. Hence, the product of voltage and current are reduced most of the time. Real power is the output power of the device, while apparent power is the product of voltage and current of the circuit or simply the power supplied to the motor wires.

When testing electric motors, one should follow some published standards. Listed below are examples of those standards from the U.S. and European Union (USDOE, 2014).

TABLE 13.1

Various Input/Output Power Formulas or Efficiency for Different Types of Electric Motors

Power Output Unit/Efficiency	DC Motor	1Φ Motors	2Φ-4 Wire Motors	3Φ Motors
Watts (W)	$E \times I \times \eta$	$E \times I \times \eta \times PF$	$E \times I \times 2 \times \eta \times PF$	$E \times I \times \eta \times PF \times 1.73$
HP	$\dfrac{E \times I \times \eta}{746}$	$\dfrac{E \times I \times \eta \times PF}{746}$	$\dfrac{E \times I \times 2 \times \eta \times PF}{746}$	$\dfrac{E \times I \times \eta \times PF \times 1.73}{746}$
Efficiency (η)	$\dfrac{Watts}{E \times I}$	$\dfrac{Watts}{E \times I \times PF}$	$\dfrac{Watts}{E \times I \times 2 \times PF}$	$\dfrac{Watts}{E \times I \times PF \times 1.73}$

HP = horsepower; E = voltage (root mean square [RMS] voltage for three-phase alternating current); I = current (also RMS for three-phase circuits), η = efficiency; PF = power factor

Unfortunately, there is no common standard acceptable worldwide akin to the those determined by the International Organization for Standardization (ISO). Each country adopts their own local standard that may differ from other countries.

Common standards for testing electric motors around the world are given below:

a. IEEE 112-2004 "Standard Test Procedures for Polyphase Induction Motors and Generators" (US)

b. IEC 60034-2-1:2007 "Standard on Efficiency Measurement Methods for Low Voltage Induction Motors" (IEC)

c. CSA C390-10 "Test Methods, Marking Requirements and Energy Efficiency Levels for Three-Phase Induction Motors" (Canadian Standards Association [CSA])

d. JEC-37 (Japanese Electrotechnical Committee [JEC])

e. BS-269 (British)

Example 13.9: Determining Motor Load Using Input Power Measurement

Determine the input power and the output power as a percentage or nameplate-rated power of an old three-phase 40 hp [29.84 kW] motor that has a reported or nameplate full-load efficiency of 92%. The average measured line voltage for each phase was 470 volts, and the average amperage on each line was measured at 38 amps with an average PF of 0.77.

SOLUTION:

a. The input power in watts is calculated using the equation in Table 13.1:

$$Input\ Power(Watts) = 470 \times 38 \times 0.77 \times 1.73 = 23,791.31\ Watts\ [31.9\ hp]$$

b. The output power may be calculated using the power at rated load and the corresponding PF as follows:

$$Power\ at\ Rated\ Load = 40\ HP \times \frac{746\ Watts}{1\ HP} \times \frac{1}{0.92} = 32,434.8\ Watts\ [43.5\ hp]$$

c. Hence, the output power as a percentage of nameplate-rated power is calculated as follows:

$$\%\ Output\ Power = \frac{23,791.31\ Watts}{32,434.8\ Watts} \times 100\% = 73.35\%$$

If one facility uses numerous electric motors, they are sometimes penalized by the utility company with a low PF. PF is easily related to overall motor efficiency as shown in Table 13.1. In fact, in the United States, the Energy Policy Act (EPAct) that became effective in 1997 requires 1 to 200 hp [746 watts to 150 kW] general purpose motors sold to meet minimum efficiency levels (referred to as EPAct levels). Motor efficiency is the ratio of the amount of work the motor performs to the electrical power it consumes. If one looks at equations for efficiencies in Table 13.1, the efficiency values go up when the PF is near unity. Of course, households do not have electric motors, and they would not be affected significantly with improved efficiency. Still, households can lower their electricity bill by investing in more efficient electric motors.

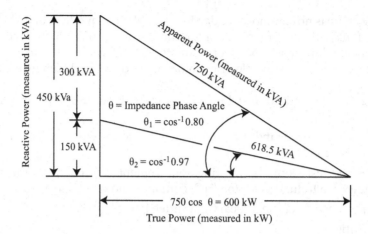

FIGURE 13.2
A typical power triangle for improving motor PF.

In AC, when voltage and current are out of phase, the cosine of the angular displacement is also called the PF. In households, the components affected by conversion efficiencies are typically appliances that run a fan, the washer and dryer, as well as illuminating the room. Appliances in the house that are directly affected by the PF are induction motors, fluorescent lamps, and water heaters. If these appliances are used simultaneously, they affect the harmonic currents, distorting their shape—instead of a perfect sine curve, the sine curve is flattened, affecting the PF and lowering the value. As a result, some industrial or commercial companies invest in performing PF corrections. Figure 13.2 shows the relationship between load power and input and output power, referred to as kVA and kW, respectively. If one takes the cosine of the angle between kVA and kW, one will get the value of PF. Example 13.10 shows an example of improving the PF with addition of a shunt capacitor that provides reactive power.

Example 13.10: Calculating Overall PF of an Electrical Setup

An electrical motor application has a 750 kVA load operating at 80% lagging PF. To improve the efficiency of the system, a 300 kVA shunt capacitor was added to act as reactive power, thereby improving the circuit and only taking 150 kVA of power from the source. Determine the new overall PF for this setup. Use Figure 13.2 for the calculation.

SOLUTION:

a. If one were to draw the power triangle for this application as depicted in Figure 13.2, these values for kVA and kW are found. The original PF is calculated below.

$$Power\ Factor\,(1) = \frac{600\ kW}{750\ kVA} = 0.80$$

b. The new PF is calculated below using the smaller triangle from the figure as follows:

$$Power\ Factor\,(2) = \frac{600\ kW}{618.5\ kVA} = 0.97$$

c. The new PF with the additional shunt capacitor has improved to 97%.

13.10 Non–Energy-Related Applications of Gravitational Field Energy

It must be noted that these topics are rarely discussed in mainstream energy or physics courses. I was inspired to write this chapter to further research these atypical applications. In 1991, as a young, newly graduated PhD with expertise in renewable energy, having just returned to the Philippines to resume my teaching and research duties, I was called to an urgent meeting with the chancellor at the University of the Philippines in Los Baños, Dr. Ruben B. Aspiras. He served as chancellor between 1991 and 1993 and was a noted microbial ecology expert. He had been informed I had returned from energy-related studies abroad and wanted my opinion on Hans Nieper's *Conversion of Gravity Field Energy: Revolution in Technology, Medicine and Society*. The book was engrossing to say the least, and much has already been discussed here; however, the second half of Nieper's book approached the medical applications of tachyon energy. As it turned out, the chancellor had cancer and was curious about the plausibility of harnessing gravitational energy to cure diseases such as cancer. Nieper was indeed trained and active as an oncologist. Nieper believed that there was also a deep connection between cancer and the annals of gravity physics. He also believed that the energy problem and the cancer problem, while very different in nature, were two of the most pressing problems in the world. While the chancellor did not ask me to solve the energy issues of the country, he wanted to know if the theories made some sense. I had no answer for the chancellor at the time, knowing full well that this theory contradicted what most in the academic field believed. We must acknowledge, anyhow, that there are unorthodox devices developed in hopes of curing illnesses that also defy our basic knowledge in medicine. Like the theory of gravitational field energy, these methods violate what we have learned in biology, chemistry, and the sciences. Nieper hoped future governments would be very open to such innovative thinking, but he was mistaken.

I continued my work on renewable energy after all these years and like most researchers have kept a discerning yet open mind. Neither gravitational field energy nor tachyon field energy has been reported to cure any of the enumerated diseases above. Likewise, we continue to hope that creative theories will advance us on the fronts of disease and energy.

13.11 Conclusions

I have included these topics here to entertain open-mindedness, looking back at the contexts in which these theories developed, and noting how little research has progressed without support and funding. Perhaps difficult magnetism and electricity theories are beyond our scope at present, but attempts to coin more appropriate terminology can help the scientific community move forward with out-of-the-box theories. We now hear of zero free energy and low-energy nuclear reactions, even though pioneering scientists have not seen their theories proven. Without funding, it is difficult to continue work on such theories, but it is also difficult to gain the support and faith of funders.

The most exciting energy frontier is the ITER, the project of 35 nations (along with the United States) to help build a massive nuclear fusion reactor in the south of France. The U.S. Congress has recently approved additional funding. The ITER is considered one of the most ambitious energy projects in the world today. However, it will still take years before the facility becomes fully operational, aiming to generate power by around 2025 (Bardi, 2017).

The main barriers to such technology are the materials used. Under fusion, high temperature is required. and materials that can withstand high temperatures over a long period of time are difficult to find. I have personally been approached by a couple of groups in the last few years, most recently at Texas A&M University. A group came to my laboratory and had me and my electrical engineer test devices that generated more power than the input. Hence, we have designed a simple experiment to measure both the input power and the output power (using power meters and an oscilloscope). True enough, the output power was greater than the input power. My proposal to the group was to test this on a long-term basis to ensure that in time the materials could withstand the rigors of power production and become sustainable. I had heard anecdotally that some parts could melt at some point. I have not heard from the group since then. These people are highly educated individuals with PhDs and should know their science well. They claim to have developed numerous prototypes but cannot proceed due to lack of funding. Investors are always skeptical and would not want to part ways with their investments until they see workable units running on a long-term basis. Hence, this group will struggle to commercialize their devices because of these simple rules that have been set by the government and the scientific community.

We are hopeful that future Nikola Teslas will emerge in this world and come up with unorthodox theories that could change the way we look at renewable energy for the growing population. Basic energy resources will continue to be very expensive for many nations in the next decade, and we should urgently seek a new source of energy to replace the diminishing fossil fuels.

13.12 Problems

13.12.1 Energy Balance in Electrolysis Setup

P13.1   Determine the difference between the energy provided to the electrolysis setup similar to that shown in Figure 13.1 from the energy generated by the medium using Equations 11.1 to 11.3. The following experimental data were measured:

Voltage = 3.5 volts

Amperage = 0.4 amps

Duration of experiment = 25 minutes

Rise in temperature = 0.50°C [0.9°F]

Amount of medium = 0.75 kg [1.65 lbs]

Specific heat of medium = 4.2 kJ/kg°C [1.0032 Btu/lbs°F]

13.12.2 Heat Capacity of Calorimeters

P13.2   A 200 Watt 0.268 hp] heater was used to heat a bomb calorimeter. The temperature rise was reported to be 48 K in 4 minutes. (a) Calculate the heat capacity of the calorimeter in units of J/K, (b) determine also the heat capacity in units of J/K-mol if water is used as medium assuming specific heat of water to be 4.184 kJ/kg K [1 Btu/lb°F], and (c) determine the number of moles of water used in this experiment.

13.12.3 Heat Released from Combustion of Chemicals

P13.3 Benzoic acid is a common material used in the calibration of a calorimeter. The reported heating value was 26.419 kJ/g [11,383 Btu/lb]. A calorimeter with a heat capacity of 1,000 J/K was used to measure the heat of combustion from a gram of benzoic acid with a chemical formula of $C_7H_6O_2$. The temperature increase was found to be 26.4 K. Compare the heat released in units of Joules when a gram of benzoic acid is combusted in a calorimeter and the amount of heat released by a gram of this chemical in units of kJ/mol.

13.12.4 Energy Balance in N-Machine or N-Generator

P13.4 The drive motor for the N-machine has the following data:

Input motor amperage	= 43.5 amperes
Input motor voltage	= 115 volts
Output drive motor current at no load	= 15 amperes
Voltage output of N generator when loaded	= 1.05 VDC
Current output of the N-generator	= 4,500 amperes
Voltage output of the N-machine at no load	= 1.5 VDC
Current output at no load	= 14,500 Amps

a. Calculate the input energy in watts and in hp.

b. Calculate the power output of the N-generator when loaded using the measured voltage and current. Compare this with the measured input power.

c. Calculate the power output of the N-generator at no load using the measured voltage and current. Compare this with the measured input power.

d. Calculate the input energy in watts and in hp.

e. Calculate the power output of the N-generator when loaded using the measured voltage and current. Compare this with the measured input power.

f. Calculate the power output of the N-generator at no load using the measured voltage and current. Compare this with the measured input power.

13.12.5 Determining Magnetic Fluxes, Voltages, and Current in Conducting Coils

P13.5 Determine the voltage and the current induced in a group of conducting coil with an area of 2 m² and a constant magnetic field of 5 Teslas perpendicular to the surface of the ring. Assume that there is only one turn or loop of the wire. Over the next 4 seconds, the magnetic field has increased to 10 Teslas. The conducting coil has a resistance of 2 ohms. Determine (a) the change on flux over this period, (b) the voltage, and (c) current induced.

13.12.6 Estimating Gravitational Forces at Various Elevations

P13.6 Determine the gravitational force between the earth and a person at sea level. Assume that the mass of earth is 5.98×10^{24} kg [1.32×10^{25} lbs] and that of the person at 100 kg [220 lbs]. The distance from the mean sea level and earth center is assumed to be 6.38×10^6 meters [20,926,400 ft]. Compare this force when the person is at the top of Mount Denali in the United States, which is 6,192 meters [20,310 ft] from mean sea level. Will the gravitational force increase or decrease with distance?

13.12.7 Calculating Acceleration due to Gravity at Various Elevations

P13.7 Determine the value of the acceleration due to gravity for a 75 kg [165 lbs] person who is of a distance equal to twice the center of the earth's distance (or 12.76×10^6 m) [41,852,800 ft] or virtually in outer space.

13.12.8 Calculating Voltages, Current, and Magnetic Fluxes in Coils

P13.8 Determine the voltage and the current induced in a group of conducting coil having an area of 25 m² with a constant magnetic field of 5 Teslas perpendicular to the surface of the ring. Assume that there are 200 turns or loop of the wire. Over the next 5 minutes, the magnetic field increases to 16 Teslas. The conducting coil has a resistance of 12 ohms. Determine (a) the change on flux over this period, (b) the voltage, and (c) current induced.

13.12.9 Calculating Input and Output Power in an Electric Motor

P13.9 Determine the input power and the output power as a percentage or nameplate-rated power of an old three-phase 40 hp [29.84 kW] motor that has a reported or nameplate full-load efficiency of 90%. The average measured line voltages for each phase was 475 volts, and the average amperage on each line was measured at 40 amps with an average PF of 0.75.

13.12.10 Improving the PF of Resistive Motors

P13.10 An electrical motor application has 150 kVA load operating at 70% lagging PF. To improve the efficiency of the system, a 77 kVA shunt capacitor was added to act as a reactive element, thereby improving the circuit and only taking 30 kVA of power from the source. Determine the new overall PF for this setup. Use Figure 13.2 for the calculation.

References

American Elements. 2017. Catalog of Materials. The Advanced Material Manufacturer, Los Angeles, CA. Available at: https://www.americanelements.com/company.html. Accessed June 10, 2019.

Assis, A. K. T. 1992. Deriving gravitation from electromagnetism. Canadian Journal of Physics 70(5): 330–340.

Bardi, J. S. 2017. The future of fusion energy: Limitless, clean energy to secure our planet's future (part 1 of a 5-part series). July 17. Inside Science: Reliable News for an Exciting Universe. Available at: https://www.insidescience.org/video/future-fusion-energy. Accessed June 10, 2019.

Bearden, T. E. 1988. Excalibur Briefing: Explaining Paranormal Phenomena, Revised and Expanded Subsequent Edition. Strawberry Hill Publishing.

Biberian, J.-P. 2007. Condensed matter nuclear science (cold fusion): An update. International Journal of Nuclear Science and Technology 3, (1): 31–42. doi:10.1504/IJNEST.2007.012439.

Casual Chemist. 2017. Cold Fusion: Not Dead Yet. Volume 1, Issue 1, September. Adapted with permission from Chemical Engineering News 96 (44): 34–39. Copyright 2016, American Chemical Society.

Cengel, Y. A. and M. A. Boles. 1989. Thermodynamics: An Engineering Approach. McGraw-Hill, New York.

Christyn, R. 2015. Incredible scientist makes free energy perpetual motion generator. Your News Wire. April 17. Available at: https://newspunch.com/incredible-scientist-makes-free-energy -perpetual-motion-generator/ Accessed June 10, 2019.

Clery, D. 2015. ITER fusion project to take at least 6 years longer than planned. Science Online. November 19. Available at: http://www.sciencemag.org/news/2015/11/iter-fusion-project -take-least-6-years-longer-planned. Accessed April 26, 2018.

De Clercq, G. 2018. ITER nuclear fusion project avoids delays as US doubles budget. Reuters Business News, March 26.

De Palma, B. 1990. On the possibility of extraction of electrical energy directly from space. British Science Journal, Speculations in Science and Technology 13 (4).

Encyclopedia Britannica. 2016. Tesla: Unit of energy measurement. Available at: https://www .britannica.com/science/tesla. Accessed April 3, 2018.

Encyclopedia Britannica. 2018. Lenz's Law. Available at: https://www.britannica.com/science/ Lenzs-law. Accessed April 3, 2018.

Lee, T. D. 1993. Gerald Feinberg. Physics Today 46 (1): 84. Available at: https://physicstoday.scitation .org/doi/10.1063/1.2808794. Accessed June 18, 2018.

Fleischmann, M., S. Pons, M. W. Anderson, L. J. Li and M. Hawkins. 1990. Calorimetry of the palladium-deuterium-heavy water system. Journal of Electroanalytical Chemistry 287: 293–348.

Fleischmann, M., S. Pons and M. Hawkins. 1989. Journal of Electroanalytical Chemistry 261 and 263: 197, 301.

Horibe, Y. and N. Ogura. 1968. Deuterium content as a parameter of water mass in the ocean. Journal of Geophysical Research 73 (4): 1239–1249.

ITER. 2016. International Thermonuclear Experimental Reactor (ITER) Organization 2016 Annual Report. ITER Organization Headquarters, St. Paul-lez-Durance Dedex, France.

Khan, H. 2009. Deuterium: From discovery to applications in synthesis. Paper presented at the Monday Night Seminar, Department of Chemistry, University of Toronto, January 9.

Krivit, S. B. 2016. The pied piper of Bologna: Andrea Rossi and the E-Cat con. A six-year retrospective of the Andrea Rossi E-Cat con. New Energy Times. November 2. LENR Reference Site, San Rafael, CA.

Lucas, J. 2016. What is Faraday's Law of induction? Live Science. January 27. Available at: https:// www.livescience.com/53509-faradays-law-induction.html. Accessed June 10, 2016.

Mills, R. L. 2016. The Grand Unified Theory of Classical Physics: Volume I. Atomic Physics. Brilliant Light Power, Cranbury, NJ.

Moray, J. E. 2011. The Sea of Energy in Which the Earth Floats. 5th Edition. Xlibris Corporation.

Nagel, D. J. 2015. Lattice-enabled nuclear reactions in the nickel and hydrogen gas systems. Current Science 108 (4): 646–652.

Nagel, D. J. and R. A. Swanson. 2015. LENR excess heat may not be entirely from nuclear reactions. Journal of Condensed Matter Nuclear Science 15: 279–287.

Nieper, H. A. 1983. Revolution in Technology, Medicine and Society: Conversion of Gravity Field Energy. MIT Verlag, Oldenburg, Germany.

Parr Instrument Company. 2016. Series 6000 oxygen bomb calorimeters catalog: Designing and building high precision bomb calorimeters for over 100 years. May. Parr Instrument Company, Moline, IL.

Park, R. L. 2016. Cold fusion day: Does Fleischmann still brew tea on hot plate? Wayback Machine, What's New with Bob Park. Archived April 25, 2012, from Wayback Machine publications by Robert Park.

Petit, C. 2009. Cold panacea: Two researchers proclaimed 20 years ago that they'd achieved cold fusion, the ultimate energy solution. The work went nowhere, but the hope remains. Science News 175 (6): 20–24. doi:10.1002/scin.2009.5591750622.

Platt, C. 1998. What if cold fusion is real? Wired Magazine. November 1. Available at: https://www.wired.com/1998/11/coldfusion/. Accessed June 10, 2019.

Puthoff, H. E., S. R. Little and M. Ibison. 2002. Engineering the zero-point field and polarized vacuum for interstellar flight. Journal of British Interplanetary Society 55: 137–144.

Ritter, S. K. 2016. Cold fusion lives: Experiments create energy when none should exist. Chemical & Engineering News, November 28.

Seike, S. 1983. The Principles of Ultra Relativity. 7th Edition. Gravity Research Laboratory, Ehime, Japan. ASIN: B0007AYNAG.

Seike, S. 1986. The Principles of Ultra Relativity. 8th Edition, Revised. Ninomiya Press, Japan and Gravity Research Laboratory, Uwajima, Japan.

Takahashi, A., A. Kitamura, K. Takahashi, R. Seto, Y. Yokose, A. Taniike and Y. Furuyama. 2016. Anomalous heat effects by interaction of nano-metals and H(D)-gas. Paper presented at the International Conference on Condensed Matter Nuclear Science (ICCF20), October 2–7, 2016, Sendai, Japan.

Tewari, P. 1984. Beyond Matter. Printwell Publishers Distributors, Jaipur, India.

Tewari, P. 2007. Universal Principles of Space and Matter: A Call for Conceptual Reorientation. Crest Publishing House.

Tewari, A. 2017. Obituary – Paramahamsa Tewari. Available at: https://www.tewari.org/. Accessed June 10, 2019.

USDOE. 2004. Report of the Review of Low Energy Nuclear Reactions. Washington, DC: U.S. Department of Energy. Archived from the original on February 26, 2008. Accessed March 8, 2018.

USDOE. 2014. Premium Efficiency Motor Selection and Application Guide: A Handbook for Industry. Energy Efficiency and Renewable Energy (EERE), Advanced Manufacturing Office. Golden, CO. With Support from Copper Development Association, Inc. (New York), and Washington State University Energy Program. Report No. DOE/GO-102014-4107. February. Available at: https://www.energy.gov/sites/prod/files/2014/04/f15/amo_motors_handbook_web.pdf. Accessed June 10, 2019.

14

Environmental and Social Cost of Renewables

Learning Objectives

Upon completion of this chapter, one should be able to:

1. Enumerate the economic, environmental, and social advantages of renewables.
2. Relate the sustainability of renewable energy technologies with one another as well as with other conventional sources.
3. Describe a range of life cycle assessment procedures for the various renewables.
4. Relate the overall social and techno-economic issues concerning all renewable energy systems discussed.

14.1 Introduction

Humans have utilized renewable energy resources for as long as they have required some form of energy to live. The discovery of fire, crude oil, and the industrial revolution are pivotal developments in fulfilling the energy needs of humankind. Energy consumption worldwide has steadily risen and will continue to rise. Many developing countries seek new forms of energy that will help their country progress regardless of cost. Affordable energy for a growing population would be the pinnacle of research.

The basic definition of a renewable energy technology remains as follows: a resource that can be replenished at the same rate as it is used (Bhattacharya, 1983). Any energy resource described in the earlier chapters of this book is not considered renewable if it is consumed more quickly than it can be replaced. Geothermal energy, despite not quite conforming to this definition, is utilized at such a minimal rate compared with available energy that for the purposes of energy research it can be considered a renewable resource. This chapter will discuss the economic, environmental, and prevailing issues concerning renewable energy technologies.

Each country should plan on the conscientious use of their renewable energy resources. The protection of the environment is a main issue. We enumerate some primary parameters for evaluating the economic, social, and sustainability merits of a renewable technology below:

a. Most renewable energy resources require large tracts of land.
b. The initial capital cost for most renewables are still quite high even with numerous technological developments.

c. The issue of balance of systems, a common feature of all renewables, will become a major issue in the future.

d. The global population growth is persistently on the rise, thereby requiring increased energy usage.

e. Numerous renewable technologies that require special materials for construction are affected by technological developments in materials engineering.

f. The transport of energy from generation to usage is a major issue to consider.

g. Timeliness of delivery and the use of smart grid systems for electrical power, as well as innovative fuel distribution systems toward various growing urban and industrial areas, are necessary.

h. There remains the need to store excess renewables for future use.

The world has heavily relied on fossil fuel carbon for both power and fuel needs. This will have to change. We can save on expensive energy and power by focusing on improving the efficiency of conversion. Nations must make a concerted effort to significantly reduce their carbon footprints and lessen the harmful effects to the environment. The environment is affected regardless of the type of energy consumed. The remaining sections of this chapter discuss various economic, environmental, and social issues surrounding renewables while summarizing innovative technological developments around each type.

14.2 Technical Advancement of Renewable Energy Technologies

Conversion efficiencies for renewable solar energy technology, such as solar photovoltaic (PV) cells, have advanced. Over the years, the solar PV cells have improved conversion efficiencies to as much as 40%, as shown in Figure 14.1 (US DOE, 2018). Today, the current practical and average conversion efficiency is around 15%, a 25% increase in overall efficiencies. It was only slightly above 10% for many years. For example, one may be able to decrease the solar collection area by 40% by an improvement in efficiency of 10% using the highest-efficiency PV cells. This is shown in Example 14.1.

Example 14.1: Reduction in Solar Collector Area with Increased Efficiency

Determine the reduction in solar collector area for a 1 MW solar PV system if the conversion efficiency is increased from 15% to 25%. Assume the average solar radiation in the area is around 5 kWh/m²/day. The average amount of daily solar radiation lasts around 12 hours.

SOLUTION:

a. The original area required for a 1 MW solar PV farm is calculated first:

$$Area(acres) = \frac{1\ MW}{0.15} \times \frac{m^2 - day}{5\ kWh} \times \frac{12\ hrs}{day} \times \frac{1,000\ kW}{1\ MW} \times \frac{hectares}{10,000\ m^2} = 1.6\ ha$$

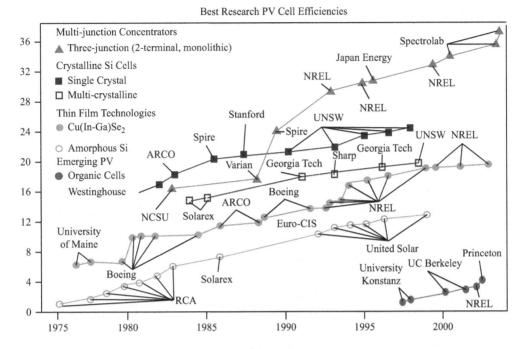

FIGURE 14.1
Best conversion efficiencies of solar PV cells as reported by the USDOE.

Source: NREL (2019a)

    b. With a new conversion efficiency, the required area will be calculated as follows:

$$Area\left(acres\right) = \frac{1\ MW}{0.25} \times \frac{m^2 - day}{5\ kWh} \times \frac{12\ hrs}{day} \times \frac{1,000\ kW}{1\ MW} \times \frac{hectares}{10,000\ m^2} = 0.96\ ha$$

    c. The difference in area with an improvement in efficiency will be about 0.64 hectares. The difference in area is what is saved in land cost with the installation of new efficiency PV systems.

The noteworthy breakthroughs in wind energy technologies are related to cost reduction (McMahon, 2017). Onshore wind machines are mature technologies with well-established key manufacturers. Developers look to extend the life of wind machines and bring down operational expenses, according to a survey conducted by Ryan Wiser published in 2016. Experts anticipate a 24% to 30% reduction in costs by 2030 and a 35% to 41% reduction by 2050 (Wiser, et al., 2016).

Biofuels research has had remarkable technological achievements over the past decade. Fuels such as biogasoline, green diesel, and aviation fuel have now been produced from biomass. ExxonMobil and Synthetic Genomics, Inc. (Ajjawi, et al., 2017) announced a breakthrough in joint research into advanced biofuels involving the modification of an algae strain, *Nannochloropsis gaditana*, doubling the oil content without inhibiting the strain's growth. They used advanced cell engineering technologies to bring oil content to 40% from the original 20% oil weight from the algae strains. Honeywell UOP

(Universal Oil Products, Chicago, IL) is now converting algal biofuel into aviation fuel. A 20% increase in oil content by weight translates to over 200 liters [50 gal] of oil per ton as shown in Example 14.2.

Example 14.2: Potential Oil Production from Algae

Determine the potential oil production in liters per metric tonne [gallons per ton] for an algae species that has oil content (by weight) of 20%. Assume an oil density of about 1.1 kg/L [9.2 lbs/gal]. How much of an increase in yield will be achieved (L/tonne or gal/ton) if there is a 20% increase in potential oil yield (by weight) of algae?

SOLUTION:

a. A 20% oil by weight in algae means that for every 100 g, there are 20 g of oil. The oil yield per tonne of material is calculated below using the density of oil:

$$Oil\ Yield\left(\frac{liters}{tonne}\right) = \frac{0.2\ tonne\ oil}{tonne\ algae} \times \frac{L}{1.1\ kg} \times \frac{1,000\ kg}{tonne} = 181.8\frac{L}{tonne}$$

$$Oil\ Yield\left(\frac{gal}{ton}\right) = \frac{0.2\ ton\ oil}{ton\ algae} \times \frac{gal}{9.2\ lbs} \times \frac{2,000\ lbs}{ton} = 43.5\frac{gal}{ton}$$

b. A 20% increase in oil yield will provide the following new calculations:

$$Oil\ Yield\left(\frac{liters}{tonne}\right) = \frac{0.2 \times 1.2\ tonne\ oil}{tonne\ algae} \times \frac{L}{1.1\ kg} \times \frac{1,000\ kg}{tonne} = 218.2\frac{L}{tonne}$$

$$Oil\ Yield\left(\frac{gal}{ton}\right) = \frac{0.2 \times 1.2\ ton\ oil}{ton\ algae} \times \frac{gal}{9.2\ lbs} \times \frac{2,000\ lbs}{ton} = 52\frac{gal}{ton}$$

c. There is also a 20% increase in yield per tonne.

Boeing, on the other hand, found a new desert plant they believed would be a more sustainable source of biofuel (Shahan, 2014). They discovered a type of plant that grows well in deserts—halophytes—which can be irrigated with salt water and do not have to take up arable lands for production. Halophytes are also considered a nuisance in many fish stocks and are in fact an aquaculture waste that can be repurposed. These halophytes are mainly cellulosic plants with sugars and lignin. Lignin is easily separated, and the sugars and cellulosic materials can be easily converted into ethanol and other liquid hydrocarbon fuels. The Boeing group has developed a production system called ISEA, the Integrated Seawater Energy and Agriculture System. This technology uses coastal seawater to raise fish and shrimp for food, whose nutrient wastewater fertilizes the plants rich in oils that can be harvested for aviation biofuel production (Shahan, 2014).

The overall costs of renewable energy technology are still not at ideal levels. We have seen reports of solar PV prices dipping below $1/Wp as well as wind machines nearing $1 million/MW. Small wind energy systems are still quite expensive, with turbines under 100 kW roughly costing between $3,000 to $8,000/kW of capacity. A 10 kW wind machine, the size appropriate for a modest household, would cost around $50,000 to

$80,000. For utility-sized turbines, the lowest quoted price is around $1.3 million/MW to $2.2 million/MW of installed capacity (Daniels, 2007). Most of the commercial-sized wind turbines that are manufactured at present have an average size of 2 MW and would cost roughly $3 million to $4 million installed. The primary reason the overall costs are rigid is the balance of system (BOS) cost. For solar energy technologies such as solar PV, costs include charge controllers, electrical wires and lines for transmission of power, batteries for energy storage and to run the inverters, and other residuals. These systems will be discussed in the following section. In most cases, the BOS accounts for about half, or 50%, of the overall cost (Fu, et al., 2018).

For solar PV systems, the USDOE (Fu, et al., 2018) estimated that prices for residential and small commercial systems (those less than 10 kW) were around $4.69/W (median value). Large commercial systems (greater than 100 kW) were reported to be around $3.89/W (median), and utility-scale systems (greater than 5MW, ground mounted) were about $3/W (weighted average). Example 14.3 shows the size of PV system needed for a typical household in the United States. It would require about at least 13 kW average power, costing more than $60,000 in initial direct capital investment. Many residents in the United States are currently taking advantage of this technology due to a 100% subsidy in capital costs in some states.

Example 14.3: PV Sizing for Households

Determine the size of PV system for an average household in the United States. The monthly average electrical consumption is around 1,000 kWh/month. Assume overall solar PV system efficiency is 85% (note that this is not the solar cell efficiency) and a climate generation factor of 3. Use Equation 2.17 for this problem. Make an estimate of this cost based on the above data of around $4.69/Watt.

SOLUTION:

a. The average daily load requirement (watts-hours) is first calculated as shown below:

$$DLR(Whr) = \frac{1,000 \ kWh}{1 \ month} \times \frac{1 \ month}{30 \ days} \times \frac{1,000 \ Watts}{kW} = 33,300 \ W-hrs$$

b. The size of solar PV system is then calculated using the conversion efficiency:

$$Size(W_p) = \frac{DLR}{CFG \times \varepsilon_s} = \frac{33,000}{3 \times 0.85} = 13,071 \ W \ or \ 13 \ kW$$

c. The average cost for this household solar PV system will be around $61,000 as shown in the calculation below:

$$Solar \ PV \ Cost(\$) = 13,071 \ Watts \times \frac{\$4.69}{Watt} = \$61,303$$

d. Note that because of the fact that solar radiation varies throughout the year, the household may need to invest in battery systems such that excess power from one day is carried onto the next so that power can be made available during cloudy days.

14.3 Balance of Systems

Each renewable energy source will have an associated balance of systems (BOS). BOS is defined as all components besides the prime component of a renewable energy technology. For solar PV systems, these are components other than the PV cell itself. For wind energy, these include any other component parts of the wind power–generating system other than the turbine itself.

For a solar PV system these include the following:

a. Wiring

b. Switches

c. Mounting system

d. Inverters

e. Charge controllers

f. Battery banks

g. Battery chargers

h. Tie grid system

i. Electrical power and energy monitoring

It gets a little complicated when the system includes advanced monitoring and tracking systems. Land is sometimes part of the BOS and could be a significant component of the cost. The levelized cost of energy can be significantly reduced by boosting the reliability of a system through improved BOS components (Dhere, 2005; Fife, 2010). Levelized cost means that all associated cost components are spread throughout the life of the facility. The initial capital cost of a renewable energy technology is still exceedingly high per MW. So, in order to compete with other non-renewable energy sources, the renewable industry must find comparable cost figures, usually reported as "levelized cost of energy" or LCOE. LCOE is defined as the price at which a renewable energy technology is valued, taking into account all the lifetime costs of the complete energy system (NREL, 2019b). The LCOE includes the initial capital cost as well as the cost of system operations, including maintenance and repair—the latter becoming a bigger portion of the total. In fact, even though some of these BOS components represent only 10% of the overall system cost, they are historically responsible for 70% of system failures and additional costs to the operational system (Fife, 2010).

For wind turbines, the BOS and soft costs for land are sometimes separate (Mone, et al., 2014), which collectively comprise over 30%, or about a third, of the overall cost. These additional components will then contribute to the overall LCOE. The USDOE has developed a new land-based wind plant BOS cost model based on developer-provided data to evaluate strategies for reducing LCOE for wind turbines. Overall conclusions for the model are as follows:

a. Costs are reduced with larger project size.

b. The electrical costs represent a significant percentage for small wind turbine sizes.

c. The combined mass of the nacelle and rotor assembly directly impacts foundation costs.

TABLE 14.1

Breakdown of Installed Cost for a Residential Solar
Photovoltaic System

	NREL 1Q 2015	GTM 1Q 2015
1. Module	$0.70	$0.75
2. Inverter	$0.29	$0.29
3. Structural BOS	$0.12	$0.18
4. Electrical BOS	$0.20	$0.19
5. Direct labor	$0.33	$0.35
6. Engineering and interconnection	$0.21	$0.25
7. Supply chain and margin	$1.24	$1.44
Total installed cost/watt	$3.09	$3.46

Source: Fu, et al., 2018.

d. Development and project management costs are substantial additions to the LCOE.

e. Transmission and interconnection costs will decrease with turbines larger in size.

f. Minimal additions to the LCOE include permanent operation and maintenance procedures as well as initial engineering costs, which include meteorological masts.

The BOS for renewable biofuels are not well defined at present. However, there are also issues with other bio-refinery components that are not directly related to the production of biofuels such as bioethanol and biodiesel. Biofuel technology is also beset with high initial capital costs and corresponding high operation and maintenance costs. Table 14.1 shows the recent USDOE NREL benchmark residential PV price breakdown in price per watt (Fu, et al., 2018). In 2015, there was a 7% drop in system prices based on the previous year. Comparative prices from a commercial manufacturer are also shown. Example 14.4 shows the percentage of PV module costs compared with other costs of the system. This example shows that the BOS and other associated costs could make up more than 50% of the total installed cost. Even if the price of PV modules decrease through the year, it will not affect the overall cost of the system due to additional costs of installation.

Example 14.4: Solar PV Cost Versus Balance of Systems

Determine the percentage of solar PV module cost compared with other associated costs, including electrical and structural balance of system (BOS) costs. Use Table 14.1 for the calculation and USDOE NREL's data.

SOLUTION:

a. The estimation of module cost is made by simply adding all associated costs excluding that of the module. The total installed price without the module is equal to $2.49/W. The percentage of module cost compared with total cost is calculated as follows:

$$\% \, Module \; Cost(\%) = \frac{\$2.49}{\$3.09} \times 100\% = 80.6\%$$

b. The solar PV module cost is just 19.4% of the total installed cost.

In reality, because of so many additional factors shown above, the overall balance of system costs can be more than 50% of the total installed cost due to direct labor, engineering, interconnection, supply chain, and margin cost for solar PV providers. These costs will further lengthen the overall payback period for solar PV systems. For example, at a nominal cost of $0.12/kWh, the households paying $1,440 per year for electricity will have a minimum payback of over 42 years without subsidy. The United States has attempted to heavily subsidize renewable energy systems so as to bring down the number of payback years to a more manageable period for consumers.

14.4 Overall Economics and Levelized Cost of Renewable Energy

Perhaps the biggest hindrance to the adoption of a renewable energy technology is its high initial capital cost. Renewable technologies are technically cheap to operate, such as solar and wind systems, which collect energy from the virtually cost-free sun and wind. However, the bulk of the expenses are from manufacturing the complete technology and the other associated costs as discussed above. Even if the overall cost has declined over the years, renewable energy technologies are still relatively pricey compared with other conventional sources of energy. In this section, we will mostly compare electrical power costs since solar and wind energy, as well as most other renewables like geothermal and hydro power, are used to produce electricity and not transport fuels. Only biomass energy resources are to be compared with conventional fuels, such as gasoline, aviation fuel, and diesel fuels for major transport.

A significant barrier to the adoption of renewable energy technologies is the intermittent nature of the resources' availability. Solar energy in only available during clear, sunny days, and wind energy is only serviceable during windy days and hydro power during rainy days. Geothermal energy is highly dependent on location. The overall cost of renewables varies between countries and is strongly affected by each country's subsidies. Generally, renewables' capital costs are higher than fossil fuel–based systems, including nuclear energy, but their production costs per kWh are competitive. Table 14.2 shows the production costs per kWh for various energy sources, demonstrating the competitiveness of these renewables. The values in the table were based on levelized costs of electrical power. That is, prices are scaled

TABLE 14.2

Levelized Cost of Electrical Power from
Various Energy Sources

Power Plant Type	Cost ($/kWh)
1. Coal	$0.11–$0.12
2. Natural gas	$0.053–$0.11
3. Nuclear	$0.096
4. Wind	$0.044–$0.20
5. Solar PV	$0.058
6. Solar thermal power	$0.184
7. Geothermal power	$0.05
8. Biomass power	$0.098
9. Hydro power plants	$0.064

Source: US EIA, (2017).

considering the overall life span of each facility. Renewable energy technologies usually last longer than many fossil fuel–based systems, and as a result, their levelized cost is typically lower. In fact, the costs of most renewables (besides solar thermal) per kWh are lower. Note that these are production costs. The retail prices are always higher. In the table, it is shown that geothermal energy, coal, and natural gas are the most economical, taking account of penalties for the use of coal and its negative effect on the environment. The cost for coal is forecast to double due to the government-imposed costs of CO_2 emissions (US EIA, 2017).

On topics related to LCOE, the most popular publication is written by Lazard (2018). Lazard's analysis is on a $/MWh basis but is complicated because of the inclusion of various U.S. subsidies on renewables, varying fuel costs across the nation, and the cost of capital. Because of the high initial capital costs for most renewables and without accounting for the levelized energy costs, they will consistently be too expensive compared to existing fossil fuel–based systems. The recent unsubsidized cost of energy comparison is depicted in Figure 14.2, showing the upper range of cost in $/MWh. Note that solar PV systems for residential units are still about the same in range as diesel reciprocating engines, whereas wind is the cheapest among all energy sources. Residential and commercial rooftop solar PV systems, solar thermal, and fuel cells remain quite expensive at this time. However, it

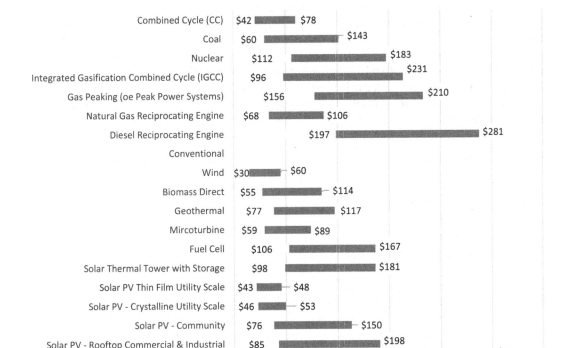

FIGURE 14.2
Unsubsidized levelized cost of energy comparison between conventional and renewable technologies.

Source: Adapted from Lazard (2018).

can be concluded that the direct use of biomass for power generation is becoming competitive with coal power plants using the LCOE approach.

As one can observe, the costs of residential rooftop solar PV systems are highest in range, as are those of diesel reciprocating engines. Along with wind power, utility-scale thin files, solar PV, and crystalline types have the lowest LCOE. According to the same graph, direct biomass power is also becoming more competitive with geothermal power.

The USDOE has an online LCOE calculator (NREL, 2019b). This software provides an easy way to evaluate both utility-scale and distributed generation (DG) systems. It compares capital costs, operating and maintenance costs, performance indicators, and fuel costs whenever needed. It does not include financing issues or depreciations. Table 14.3 shows an example of results given by this online calculator with varying performance values. In the table, the levelized cost of renewable energy goes down even if the capital cost increases, with a corresponding decrease in the electricity price.

One may estimate variations in the costs and performance of renewables by sliding the costs for items listed in the table. The discount rate used may be nominal or real based on inflation rates such that the time value of money is accounted for. Related to this discount factor is the capital recovery factor (CRF). The CFR is the ratio of a constant annuity to the present value of receiving annuity for a given length of time. The project assumes that money will be borrowed toward the initial capital cost of the system. Another important feature of the calculator is the capacity factor. The capacity factor is a value between 0 and 1 and represents the portion of the year that the power plant actually generates power— recall this definition from the wind energy chapter used in association with availability of wind. All other costs are associated with related costs of fuel for conventional systems such as natural gas or coal. Therefore, note that the only way for the levelized cost of energy to go down is with a higher selling cost of the electricity and/or by lowering the initial capital cost for the system. The selling price is always negotiated through the local electrical utilities provider. Simply, the levelized cost of energy is the minimum price at which energy must be sold for any particular energy project to break even.

TABLE 14.3

LCOE Calculator Results for Various Factors

Given/Encoded Items	Run 1	Run 2	Run 3
Period (Years)	20	20	20
Discount rate (%)	3%	3%	3%
Cost and Performance Values			
Capital cost ($/kW)	$1.050	$1,520	$1,990
Capacity factor (%)	43.6	43.6	43.6
Fixed O&M cost ($/kW-yr)	25	32	35
Variable O&M cost ($/kWh)	0.002	0.003	0.003
Heat rate (Btu/kWh)	10,000	10,000	10,000
Fuel cost ($/MMBtu)	8	8	8
Current Utility Electricity Cost			
Electricity price (cents/kWh)	12	10	8.1
Cost escalation rate (%)	3.0	3.0	3.0
Results			
Levelized cost of utility electricity (cents/kWh)	16.1	13.4	10.9
Simple levelized cost of energy (cents/kWh)	10.7	11.8	12.7

14.5 Life Cycle Analyses of Renewables

In the late 1990s, the USDOE initiated studies to compare the life cycle analysis (LCA) of coal conversion into electricity with that of biomass-based power production systems (Mann and Spath, 1999; Spath, et al., 1999). Results showed that the CO_2 emissions of coal power plants total to about 1,022 g CO_2/kWh of electricity produced. If biomass is used to generate electrical power, the net CO_2 emissions total only about 46 g CO_2/kWh. LCAs for other renewables have also been established of late, particularly for solar PV systems, wind, and other major energy technologies. There is no generalized LCA for renewables. Analyses will have to be estimated on case-by-case basis. Even within one type of technology, there will be numerous variations depending on design. For example, there are four solar PV technologies that are now considered mature, and there are numerous LCA analyses completed on them. Enumerated below are the four major solar PV designs:

 a. Mono-crystalline Si
 b. Multi-crystalline Si
 c. Cadmium telluride (CdTe)
 d. High concentration PV (HCPV) using III/V cells

In an LCA for a renewable, the indicators used include (a) energy payback times (EPBT), (b) greenhouse gas emissions (GHG), (c) criteria pollutant emissions, and (d) heavy metal emissions. One major hurdle against the accuracy of these LCA calculations is the lack of availability of real and actual data to develop a comprehensive method of quantifying the material and energy flows and impacts associated with the manufacturing process. In many instances, much of the required data is missing, and evaluators must resort to making scientific assumptions.

Figure 14.3 shows a typical LCA flow chart for a solar PV system (IEA, 2011). Other renewable types each have their own LCA flow chart. For example, in biofuels, the flow chart is divided into two or three components or steps: biomass production, biofuels

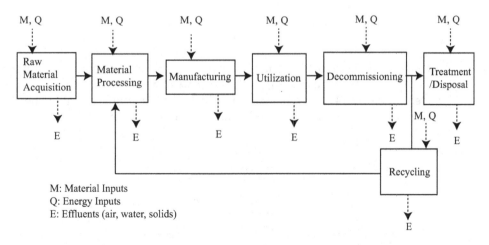

FIGURE 14.3
Life cycle analysis (LCA) flow chart for solar PV systems.

TABLE 14.4

Early LCA Studies on Various Solar PV Systems

PV Type	Primary Energy Consumption MJ/m²	EPBT Years	GHG Emissions g CO₂-eq/kWh	Efficiency %
Mono-crystalline	5,700	3.1	63	13
Multi-crystalline	4,200	2.5	46	14

production, and fuel use in transport vehicles. Also note that across countries, there will be variations for even the same technologies. We may separate regions into the United States, Europe, and Asia. The primary reason flow charts differ is the varied irradiation in these three regions.

Table 14.4 shows the various parameters used to describe the life cycle of a particular PV technology in earlier studies. These values were considered older LCA numbers from reports by Alsema (2000). Only the multi-crystalline and mono-crystalline Si were compared. Newer studies show that the values have gone down because of improvements in manufacturing processes. Table 14.5 shows more recent data on various solar PV technologies (Fthenakis and Alsema, 2009). Note that the data are applicable to the Southern European condition of around 1,700 kWh/m²/year insolation [4.7 kWh/m²/day]. The average value of U.S. insolation is around 1,800 kWh/m²/year insolation [4.9 kWh/m²/day]. The values could be further reduced as newer PV systems and studies are reported. Note also the similarities with GHG emissions of around 46 g CO₂-eq/kWh of electrical power produced from biomass systems analyzed several decades ago.

More recent data are shown and reproduced in Table 14.6 based on newest studies and reports by US EIA (2019) and Fthenakis and Kim (2012) for the four common PV systems. To reiterate, the basic definition of energy payback time is the period required for a renewable energy system to generate the same amount of energy (either primary or kWh equivalent) that was used to produce the system itself. It would be considered sustainable if the resulting number is less than 1, as shown in the table as the expected future values of these technologies.

The primary equation for the EPBT is shown in equation 14.1.

$$EPBT\left(years\right) = \frac{E_{mat} + E_{manuf} + E_{trans} + E_{inst} + E_{EOL}}{E_{agen} - E_{aoper}} \qquad (14.1)$$

In this equation, the numerator includes the energy demand to produce the materials used—for manufacture, for transport, installation, and end of life management of the system (depicted in Figure 14.3). The denominator includes the annual energy generation minus the energy demand for operation and maintenance of the system.

TABLE 14.5

Revised LCA Studies on Various Solar PV Systems

PV Type	Primary Energy Consumption MJ/m²	EPBT Years	GHG Emissions g CO₂-eq/kWh
Mono-crystalline	4,200	2.7	45
Multi-crystalline	3,700	2.2	37

TABLE 14.6

Recent LCA Studies on Various Solar PV Systems.

PV Type	EPBT (Years)		GHG Emissions (g CO$_2$-eq/kWh)	
	Current	Future	Current	Future
Mono-crystalline	2.7	0.8	37	12
Multi-crystalline	2.2	0.6	29	11
Ribbon-Si	1.7	0.7	22	12
CdTe	1.0–1.1	0.4	19	10

Source: Fthenakis and Kim, 2011.

The life cycle analysis for a wind machine is done similarly to the solar PV systems, where all the energy used to manufacture the complete system is accounted for and compared with the energy produced, subtracting its operating and maintenance demand. One study reported an EBPT for a 2 MW wind machine ranging between 5.94 to 9.27 months, where the lower value is when the wind machine parts are recycled after a 25-year life span (Ghenai, 2012). However, the values reported on GHG emissions ranged from 216–260 g CO$_2$/kWh whether the wind machine was recycled or not. The wind values are quite high compared with solar PV and biomass power. GHG emission values for a wind machine in China (Tang, et al., 2010) show the value to be around 170 g CO$_2$-eq/kWh. A study in Norway brings down the value of GHG emissions to between 19±13 g CO2-eq/kWh, suggesting better performance than other fossil-fueled power generation systems (Arvesen and Hertwich, 2012). This report also summarized the various GHG emissions, shown in Table 14.7.

The values reported above clearly illustrate variations across countries. These deviations are due to the parameters used in each analysis as well as the available renewable energy resource conditions of each place. For wind machines, the major factors are the assumed capacity factor and lifetime. For solar PV systems, the major factors affecting the LCA values are the insolation, type of manufacturing processes used, and efficiencies of conversions.

TABLE 14.7

GHG Emissions for Various Renewables and Other Conventional Power Sources

Energy Source	GHG Emission Range (g CO$_2$-eq/kWh	GHG Emissions Median (g CO$_2$-eq/kWh)
Wind	8–20	12
Nuclear	8–45	16
Hydro	3–7	4
Concentrating solar	14–32	22
Solar PV	29–80	46
Coal power[*]	180–220	200
Coal power[**]	1,000	1,000
Natural gas[*]	140–160	150
Natural gas[**]	500–600	550

[*]With carbon capture and storage (CCS).
[**]Without carbon capture and storage.

14.6 Pollutant Emissions of Some Renewable Energy Technologies

There are a few renewable energy technologies that may generate pollutant emissions. We have shown that solar and wind technologies have basically no pollutant emissions besides those made during the manufacturing of their component parts; however, biomass technologies do have issues with pollutant emissions. During the process of acquiring permits, biomass technologies may not be able to pass a state's local permitting codes. In the future, the widespread use of coal to generate electrical power will hopefully lessen and be slowly replaced with renewable technologies with reduced pollutant emissions. In this section, we will compare biomass technology with traditional coal power and evaluate their relative pollutant emissions production based on major criteria pollutants.

The US Environmental Protection Agency (EPA) has established federal regulations that limit the amount of emissions allowed for various sources. Specific regulations for each industrial category are found in the Code of Federal Regulations (e.g., 40 CFR Part 50 or Part 60). This document lists emission limits, performance standards, permitting procedures, testing and monitoring requirements, record-keeping requirements, and other local reporting. Each state in the United States is bound by these U.S. laws established by the federal government. States may create their own standards but may only be more stringent than the national laws. The permit process for each technology enforces these laws.

The National Ambient Air Quality Standards (NAAQS) provides legal air quality parameters. Table 14.8 lists the six major pollutants that are being regulated on a nationwide scale. Coal and biomass technologies both make use of some level of nitrogen and sulfur and must comply with the standards. There is also a separate national New Source Performance Standards (NSPS) resource for the particulates NOx and SO_2. Examples 14.5 and 14.6 show some comparative calculations of these standards. In Example 14.5, the emission rate is based on a certain scrubber efficiency in percent, thus complicating the emission rate calculations. This may be easily calculated using a monograph developed by Molburg (1980), shown in Figure 14.4.

Most of the emissions calculations involve the use of mixed units such as parts per million (ppm) and concentrations in $\mu g/m^3$. Example 14.7 shows the simple conversion of units from ppm to $\mu g/m^3$ [or mg/L]. The definition of ppm is shown in Equation 14.2 below, and the conversion from ppm to $\mu g/m^3$ is shown in Equation 14.3 at standard conditions:

$$ppm = \frac{Volume\ of\ Pollutant\ Gas}{Total\ Volume\ of\ Gas\ Mixture} \times 10^6 \tag{14.2}$$

$$C_{mass\left(\frac{\mu g}{m^3}\right)} = \frac{1,000 \times C_{ppm} \times MW_p}{24.45} \tag{14.3}$$

The denominator in Equation 14.3 assumes that the conditions have a standard temperature of 25°C [298 K] and 1 atmosphere. That is, the volume per mole of an ideal gas (the V/n term in Equation 14.4) has the value 22.45 at T = 25°C and 1 atmosphere or 22.4 L/gmol at T = 0°C [273 K] and 1 atmosphere of pressure. One would use the ideal gas law shown in Equation 14.4 to convert this factor if a different temperature is used:

$$V = \frac{nRT}{p} \tag{14.4}$$

TABLE 14.8

The National Ambient Air Quality Standards (NAAQS)

Pollutant	Primary (1°) and Secondary (2°)	Averaging Time	Level	Notes
Carbon Monoxide	(1°)	8 hours	9 ppm	(a)
		1 hour	35 ppm	
Lead (Pb)	(1°) and (2°)	Rolling 3-month average	$0.15\ \mu g/m^3$	(b)
Nitrogen Dioxide (NO_2)	(1°)	1 hour	100 ppb	(c)
	(1°) and (2°)	1 year	53 ppb	(d)
Ozone (O_3)	(1°) and (2°)	8 hours	0.070 ppm	(e)
$PM_{2.5}$	(1°)	1 year	$12.0\ \mu g/m^3$	(f)
	(2°)	1 year	$15.0\ \mu g/m^3$	(f)
	(1°) and (2°)	24 hours	$35\ \mu g/m^3$	(g)
PM_{10}	(1°) and (2°)	24 hours	$150\ \mu g/m^3$	(h)
Sulfur Dioxide (SO_2)	(1°)	1 hour	75 ppb	(i)
	(2°)	3 hours	0.5 ppm	(j)

Source: Adapted from USEPA website (https://www.epa.gov/criteria-air-pollutants/naaqs-table).
Notes:
a. Not to be exceeded more than once per year.
b. Not to be exceeded.
c. 98th percentile of 1-hour daily maximum concentrations, averaged over 3 years.
d. Annual mean.
e. Annual fourth-highest daily maximum 8-hour concentration, averaged over 3 years.
f. Annual mean, averaged over 3 years.
g. 98th percentile, averaged over 3 years.
h. Not to be exceeded more than once per year on average over 3 years.
i. 99th percentile of 1-hour daily maximum concentrations, averaged over 3 years
j. Not to be exceeded more than once per year.

where
V = volume of air (L)
n = gmol
R = 0.082 L-atm/gmol-K
T = K

Example 14.5: Calculating CO Concentrations in Ambient Air

A cubic meter of air was used to determine the concentrations of carbon monoxide. The gas analysis showed that 9 mL of carbon monoxide was found in this sample of air. Calculate the resulting concentration of CO in the air in units of ppm. Recalculate the CO concentrations in units of $\mu g/m^3$.

SOLUTION:

a. The concentration of CO in air in ppm is calculated using Equation 14.2 as follows:

$$C_{ppm} = \frac{9\ mL}{1\ m^3} \times \frac{1\ m^3}{1,000\ L} \times \frac{L}{1,000\ mL} \times 10^6 = 9\ ppm$$

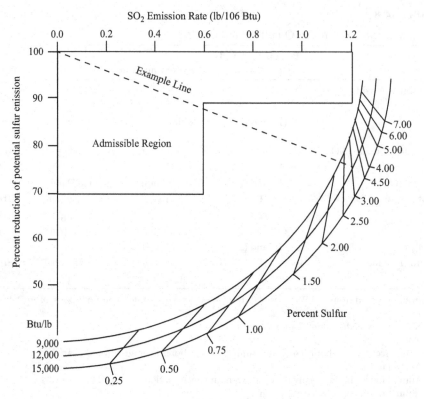

FIGURE 14.4
Graphical estimation of National Source Performance Standards (NSPS) for SO_2 emissions from coal-powered plants.

Source: Adapted from Molburg (1980).

    b. To convert this into units of $\mu g/m^3$, one will need the MW of carbon monoxide [12 + 16 = 23], which is 28, and the use of Equation 14.3 above:

$$C_{\mu g/m^3} = \frac{1,000 \times 9 \; ppm \times 28}{24.45} = 10,307 \; \frac{\mu g}{m^3}$$

    c. This value is within the maximum prolonged exposure limit of CO for humans based on ASHRAE standards.

Example 14.6: Calculating Daily SO_2 Emission Rates for a Coal-Fired Power Plant

Calculate the daily emissions rate (ER) of particulates and SO_2 from a new 1,000 MW coal-fired power plant that must meet the performance standards of 0.60 lb SO_2 per million Btu heat input at approximately 87% scrubber efficiency. Assume that coal has a heating value of 12,000 Btu/lb and contains 3.0% sulfur. The plant is 40% efficient.

SOLUTION:

    a. First, one should calculate the heat input at a rate of 40% efficiency:

$$E_{in} = \frac{1,000 \; MW}{0.40} \times \frac{1,000 \; kW}{1 \; MW} \times \frac{24 \; hrs}{day} \times \frac{3,412 \; Btu}{kWh} = 2.05 \times 10^{11} \frac{Btu}{day}$$

b. The amount of particulates emitted is then estimated as follows:

$$SO_2\left(\frac{tons}{day}\right) = \frac{0.03 \; lbs \; SO_2}{1 \times 10^6 \; Btu} \times \frac{2.05 \times 10^{11} \; Btu}{day} \times \frac{1 \; ton}{2,000 \; lbs} = 3.075 \frac{tons}{day}$$

c. From Figure 14.4, using the dotted line, the emission rate is about 0.6 lb SO_2 per million Btu heat input. This is at about 87% scrubber efficiency. The emitted SO_2 in tons per day will be calculated as follows:

$$SO_2 \; Emissions\left(\frac{tons}{day}\right) = \frac{0.60 \; lb \; SO_2}{10^6 \; Btu} \times \frac{2.05 \times 10^{11} \; Btu}{day} \times \frac{1 \; ton}{2,000 \; lbs} = 61.5 \frac{tons}{day}$$

Example 14.7: Calculating Daily SO_2 Emission Rates for a Biomass-Fueled Power Plant

Calculate the daily emissions rate (ER) of particulates and SO_2 from a new 1,000 MW biomass-fired power plant that must meet the performance standard of 0.60 lb SO_2 per million Btu heat input at approximately 70% scrubber efficiency. Assume that the biomass has a heating value of 9,050 Btu/lb [21 MJ/kg] and contains 0.25% sulfur. The plant is 50% efficient.

SOLUTION:

a. First, one should calculate the heat input at a rate of 40% efficiency:

$$E_{in} = \frac{1,000 \; MW}{0.50} \times \frac{1,000 \; kW}{1 \; MW} \times \frac{24 \; hrs}{day} \times \frac{3,412 \; Btu}{kWh} = 1.64 \times 10^{11} \frac{Btu}{day}$$

b. The amount of emitted particulates is then estimated:

$$SO_2\left(\frac{tons}{day}\right) = \frac{0.001 \; lbs \; SO_2}{1 \times 10^6 \; Btu} \times \frac{1.64 \times 10^{11} \; Btu}{day} \times \frac{1 \; ton}{2,000 \; lbs} = 0.082 \frac{tons}{day}$$

c. From Figure 14.4, using the dotted line, the emission rate is about 0.18 lb SO_2 per million Btu heat input. This is at about 70% scrubber efficiency. The emitted SO_2 in tons per day will be calculated as follows:

$$SO_2 \; Emissions\left(\frac{tons}{day}\right) = \frac{0.18 \; lb \; SO_2}{10^6 \; Btu} \times \frac{1.64 \times 10^{11} \; Btu}{day} \times \frac{1 \; ton}{2,000 \; lbs} = 14.8 \frac{tons}{day}$$

d. The SO_2 emissions are significantly lower than those of the coal power plant—about 25% of the emissions of a coal power plant.

Example 14.8: Calculating CO Concentrations in Ambient Air

A cubic meter of air was used to determine the concentration of carbon monoxide. The gas analysis showed that 9 mL of carbon monoxide was found in this sample of air. Calculate the resulting concentration of CO in air in units of ppm. Recalculate the CO concentration in units of µg/m³.

SOLUTION:

a. The concentration of CO in air in ppm is calculated using Equation 14.2:

$$C_{ppm} = \frac{9\ mL}{1\ m^3} \times \frac{1\ m^3}{1,000\ L} \times \frac{L}{1,000\ mL} \times 10^6 = 9\ ppm$$

b. To convert this into units of $\mu g/m^3$, one will need the MW of carbon monoxide [12 + 16 = 28], which is 28, and use Equation 14.3 above:

$$C_{\mu g/m^3} = \frac{1,000 \times 9\ ppm \times 28}{24.45} = 10,307\ \frac{\mu g}{m^3}$$

c. This value is within the maximum prolonged exposure limit of CO for humans based on ASHRAE standards.

14.7 Sustainability Issues of Renewables

It has been said that renewable energy, despite being renewable, needs to be sustainable (Helder, 2015). Recall that in future projections of solar PV technologies, the EBPT must be less than 1 or must be recovered in one year such that the following year it would not be necessary to pay for excess energy to produce the solar PV material (that is, EBPT numbers are less than a unity). If not, this renewable energy resource would not be sustainable for a given year.

Some authors also define sustainability in the contexts of environmental, social, and economic parameters (Helder, 2017). So far, we have discussed the economic aspects of renewable energy and whether or not a project may be able to pay for itself throughout its lifespan. Environmental sustainability simply means the technology is no harm to the environment. We will devote another section to the issue of social sustainability.

Researchers must also consider climate change. There is widespread correlation between greenhouse gas emissions of CO_2 by fossil fuel–based systems and increased global temperature. However, climate change is complicated. There have been drastic changes in the weather across the globe, and scientists are still in the process of recognizing the causes and coming up with solutions. We will not delve into climate research in this book.

There are numerous ways of evaluating the sustainability of a renewable energy technology. For example, in biofuels, the two most common parameters used are the net energy ratio (NER) and net energy balance (NEB). These parameters are shown in Equations 14.5 and 14.6, respectively. Examples 14.5 and 14.6 show the values of NER and NEB for making bioethanol. NER is the amount of energy required to produce a biofuel divided by the energy content of the biofuels produced (Shapouri, et al., 2003). A value greater than 1 means that there is more energy from the fuel produced than the energy used to make this biofuel. There must be a comprehensive analysis of the energy inputs from the moment the biomass is planted up to the distribution of the fuel prior to its use in engines. The NER of biofuel from corn via the dry method process has been reported to be 1.10 (Shapouri, et al., 2006):

$$NER = \frac{Energy\ Content\ of\ Biofuel\,(MJ)}{Energy\ Required\ to\ Produce\ the\ Biofuel\,(MJ)} \tag{14.5}$$

$$NEB = HV\ of\ Biofuel - Energy\ Required\ to\ Produce\ the\ Biofuel \tag{14.6}$$

Example 14.9: Calculating Net Energy Ratio (NER) for Bioethanol Production

Determine the net energy ratio (NER) for making bioethanol from corn. Data show that the energy used to produce the ethanol via the dry mill process is as follows:

   a. Corn production = 5.261 MJ/L
   b. Corn transport = 0.596 MJ/L
   c. Ethanol conversion = 13.133
   d. Ethanol distribution = 0.414

The heating value of bioethanol produced was reported to be 21.28 MJ/L.

SOLUTION:

   a. The first step is to calculate the total energy required per liter to produce the fuel:

   Energy Required to Produce Fuel = 5.261 + 0.596 + 13.133 + 0.414 = 19.404 *MJ*

   b. The NER is calculated based on Equation 14.2 as follows:

$$NER = \frac{21.28 \ MJ}{19.404 \ MJ} = 1.10$$

   c. The NER of making bioethanol via the dry milling process was estimated to be 1.10—a positive number— and we can conclude that there is more energy from the fuel produced than all the energy associated with making the fuel. This was a sustainable process.

Example 14.10: Calculating Net Energy Balance (NEB) for Bioethanol Production

Determine the net energy balance (NEB) for making bioethanol from corn via a dry milling process with co-products allocation or recovery. Data shows that the energy used to produce the ethanol via the dry mill process with recovery of co-products is as follows:

   a. Corn production = 4.314 MJ/L
   b. Corn transport = 0.489 MJ/L
   c. Ethanol conversion = 10.769
   d. Ethanol distribution = 0.414

The heating value of bioethanol produced was reported to be 21.28 MJ/L.

SOLUTION:

   a. The first step is to calculate the total energy per liter required to produce the fuel:

   Energy Required to Produce Fuel = 4.314 + 0.489 + 10.769 + 0.414 = 15.986 *MJ*

   b. The NEB is calculated based on Equation 14.3 as follows:

$$NEB = 21.28 - 15.986 = 5.294$$

c. The NEB of making bioethanol via the dry milling process with co-products recovery was estimated to be 5.294—a positive number—and we can conclude that there is more energy from the fuel produced than all the energy associated with making the fuel. This was a sustainable process.

The same principle of NER and NEB may be applied to other renewables. However, for example, for solar PV cells, we have shown that it will take 4,200 MJ of energy to produce a square meter of solar PV cell/m^2 of solar cell area (Table 14.5) for a mono-crystalline type. We cannot simply take this energy production value and divide or subtract the instantaneous energy produced by the solar PV cell because the true value will be significantly less. However, when levelized to the life of the PV cell, the NER and NEB will be significantly positive. The reader should make this very simple calculation based on the average solar radiation received in the area. For example, in Texas, the average solar radiation is around 12.51 MJ/m^2. If we take these ratios and account for conversion efficiencies, the NER and NEB values will be less than 1, or negative, based on instantaneous output. The output must be levelized over the life of the unit for the numbers to become within sustainable ranges. In short, it will take more days to recover the energy used to produce the solar cell. Example 14.7 shows the calculation of recovering the energy used to produce the solar cell on a unit basis.

Example 14.11: Estimating the Time it Takes to Recover Energy Usage for PV Systems

Determine the time it will take for a mono-crystalline solar cell to recover the energy used for its production. The data show that it will take 4,200 MJ of energy to produce a square meter of mono-crystalline PV cell. The overall conversion efficiency from solar energy to electrical power was reported to be 13%. The average solar energy per square meter on-site per day is around 12.51 MJ. Report the energy recovery period in days.

SOLUTION:

a. The first step is to determine the solar energy output per day per square meter using the insolation given:

$$Output\ Energy\,(MJ) = 12.51\ MJ \times 0.13 = 1.63\ MJ$$

b. The number of days required to recover the energy used to produce the solar cell is simply calculated:

$$Number\ of\ Days = \frac{4,200\ MJ}{1.63\ MJ/day} = 2,577\ days$$

c. Surprisingly, it will take several years to recover the energy used to produce the solar PV cell. In this example, it would take a little more than seven years.

14.8 The Social Costs of Renewables

The social costs of installing renewable energy technologies include the healthy circumstances of workers, decent wages, and local welfare. For example, in some parts of the Philippines where numerous wind farms are installed, some communities benefit from low power costs by taking advantage of an option for users to purchase electricity directly from renewable energy facilities—what they call "open access" (RA 9512, Renewable Energy Act of 2008, De Los Santos, 2017). However, other locations even within the Philippines may not adopt this benefit. As written in the act, power may already be contracted to a different region and local families cannot benefit (IRENA, 2017).

We have learned that most renewable energy technologies are still expensive from a business perspective, without including subsidies. However, the social cost and benefits of renewable energy should also be considered in future analyses, as these benefits are usually not accounted for. There is always a need to provide incentives for renewables so that consumers can take advantage of social and environmental benefits. Tax exemptions and subsidies fall under programs that have immediate social benefits to consumers. Lately, carbon credits and carbon footprint reductions have been employed as incentives.

The social benefits of renewable energy are typically related to the negative external effects of using fossil energy. There is now a great deal of knowledge of global warming as a result of excessive CO_2 emissions from cars running on fossil fuels. Many facilities that utilize fossil fuels, such as coal power plants, contribute to smog, pollute the area, and affect soil and water where agricultural produce is grown. The water-energy and food nexus group (WEF) is active in promoting the interconnecting influences of the three major requirements for a sustainable society: sustainable water, energy, and food. It should be no surprise that these groups encourage renewable energy technologies over fossil-based energy resources.

There are also research groups who assert that the use of renewable energy helps to reduce external effects to communities and thereby reduces social costs of energy production (Wiesmeth and Golde, 2016). The economic improvement as a result of installation of a renewable energy technology is projected to yield social improvements and alleviate poverty. The local community must be in conversation with the installers and will have to be educated on the consequences of renewables and how they affect the local environment, water sources, and other issues. The local population may be employed for additional labor and necessary manpower during installation and operation. Many social scientists are also documenting the effect of renewables on the local community. These groups use different parameters to determine whether or not renewable energy can flourish in the community. There will always be negative perceptions and reluctance toward change. In an article by Kumarankandath (2017) from India, the government was said to have released a report stating that investing in renewable energy would have a social cost three times that of coal in terms of dollars per unit of electricity produced. There are arguments on the basis of these data, such as opportunity cost of land, social cost of carbon, health costs, as well as costs of stranded assets. The government report mainly states that investing in solar and wind energy would reduce the operation of coal power plants, leading to job losses and forfeiture of loans by coal companies, which is arguably bad for the country's economy. Kumarankandath claims the data used for the projections were erroneous but it was enough to make people doubtful of renewables (Kumarankandath, 2017).

Engineers will have to start working with social scientists to link renewable energy to its social costs and benefits to local communities. Researchers may document the effects of renewables to communities with regard to education, a clean environment, health benefits, and other intangibles or non-financial indicators to advocate for widespread adoption of renewable energy technologies.

14.9 Conclusion

In this chapter, we discussed the economics and sustainability issues of renewable energy resources, resources that are naturally replenished at the same rate as they are used. However, it is understandable that some renewable energy resources may not be fully sustainable and may inadvertently harm or be of no benefit to the environment. The primary goal of a renewable technology is to have the least negative impact on the environment. Sustainability is also a broader term which describes how the use of a certain energy type affects or pollutes the environment. For example, some manufacturing processes for component parts of a renewable energy technology can cause harm to the environment with widespread use, such as mining the component parts (e.g., gallium and indium and other rare-earth metals) of a solar PV. These rare-earth metals have to be replaced with common metals, such as zinc and copper, in order to make the component parts cheaper and more sustainable (Woods and Bernstein, 2012).

The same is true for fuel cells. We have shown that it is easy to convert hydrogen gas into electricity, but the process would require a platinum electrode that is even rarer than indium. Indium is priced at $1,000/kg, and platinum is priced in $/gram and not $/kg (Barras, 2009). If around 500 million electric vehicles are put to use today fitted with fuel cells that use platinum, the world's platinum would be used up in 15 years (as stated in the same report by Barras). We would have to find ways to search for cheaper alternatives that are also sustainable.

We have learned that the capital costs of most renewable energy technologies are high but that their levelized cost of energy (LCOE) values are very competitive. This is because most renewable technologies could last for more than a decade, and these capital costs spread over many years make the yearly average cost much lower.

The sustainability of a renewable energy technology should be assessed from the moment a material is sourced to the final use of the product. We have shown in this chapter that the major renewable energy technologies have positive net energy balances (NEB) or net energy ratios (NER). Most are very close to 1, and there is still room for improvement. For some renewable component parts, it may take more than five years to recover the energy used to manufacture the device. Still, most energy used to manufacture a renewable component comes from fossil fuel–based input. These types of production processes would have to change.

Mark Jacobson, a professor at Stanford University, listed the top seven alternative energy systems that have better global warming air pollution energy security potential indicators. Wind power tops the list, followed by concentrated solar power, geothermal energy, tidal energy, solar PV panels, wave energy, and finally hydroelectric power (Brahic, 2009). Biofuels do not make the list due to their more demanding requirement of land, water, and sunlight. Also, most biofuels would produce more greenhouse gases than wind power.

The ideal situation, as espoused by Jacobson, is to put people to work in building wind turbines, solar plants, geothermal plants, and other renewables not only to create newer jobs but also to reduce the cost of health care, land use, crop damage, pollution, and climate change (Jacobson, 2009). There must be a concerted effort among nations, industries, and governments to find sustainable sources of energy that are cost effective and do not harm the environment.

14.10 Problems

14.10.1 Area Required for Solar PV Systems

P14.1 Determine the reduction in solar collector area for a 1 MW solar PV system if the overall conversion efficiency is increased from 11% to 30%. Assume the average solar radiation in the area is around 5 kWh/m²/day. The average daily solar radiation period is around 12 hours.

14.10.2 Algal Oil Production and Yield Calculations

P14.2 Determine the potential oil production in liters per metric tonne [gallons per ton] for an algae species that has oil content by weight of 30%. Assume an oil density of about 1.2 kg/L [10 lbs/gal]. How much increase in yield will be achieved (L/tonne or gal/ton) if there is a 20% increase in potential oil yield by weight of algae? Determine also the percentage increase in yield.

14.10.3 Size and Cost of PV Systems for Large Commercial Applications

P14.3 Determine the size and cost of a PV system for a large commercial application in the United States. Assume a monthly average electrical consumption of around 7,650 kWh/month. Assume overall solar PV system efficiency of 85% and a climate generation factor of 3. Use Equation 2.17. Make an estimate of this cost based on discussions of the above data of around $3.89/watt.

14.10.4 Balance of System Cost as Percentage of PV Cost

P14.4 Determine the percentage of cost of a solar PV module compared with other associated costs, including electrical and structural balance of system (BOS) costs. Use Table 14.1 and the GTM 2015 data for the calculation.

14.10.5 SO₂ Daily Emissions Rate for Coal Power Plants

P14.5 Calculate the daily emissions rate (ER) of particulates and SO_2 from a new 1,000 MW coal-fired power plant that must meet the performance standard of 0.60 lb SO_2 per million Btu heat input at approximately 87% scrubber efficiency. Assume that the coal is of semi-anthracite type with a heating value of 17,700 Btu/lb, containing 3.8% sulfur. The plant is 42.5% efficient.

14.10.6 SO₂ Daily Emissions Rate for Biomass Power Plants

P14.6 Calculate the daily emissions rate (ER) of particulates and SO_2 from a new 1,000 MW MSW-fired power plant that must meet the performance standard of 0.60 lb SO_2 per million Btu heat input at approximately 70% scrubber efficiency. Assume that the MSW has a heating value of 9,000 Btu/lb [20.9 MJ/kg] and contains 0.5% sulfur. The plant is 42.5% efficient.

14.10.7 Ozone and SO₂ Concentration Units from NAAQS Standards

P14.7 a. Determine the eight-hour ozone standard (NAAQS) in units of $\mu g/m^3$. b. Determine also the concentration of the one-hour NAAQS standard for sulfur dioxide in units of $\mu g/m^3$. c. If the temperature of SO_2 emissions for the one-hour standard is raised to 150°C, what is the concentration of SO_2 in units of ppm and $\mu g/m^3$?

14.10.8 Net Energy Ratio (NER) for Biofuels

P14.8 Determine the net energy ratio (NER) for making bioethanol from corn via the wet milling process. Data show that the energy used to produce the ethanol via the wet mill process is as follows:

 a. Corn production = 5.171 MJ/L
 b. Corn transport = 0.586 MJ/L
 c. Ethanol conversion = 14.591
 d. Ethanol distribution = 0.414

The heating value of bioethanol produced was reported to be 21.28 MJ/L.

14.10.9 Net Energy Balance (NEB) for Biofuels

P14.9 Determine the net energy balance (NEB) for making bioethanol from corn via wet milling with co-products allocation or recovery. Data show that the energy used to produce the ethanol via the wet mill process (with recovery of co-products) is as follows:

 a. Corn production = 4.188 MJ/L
 b. Corn transport = 0.474 MJ/L
 c. Ethanol conversion = 11.819
 d. Ethanol distribution = 0.414

The heating value of bioethanol produced was reported to be 21.28 MJ/L.

14.10.10 Return on Investment for the Production Cost of Solar PV Systems

P14.10 Determine the number of years it will take for a multi-crystalline solar cell to recover the energy used for its production. The data show that it will take 3,700 MJ of energy to produce a square meter of multi-crystalline PV cell. The overall conversion efficiency from solar energy to electrical power was reported to be 14%. The average solar energy per square meter on the site per day is around 5 kWh.

References

Ajjawi, I., J. Verruto, M. Aqui, L. B. Soriaga, J. Coppersmith, K. Kwok, L. Peach, E. Orchard, R. Kalb, W. Xu, T. J. Carlson, K. Francis, K. Konigsfeld, J. Bartalis, A. Schultz, W. Lambert, A. Schwartz, R. Brown and E. Moellering. 2017. Lipid production in *Nannochloropsis gaditana* is doubled by decreasing expression of a single transcriptional regulator. Nature Biotechnology 35: 647–652. doi10.1038/nbt.3865.

Alsema, E. A. 2000. Energy payback time and CO_2 emissions of PV systems. Progress in Photovoltaics: Research and Applications 8: 17–25.

Alsema, E. and M. de wild-Scholten. 2005. Environmental impact of crystalline silicon photovoltaic module production. In: material Research Society Fall Meeting, Symposium G: Life Cycle Analysis Tools for "Green" Materials and Process Selection, Boston, MA.

Arvesen, A. and E. G. Hertwich. 2012. Assessing the life cycle environmental impacts of wind power: A review of present knowledge and research needs. Renewable and Sustainable Energy Reviews 16: 5994–6006.

Barras, C. 2009. Why sustainable power is unsustainable. New Scientist, February 6. Available at: https://www.newscientist.com/article/dn16550-why-sustainable-power-is-unsustainable/. Accessed June 9, 2019.

Bhattacharya, S. C. 1983. Lecture Notes in "Renewable Energy Conversion" Class. Asian Institute of Technology, Bangkok, Thailand.

Brahic, C. 2009. Top 7 alternative energies listed. New Scientist, Daily News. January 14. Available at: https://www.newscientist.com/article/dn16419-top-7-alternative-energies-listed. Accessed June 10, 2019.

Chaouki, G. 2012. Life Cycle Analysis of Wind Turbine, Sustainable Development—Energy, Engineering and Technologies—Manufacturing and Environment, Prof. Chaouki Ghenai (Ed.), InTech. Available at: https://www.intechopen.com/books/sustainable-development -energy-engineering-and-technologies-manufacturing-and-environment/life-cycle-analysis -of-wind-turbine. Accessed June 10, 2019.

Daniels, L. 2007. Community Wind Handbook. Chapter 8: Costs. Developed by Wind Industry, Rural Minnesota Energy Board, December 15, 2006.

De Los Santos, A. S. A. 2017. Renewable Energy in the Philippines. Renewable Energy Management Bureau (REMB) of the Philippines Department of Energy (PDOE), Bicutan, Manila, Philippines.

Dhere, N. G. 2005. Reliability of PV modules and balance of system components, Photovoltaic Specialists Conference, 2005. Conference Record of Thirty-First IEEE 1 (1): 1570–1576.

Feldman, D., G. Barbose, R. Margolis, T. James, S. Weaver, N. Darghouth, R. Fu, C. Davidson, S. Booth and R. Wiser. 2014. Photovoltaic system pricing trends. Sun Shot, U.S. Department of Energy, National Renewable Energy Laboratory (NREL), and Lawrence Berkeley National Laboratory Project Update. NREL/PR-6A20-62558 Report. September 22. Available at: http://www.nrel.gov/docs/fy14osti/62558.pdf.

Fife, M. 2010. Solar power reliability and balance of system designs. Renewable Energy World. October 14. Available at: https://www.renewableenergyworld.com/articles/print/rewna/volume-2/issue-5/solar-energy/solar-power-reliability-and-balance-of-system-designs.html. Accessed June 10, 2019.

Fthenakis, V., H. C. Kim, R. Frischknecht, M. Raugei, P. Sinha and M. Stucki. 2011. Life cycle inventories and life cycle assessment of photovoltaic systems. International Energy Agency (IEA) PVPS Task 12, Report T12-02:2011.

Fthenakis, V. and E. Alsema. 2006. Photovoltaic energy payback times, greenhouse gas emissions and external costs: 2004-early 2005 status. Progress in Photovoltaics: Research and Applications 14 (3): 275–280.

Fthenakis, V. M. and H. C. Kim. 2011. Photovoltaics: Life cycle analyses. Solar Energy 85: 1609–1628.

Fu, Ran, D. Feldman and R. Margolis. 2018. U.S. Solar Photovoltaic System Cost Benchmark: Q1 2018. Golden, CO: National Renewable Energy Laboratory. NREL/TP-6A20-72399. https://www .nrel.gov/docs/fy19osti/72399.pdf. Accessed June 10, 2018.

Helder, M. 2015. Renewable energy is not enough: It needs to be sustainable. World Economic Forum. September 2. Environment and Natural Resource Security, the Future of Energy.

Hertwich, E. G., T. Gibon, E. A. Bouman, A. Arvesen, S. Suh, G. A. Heath, J. D. Bergesen, A. Ramirez, M. I. Vega and L. Shi. 2015. Integrated life-cycle assessment of electricity-supply scenarios confirms global environmental benefit of low-carbon technologies. Proceedings of the National Academy of Sciences of the United States of America 112 (20): 6277–6282.

International Energy Agency (IEA). 2011. Life cycle inventories and life cycle assessments of photovoltaic systems. IEA PVPS Task 12, Subtask 20, LCA Report IEA-PVPS T12-02:2011. October. Photovoltaic Power Systems Program.

International Renewable Energy Agency (IRENA). 2017. Renewables readiness assessment: The Philippines. Abu Dhabi, United Arab Emirates. Copyright by IRENA 2017.

Jacobson, M. 2009. Review of solutions to global warming, air pollution and energy security. Energy and Environmental Science Journal 2:148–173. Copyright by the Royal Society of Chemistry, UK.

Kumarankandath, A. 2017. The false burden of social cost on renewable energy. The Wire. September 8. Available at: https://thewire.in/174632/renewable-energy-and-social-costs-discrepancies/. Accessed June 10, 2019.

Lazard. 2018. Annual Levelized Costs of Electricity for 2017, Version 11. Publication by Energy Innovation Policy and Technology, LLC. 98 Battery Street, San Francisco, CA. January 22.

Mann, M. K. and P. L. Spath. 1999. The net CO_2 emissions and energy balances of biomass and coal-fired power systems. NREL, USDOE Project Report NEL/TP-430-23076 and NREL/TP-570-25119. National Renewable Energy Laboratory (NREL) of the USDOE, Golden, CO.

May, A. 2017, Renewable energy, what is the cost? Green Tech, What's Up With That. March 13. Available at: https://wattsupwiththat.com/2017/03/13/renewable-energy-what-is-the-cost/. Accessed June 10, 2019.

McMahon, J. 2017. Wind energy will see more tech breakthroughs, falling costs, expert predicts. Forbes. February 27, 2017.

Molburg, J. 1980. A graphical representation of the new NSPS for sulfur dioxide. Journal of the Air Pollution Control Association 30 (2). Available at: https://www.osti.gov/biblio/5459755-graphical-representation-new-nsps-sulfur-dioxide.

Mone, C., B. Maples and M. Hand. 2014. Land-based wind plant balance of system cost drivers and sensitivities. National Renewable Energy Laboratory Publications. From Proceedings of the AWEA Windpower Conference held in Las Vegas, May 5–8. Report No.: NREL/PO-6A20-61546, Golden CO.

National Renewable Energy Laboratory (NREL). 2019a. Best research-cell efficiency charts. USDOE, Golden, Colorado. Available at: https://www.nrel.gov/pv/cell-efficiency.html. Accessed June 10, 2019.

National Renewable Energy Laboratory (NREL). 2019b. Levelized cost of energy calculator. USDOE, Golden, Colorado. Available at: https://www.nrel.gov/analysis/tech-lcoe.html. Accessed June 10, 2019.

Shahan, Z. 2014. Boeing biofuel breakthrough—This is BIG deal (interview with Boeing's biofuel director). Clean Technica. January 27. Available at: https://cleantechnica.com/2014/01/27/boeing-biofuel-breakthrough-big-deal/. Accessed June 10, 2019.

Shapouri, H., A. Duffield and M. Wang. 2003. The energy balance of corn ethanol revisited. Transactions of the ASABE 46 (4): 959–968.

Shapouri, H., M. Wang and A. Duffield. 2006. Net energy balancing and fuel-cycle analysis: In: Renewable-Based Technology Sustainability Assessment (J. Dewulf and H. Van Langenhove, eds.). John Wiley and Sons, New York.

Songlin, Tang, Z. Xiliang and W. Licheng. 2011. Life cycle analysis of wind power: A case of Fuzhou. Energy Procedia 5: 1847–1851.

Spath, P. L., M. K. Mann and D. R. Kerr. 1999. Life cycle assessment of coal-fired power production. NREL, USDOE Project Report NREL/TP-570-25119. June. National Renewable Energy Laboratory (NREL) of the USDOE, Golden, CO. Under Contract No. DE-AC-36-98-GO10337.

US Energy Information Administration (US EIA). 2017. Levelized cost and levelized avoided cost of new generation resources in the Annual Energy Outlook 2017. Available at: http://www.dnrec.delaware.gov/energy/Documents/Offshore%20Wind%20Working%20Group/Briefing%20Materials/2017_EIA_Levelilzed%20Cost%20and%20Levelized%20Avoided%20Cost%20of%20New%20Generation%20Resources_Annual%20Energy%20Outlook%202017.pdf. Accessed June 10, 2019.

US Energy Information Administration (US EIA). 2019. Levelized cost and levelized avoided cost of new generation resources in the Annual Energy Outlook 2019. Available at: https://www.eia.gov/outlooks/aeo/pdf/electricity_generation.pdf. Accesses June 10, 2019.

Wiesmeth, H. and M. Golde. 2016. Social-economic benefits of renewable energy. Seminar presented for SEED Engineering (Sustainability by Engineering, Environment by Design), Brisbane, Australia, by faculties from Technical University Dresden (TUD), Germany. January 7. Available at: http://www.seedengr.com/Socio-economic%20benefits%20of%20Renewable%20Energy.pdf. Accessed June 10, 2019.

Wiser, R., K. Jenni, J. Seel, E. Baker, M. Hand, E. Lantz and A. Smith. 2016. Expert elicitation survey on future wind energy cost. Nature Energy. September 12. Article No. 16135. doi:10.1038/NEENERGY.2016.135.

Woods, M. and M. Bernstein. 2012. New solar panels made with more common metals could be cheaper and more sustainable. American Chemical Society Bulletin, August 21.

Appendix A

Table of Conversion Units

Length

1 meter	=	3.28 feet	1 mile	=	5,280 ft
1 mile	=	1.609 km	1 inch	=	2.54 cm

Weight

1 tonne	=	1.1 ton	1,000 kg	=	tonne (metric)
1 kg	=	2.2 lbs	1 ton	=	0.909 tonne
1 ton (short)	=	2,000 lbs	1 ounce (oz)	=	28.35 g
1 kg	=	1,000 g	16 ounces	=	1 lb

Volume

1 gallon	=	3.785 Liters	1 ft^3	=	7.48 gallons
1 barrel	=	42 U.S. gallons	1 m^3	=	35.28 ft^3
1 m^3	=	1,000 Liters	1 ft^3	=	1,728 in^3

Time

1 min	=	60 sec	1 day	=	24 hrs
1 hr	=	3600 s	1 year	=	365 days

Area

1 hectare	=	2.47 acres	1 m^2	=	10.76 ft^2
1 hectare	=	10,000 m^2	1 square mile	=	2.589 square km

Pressure

1 psi	=	6,894.75 Pa	1 psi	=	6.89 kPa
1 bar	=	14.5 psi	1 atm	=	14.7 psi

Power

1 hp	=	746 Watts	10^6 bbl crude oil/day	~	2.12 Quad/yr
1 MW	=	1×10^6 Watts	1 kW	=	1,000 Watts
1 Watt	=	1 J/s	1 hp	=	33,00 ft-lb/min

Energy

1 Btu	=	1055 J	1 kWh	=	3,412 Btu
1 MJ	=	1×10^6 J	1 kWh	=	3.6×10^6 J
1 calorie	=	4.184 J	1 Therm	=	100,000 Btu
1 Btu	=	251.9958 cal	1 electron-volt	=	4.45×10^{-26} kWh
1 kJ	=	1,000 J	1 MJ	=	238.85 kcal

Large Energy Units

1 Quad	=	1×10^9 MBtu	1 TWyr	=	8.76×10^{12} kWh
1 Quad	=	1×10^{15} Btu	1 Quad	=	1.055 EJ
1 EJ	=	1×10^{18} J	1 Quad	=	11 GWyr
1 TWyr	=	31.54 EJ	1 TWyr	=	29.89 Quad
1 GWyr	=	8.76×109 kWh	1 TWh	=	1×10^6 MWh

(Continued)

Force

1 lb$_f$	=	4.4485 N	1 Newton	=	1 kg/m²s
1 barrel of oil equivalent	=	5.80 MBtu	1 U.S. dry barrel	=	7,056 in³
1 barrels of oil per day (Mbd)	=	2.12 Quad/yr	1 U.S. dry barrel	=	115.6 L

Temperature

°F	=	°C × 9/5 + 3	°R	=	°F + 460.67
K	=	°C + 274.15	°R	=	1.8 × K

Specific Heats

1 kJ/kg°C	=	0.24 Btu/lb°F	1 kcal/kg°C	=	4.186 kJ/kg°C

Index

Note: Page numbers in **bold** indicate tables and those in *italics* indicate figures.

A

Absorbents, 3
Acid-producing microbes, 104
Acoustic resonance anemometer, 76
AEFC, *see* Alkaline electrolyte fuel cell (AEFC)
Aerial photography, 156
Aerospace industry, 228
AFR ratio, *see* Air-to-fuel (AFR) ratio
Afsluitdijk, 200
Agus 1 to 7, 145, 146
AHE, *see* Anomalous heat effect (AHE)
AI, *see* Alkali index (AI)
Air density, 64
Air pollution, 184, 390–393; *see also* Greenhouse gas (GHG) emissions
Air-to-fuel (AFR) ratio, 109
Albatern WaveNet, 289
Alkali index (AI), 110, 111–112
Alkaline electrolyte fuel cell (AEFC), 217–218, **225**
American Wind Energy Association (AWEA), 69, 88
American wire gauge (AWG), 56
Ammonia, 316
Amorphous silicon solar PV system, 50
Anaconda Wave Energy, 289
Anaerobic digestion, 97, 104–106, 181
Anaerobic digestion reactor, 104, *105*
Angle of incidence, 37–38
Animal manure, 96–97
Animal power, 330–334
    advantages/disadvantages, 330–331
    characteristics of draft animals that affect performance, 331
    cows, 334
    draft animal performance vs. tractors, 331
    draft animals, 330
    draft horsepower capability of various animals, 331–333
    oxen, 333
    sustainable power of various work animals, **328, 329**
    unique perspectives of, 333–334
    work horse, 328, 332–333

Anomalous heat effect (AHE), 355
Antifreeze loop and heat exchanger system, *50*
Aquaculture industry, 181
Aquatic biomass, 98
Archimedes wave swing, 279
Argonne National Laboratory (ANL), 116, 233
Asian Phoenix Resources Ltd of Canada, 11
Aspergillus awamori, 102
Aspergillus niger, 102
Aspira, Ruben B., 371
ASTM Standard B258, 56
Atargis Energy Corporation, 290
Australia, wave energy projects, 289
Automotive industry, 343
Availability/availability factor, 88
Aviation fuel, 9, 379
AWEA, *see* American Wind Energy Association (AWEA)
AWG, *see* American wire gauge (AWG)
Azimuth angle, 35, *36*

B

B5, 101
Bacon, Thomas, 216, 217
Bacon cell, 216
BAI, *see* Biomass agglomeration index (BAI)
Balance of systems (BOS), 381, 382–384
Balanced equation for combustion, 108, 109
Ballard PEMFC commercial fuel cell, 231, *231*
Ballard Power Systems, 231
Banchik, Leonardo, 198
Banki turbine, 131
Barium-strontium-titanate (BST), 366
Base load hydro power plant, 141
Base-to-acid ratio, 110, 111
Bay of Fundy, 17, 244, 258
Beam/direct radiation, 31, *32*, 42
Bearden, Tom, 368
Bed agglomeration, 110, 111
Bedini, John, 368
Belgium, wave energy projects, 290
Belt HaArava solar pond, 193
Benzoic acid, 373
Betz, Albert, 69
Betz coefficient, 69, *70*, 72
Bhumibol Dam, 128, *129*
Billabong Rams, 149

Binary fluid power generation cycle, *176*, 178–179
Binomial distribution, 64
Biochar, 107
Biodiesel, 8
Biodiesel economics, 113–114
Biodiesel production, 99–102
Bioethanol, 8, 114
Bioethanol production, 102–104, 395–396
Biogas, 102
Biogas production, 104–106, 181
Biogasoline, 379
Biological fuel cell, 223
Biomass agglomeration index (BAI), 110, 111
Biomass energy, 7–9, 93–119
    anaerobic digestion, 104–106
    animal manure, 96–97
    aquatic biomass, 98
    balance of systems (BOS), 383
    biodiesel economics, 113–114
    biodiesel production, 99–102
    bioethanol production, 102–104, 395–396
    biogas production, 104–106
    biological conversion process, 102–106
    combustion processes, 112–113
    deforestation, 93
    ethanol economics, 114
    ethanol fermentation, 102–104
    eutectic point of biomass, 110–112
    fuel wood, 98, 116
    gasification, 108–110
    GREET software, 116
    ligno-cellulosic crop residues, 97
    municipal sewage sludge, 96
    municipal solid waste (MSW), 95–96
    oil crops, 97–98
    physico-chemical conversion process, 99–102
    potential oil production from algae, 380
    pyrolysis, 107–108
    reforestation, 93, 116
    required area, 93–95
    slagging and fouling, 109–112
    sugar and starchy crops, 98
    sulfur dioxide emissions, 393
    sustainability issues, 115–116, 394–396
    technological advancements, 379–380
    thermal conversion process, 107–113
    torrefaction, 107
    wood power, 112–113
Biomass plants, 5
Biooil, 107
Blake Rams, 149
Boeing, 380

Bomb calorimeter, 353, 355
BOS, *see* Balance of systems (BOS)
Breast water wheel, 130
Bright Source Energy, 55
Brilliant Light Power (BLP), 355, 356
Brine, 198
Brown and Sharpe wire gauge, 56
Bruce Foods Corporation, 190, 193
BS-269, 369
BST, *see* Barium-strontium-titanate (BST)

C

C4 photosynthesis plants, 23
Cadmium telluride (CdTe) PV system, 55
CAFO, *see* Concentrated animal feeding operation (CAFO)
Calorimetry, 353–355
Campbell-Stokes sunshine recorder, 32
Capacity factor
    online LCOE calculator, 386
    tidal energy, 260
    wind energy, 87–88
Capital cost, *see* Economics
Capital recovery factor (CRF), 386
Carbon credits, 184, 397
Carbon dioxide production from glucose conversion, 6
Carbon emission reduction (CER), 184
Carbon footprint reductions, 397
Carbon monoxide (CO), **391**, 391–392
Carbon-to-nitrogen ratio (C:N ratio), 105
Carnot cycle efficiency
    fuel cells, 211
    OTEC systems, 301
    salinity gradient energy, 191–192, 192–193
Carnot refrigeration cycle, 168
CCel Company, 291–292
CDF, *see* Cumulative density function (CDF)
Cecoco Rams, 149
Cell phones, 342–343
CER, *see* Carbon emission reduction (CER)
Ceres Power, 219
Cerium gadolinium oxide (CGO), 219
Cerro Prieto, Mexico, 158
CETO wave power device, 289
CGF, *see* Climate generation factor (CGF)
CGO, *see* Cerium gadolinium oxide (CGO)
Charcoal igniter, 339
Charging cell phones, 342–343
Checkmate Sea Energy Group, 289
Cherokee, 358
China, OTEC systems, 319

Chrysanthemum, 40
Cigarette lighter, 339
Clack valve, 134
Claude, Georges, 19, 298
Clearness index, 43–44
Climate generation factor (CGF), 56
Clock time, 34
Closed-cycle OTEC system, 309–315
Clostridium acetobutylicum, 223
CMNS, *see* Condensed matter nuclear
    science (CMNS)
C:N ratio, *see* Carbon-to-nitrogen ratio
    (C:N ratio)
CO, *see* Carbon monoxide (CO)
Coal-fired power plant, *392,* 392–393
Cockerell wave contouring raft, 276, *276*
Coefficient of performance (COP), 168, 169,
    171, 174
Cold fusion and gravitational energy, 21–22,
    347–376
    Brilliant Light Power (BLP), 356
    calorimetry, 353–355
    cold fusion theory, 350–353
    Energy Catalyzer (E-Cat), 357–358
    gravitation energy, *see* Gravitational
        field energy
    hydrino, 355–356
    International Thermonuclear Experimental
        Reactor (ITER), 358
    key figures in fusion energy research,
        356–358
    lattice-enabled nuclear reactions, 357
    low-energy nuclear reactions (LENR),
        355–357
    other names for cold fusion, 355–356
Collares-Pereira and Rabl correlation, 42
Combustion processes, 112–113
Commercial gasoline, 9
Concentrated animal feeding operation
    (CAFO), 96
Condensed matter nuclear science (CMNS), 355
Conventional hydroelectric power plant, 142
Conversion from ppm to $\mu g/m^3$, 390
*Conversion of Gravity Field Energy: Revolution
    in Technology, Medicine and Society*
    (Nieper), 371
Converting units of measure, 405–406
COP, *see* Coefficient of performance (COP)
Copper indium gallium selenide (CuInGaSe)
    PV system, 55
Coriolis effect, 241
Corn, 114
Corn residues, 97

Cost, *see* Economics
Cotton gins, 141
Cotton hulls, 109
Cows, 334
CRF, *see* Capital recovery factor (CRF)
Cross flow turbine, 131
Crosswind Savonius windmill, 75
Crown Iron Works, 114
Crysanthemum moridolium, 40
CSA C390-10, 369
CT scanner, 344
Cumulative density function (CDF), 77, 81
Cup anemometer, 75, *76*
Curie, Jacques and Pierre, 339
Cut-in wind speed, 71, 84, *85*
Cut-out wind speed, 71, 84, *85*

D

D2Fusion, Inc., 356
DAFC, *see* Direct alcohol fuel cell (DAFC)
Daily load requirement (DLR), 55
Dams, *see* Hydro power
Dance floors, 341, 344
DARPA, *see* Defense Advanced Research
    Projects Agency (DARPA)
Darrieus windmill, 71, 74
d'Arsonval, Jacques Arsene, 19, 297
D'Aubuisson's efficiency, 136
Daylight hours, 39–40
DDG, *see* Dry distiller's grain (DDG)
De Palma, Bruce, 359, 366–367
Declination angle, 35–37, 39
Deep-cycle battery, 55
Defense Advanced Research Projects Agency
    (DARPA), 340
Deforestation, 93
Denmark, wave energy projects, 289–290
DePalma, Bruce, 21, 22
Desalination treatment plant, 198
Desmet Ballestra, 114
Destructive distillation, 107
Deuterium, 349
D2Fusion, Inc., 356
Diesel fuel, 9
Diffuse solar radiation, 31, *32,* 42
Direct alcohol fuel cell (DAFC), 233
Direct/beam radiation, 31, *32,* 42
Direct methanol fuel cell (DMFC), 216–217, **225**
Directional windmill, 75
Diversion canal hydro power plant, 147, *148*
DLR, *see* Daily load requirement (DLR)
DMFC, *see* Direct methanol fuel cell (DMFC)

Double-basin continuous power generation
system, 253–256
Double flash power generation cycle, *176,*
177–178
Downwind windmills, 74
Draft animal power, 327
Draft Animal Power Network (DAPNet), 327
Draft animals, 330; *see also* Animal power
Dry distiller's grain (DDG), 102
DTP, *see* Dynamic tidal power (DTP)
Dutch windmill, *71*
Dynamic tidal power (DTP), 257–258

E

E-Cat, *see* Energy Catalyzer (E-Cat)
Earth's attraction, 363
Economics
biodiesel, 113–114
BOS cost, 381, 382
capital recovery factor (CRF), 386
energy payback times (EPBT), 387, 388, **388**
ethanol, 114
fuel cells, 215, 234
geothermal energy, *182,* 182–183
hydro power, 149–150
levelized cost of energy (LCOE), 182, *182,*
382, 384–386
life cycle analysis (LCA), 387–389
online LCOE calculator, 386
OTEC system, 320–321
salinity gradient energy, 203–204
solar energy, 57–59
tidal energy, 258, **259**
wave energy, 287–288
wind energy, 88–89, 380–381, 382–383
EIA, *see* Environmental impact
assessment (EIA)
El Paso solar pond, 190, *190,* 193
Electric charges, 338; *see also* Piezoelectric
power
Electric motors, 368–370
Electric power consumption per capita
(selected countries), **306**
Electric Power Research Institute (EPRI), 153
Electroactive polymer artificial muscle, 290
Electrolux kerosene refrigerator, 42
Electrolysis, 5
Electromotive force (EMF), 364
Elmore, G. V., 218
EMF, *see* Electromotive force (EMF)
Energy Catalyzer (E-Cat), 357–358
"Energy Harvesting," 340

Energy payback times (EPBT), 387, 388, **388**
Energy Policy Act (1997), 369
Energy Policy Act (2005), 233
Energy units of measure, 405
Enhanced coercive flow, 154, 156
Entropy, 163
Environmental concerns
air pollution, 184, 390–393
alternative energy systems with better
pollution rates, 398
EIA, *see* Environmental impact
assessment (EIA)
geothermal energy, 183–184
GHG emissions, *see* Greenhouse gas (GHG)
emissions
hydro power, 149–150
salinity gradient energy, 205–206
sustainability issues, 394–396, 398
tidal energy, 259–260
wave energy, 286
Environmental impact assessment (EIA)
geothermal energy, 183
hydro power, 149
tidal energy, 260
Environmental Protection Agency (EPA), 390
EPA, *see* Environmental Protection
Agency (EPA)
EPAct levels, 369
EPBT, *see* Energy payback times (EPBT)
EPRI, *see* Electric Power Research Institute
(EPRI)
Ester linkage, 100, *101*
Ethanol, 8
Ethanol economics, 114
Ethanol fermentation, 102–104
Ethanol production process flowchart, *103*
European Commission Project, 278
European Marine Energy Center (EMEC), 291
European Salt Company, 200
Eutectic point of biomass, 110–112
Excess heat, 350
Extraterrestrial solar radiation, 30, 40–42
ExxonMobil, 379

F

FAA, *see* Federal Aviation Administration
(FAA)
Faraday disc, 359
Faraday's law, 363–364
Farm power, *see* Human, animal,
and piezoelectric power
Federal Aviation Administration (FAA), 87

Feinberg, Gerald, 361, 363
Financial concerns, *see* Economics
First law analysis, 163–168
"First Plasma," 358
Five percent biodiesel (B5), 101
Fixed solar collector, 35, *36*
Flansea Corporation, 290
Flash drum, 177
Flash evaporation, 177
Flash vessel, 177
Fleischman, Martin, 21, 55, 350, *351*, 352, 353, 356
Fluidized bed pyrolyzer, 107–108
Force cancellation, 290
Force units of measure, 406
Forced circulation solar water heater, 49, *50*
Fouling and slagging, 109–112
Fouling factor, 110
Four-bladed windmill, 75
France, OTEC projects, 318–319
Francis turbine, 130, 132
Frederikshawn, Denmark, 289
Free energy, 367–368
Free energy converter, 368
Fremantle, Western Australia, 289
Fuel Cell Energy Company, 221
Fuel cells, 15–16, 211–237
    advantages/disadvantages, 226–227
    aerospace applications, 228
    alkaline electrolyte fuel cell (AEFC), 217–218, **225**
    basic schematic of hydrogen-oxygen fuel cell, *213*
    Carnot cycle efficiency, 211
    components, 227–228
    cost of input feedstock, 215, 234
    direct alcohol fuel cell (DAFC), 233
    direct methanol fuel cell (DMFC), 216–217, **225**
    environmental damage, 398
    existing and emerging markets, 228–232
    future projects, 233
    heat energy losses, 215
    Heliocentris fuel cell system, 232
    high-temperature proton exchange fuel cell (HT-PEM), 216
    illustration of fuel cell reactions, *214*
    intercoolers, 227
    microbial fuel cell (MFC), 223–224
    molten carbonate fuel cell (MCFC), 220–221, **225**
    NASA Helios unmanned aviation vehicle (UAV), 229, *229*
    other fuel cells (formic acid, redox flow, etc.), 224
    PEMFC commercial fuel cell module by Ballard, 231, *231*
    phosphoric acid fuel cell (PAFC), 218–219, **225**
    practical fuel cell conversion efficiency, 213–214
    primary fuels used, 225
    propane-based fuel cell, 225–226
    proton exchange membrane (PEM) fuel cell, 215–216, **225**
    regenerative fuel cell, 221–222
    solid acid fuel cell (SAFC), 220
    solid oxide fuel cell (SOFC), 219–220, **225**
    solid polymer fuel cell (SPFC), 222
    Solid State Energy Conversion Alliance (SECA), 233
    Spider Lion, 229–230
    standby power, 228
    theoretical fuel cell conversion efficiency, 214–215
    typical design, *212*
    U.S. DOE national labs, 233
    well-conducting electrolyte, 212
    zinc-air fuel cell (ZAFC), 222–223
Fuel wood, 98, 112, 116
Fuji Films, 200
Fusion energy research, 356–358; *see also* Cold fusion and gravitational energy

G

Gallium arsenide (GaAs) PV system, 55
Gasification, 108–110
Gauss (unit of measure), 361
GE Energy, 85
Gemini spacecraft, 222
Geo-pressure geothermal power plant, 159
Geothermal energy, 1, 11–14, 153–188
    agricultural applications, 181
    air pollution problems, 184
    applications, 180–181
    binary fluid cycle, *176*, 178–179
    by-products of geothermal wells, 181
    coefficient of performance (COP), 168, 169, 171, 174
    double flash system, *176*, 177–178
    economics/cost comparisons, *182*, 182–183
    environmental effects, 183–184
    environmental impact assessment, 183
    estimating number of households serviced by geothermal facility, 179–180

first law analysis, 163–168
geo-pressure system, 159
geothermal heat pump (GHP), 168–174
geothermal power cycle, 161–168, 174–180
hot dry rock system, 159
indirect condensing cycle, *175*, 177
industrial applications, 181
inefficiencies, 166
liquid-dominated system, 158–159
low-boiling-point refrigerant, 184
non-condensing cycle, 174–175, *175*
P-V diagram, *162*, *170*
Philippines, 178–180
Rankine cycle, 161, 162, *162*, **167**
return on investment (ROI), 181–182
single flash system, *176*, 177
space heating and cooling, 181
straight condensing cycle, *175*, 175–177
system efficiency, 166, 167–168
T-s diagram, *162*, *163*, *170*
temperature profile in Earth's core, 154–158
thermodynamic cycle, 162–163, *166*
vapor-dominated system, 159
vapor refrigeration system, 168, *169*
well selection, 157–158
world's geothermal provinces, *12*,
    154–158, *155*
Geothermal heat pump (GHP), 168–174
Geothermal power cycle, 161–168, 174–180
Germanium triode, 365
GHG emissions, *see* Greenhouse gas (GHG)
    emissions
GHP, *see* Geothermal heat pump (GHP)
Giant kelp, 98
Gibbs free energy, 16, 213
Girard, Monsieur, 269
Glauber's salt, 4
Global solar radiation, 42
Glycerin, 114
Grain sorghum, 98
Grand Coulee Power Station, 122
Gravitational acceleration, 361
Gravitational attraction, 363
Gravitational constant, 10
Gravitational field energy, 21, 359–371
    calculating gravitational force between
        earth and a person, 362
    calculating gravitational force exerted
        by a person on earth, 363
    calculating voltage and current in
        electrical coils, 364
    De Palma, Bruce, 366–367
    electric motors, 368–370

Faraday's law, 363–364
    free energy, 367–368
    Klein coils, 365
    Len's law, 364
    Moray device, 365
    N-machine, 359–360, 366–367
    non-energy-related applications, 371
    nuclear electrical resonance (NER), 366
    power factor (PF), 368–370
    Seike, Shinichi, 365–366
    space power generator (SPG), 367
    tachyon field energy, 361–362
    ultra relativity, 365
    zero point energy, 367, 368
Gravity gravitons, 363
Green diesel, 379
Green Wave, 289
Greenhouse effect, 23
Greenhouse gas (GHG) emissions
    biological microbial growths, 24
    biomass systems, 388
    solar PV system, **388**, **389**
    wind machine, 389
GREET software, 116
Grid parity, 58
Gujarat solar pond, 193
GW, 121

H

Haematococcus pluvialis, 97
Hainan Island, 319
Halocline, 190
Halophytes, 380
Hanstholm, 290
Heat of evaporation calculations, 316–317
Heaving buoy device, 280
Heck, Jon and Kaley, 181
Heliocentris, 232
Heliocentris fuel cell system, 232
High-capacity hydro power plant, 142
High conductive gradient, 154, *157*
High head diversion hydro power plant, 147,
    148
High head hydro power plant, 142
High-speed windmills, 73
High-temperature proton exchange fuel cell
    (HT-PEM), 216
High-temperature solar heat engine, 54–55
High-tonnage sorghum biomass, 97
Highways, runways, racetracks, 341
Homopolar generator, 366
Honeywell UOP, 379–380

Hoover Dam, 128
Horizontal axis windmills, 74–75, 84–85
Hose pump, 279–280
Hot dry rock geothermal power plant, 159
Hour angle, 36
HT-PEM, *see* High-temperature proton
      exchange fuel cell (HT-PEM)
Huldbergen, Kees, 257
Human, animal, and piezoelectric power,
      20–21, 327–348
  animal power, *see* Animal power
  conclusion, 345
  human power, *see* Human power
  introduction, 327–330
  piezoelectrics, *see* Piezoelectric power
Human power, 334–338
  advantages/disadvantages, 335–336
  body movement, 337
  calculating amount of power generated
      by humans, 329, 335
  drudgery from full day of work, 345
  ergonomic factor, 337
  estimating output leg power generated
      by humans, 336
  heartbeat ratio, 337
  human-machine interfaces, 337
  man-hour requirements of various
      farm activities, **338**
  power of draft animals vs. human
      power, **329**
Hybrid OTEC system, 316
Hybrid tidal system, 258
HYDRA, 218
Hydraulic air compressor, 129
Hydraulic ram, 134–141
  basic components, 134, *135*
  D'Aubuisson's efficiency, 136
  delivery valve, 134
  energy efficiency, 136–138
  nomenclatures, *137*
  overall efficiency, 137
  principles of operation, 134
  Rife ram, 138–139
  specifying drive pipe sizes and lengths,
      139–140
  starting procedure, 140
  troubleshooting, 141
  typical installation, *135*, 135–136
  volumetric efficiency, 136, 137
  waste valve, 134
  water use efficiencies, 136
Hydrino, 355–356
Hydro power, 9–11, 121–152

base load plant, 141
  components of hydro power plant, *127*,
      127–128
  conventional hydroelectric power plant, 142
  cross flow turbine, 131
  diversion canal power plant, 147, *148*
  energy losses, 125–126
  environmental and economic issues, 149–150
  environmental impact assessment, 149
  Francis turbine, 130
  high head diversion power plant, 147, *148*
  high head power plant, 142
  hydraulic ram, *see* Hydraulic ram
  hydro power plant calculations, 128–129
  hydrologic cycle, 121, *122*
  impulse turbine, 130
  inefficiencies, 125–126
  Kaplan runner, 130
  low head power plant, 142
  overflowing of a dam, 149
  peak load plant, 141–142
  Pelton runner, 130–131
  plant capacity, 142
  pumped storage system, 142–147
  reaction turbine, 130
  run-of-river power plant, 147, *148*
  siltation, 149, 150
  specific speed for pumps/turbines, 131–132
  theoretical and actual power derived from
      water stream, 10, 124
  top hydro power-producing states,
      122–123, *123*
  tub wheels, 130
  turbine application chart, *133*
  turbines, 130–133
  Turgo wheel, 131, *131*
  types of plants, 141–148
  useful equations, 124–125
  valley dam power plant, 147, *148*
  water power-generating devices, 129–133
  water quality, 149, 150
  water wheels, 130
Hydro turbines, 130–133
Hydrogen-oxygen fuel cell, 212, *213*, 215;
      see also Fuel cells
Hydrologic cycle, 121, *122*

I

Ideal Carnot cycle efficiency, *see* Carnot
      cycle efficiency
Ideal gas law, 390
Ideal Rankine cycle, *see* Rankine cycle

Ideal refrigeration cycle, 170
IEC 60034-2-1:2007, 369
IEEE 112-2004, 369
IHI Corporation, 318
Impulse turbine, 130
Incidence angle, 37–38
Indirect condensing geothermal power cycle,
        175, 177
Industrial Heat, 358
Integrated seawater energy and agriculture
        system (ISEA), 380
International Thermonuclear Experimental
        Reactor (ITER), 358
Ireland, wave energy projects, 290–291
Isaac-Seymour wave power system, 278,
        278–279
ISEA, see Integrated seawater energy and
        agriculture system (ISEA)
Isentropic process, 163
Israel, wave energy projects, 291
Israel Institute of Technology, 341
Itaipu Dam, 9
ITER, see International Thermonuclear
        Experimental Reactor (ITER)
Ivanpah Project, 55

J

Jacobson, Mark, 398
Jameel, Abdul Latif, 198
Japan, OTEC systems, 318
JEC-37, 369
Joules to Btu conversion, 7

K

Kalayaan pumped storage hydro power
        plant, 145
Kaplan runner, 130, 132
Kerosene absorption refrigerator, 42
Kincheloe, Robert, 366
Kinetic energy equation, 6, 65
KIOST, see Korea Institute of Ocean Science and
        Technology (KIOST)
Kishorn Port, 289
Klein coils, 365
Knock sensors, 343
Korea Institute of Ocean Science and
        Technology (KIOST), 319
Kromrey converter, 368
Kume Island, 318
Kumejima, 318

L

La Rance, France, 256, 258, **259**
Lake Caliraya, 143
Lanao Del Sur, 145
Lanoa Del Norte, 145
Las Gaviotas, 149–150
Latent heat of vaporization, 316
Lattice-enabled nuclear reactions, 357
Law of universal gravitation, 361
Lawrence Berkeley National Laboratory
        (LBNL), 233
LCA, see Life cycle analysis (LCA)
LCOE, see Levelized cost of energy (LCOE)
Lead zirconate titanate crystal, 20
LENR, see Low-energy nuclear reactions
        (LENR)
Len's law, 364
Leonardo Corporation, 357
Levelized cost of energy (LCOE), 182, 182, 382,
        384–386
Lienhard, John, 198
Life cycle analysis (LCA), 387–389
Ligno-cellulosic crop residues, 97
Liquid-dominated geothermal power plant,
        158–159
Load factor, 87
Lockheed Martin, 319
Loeb, Sidney, 189
Lorentz force, 366
Los Baños, 178
Low-boiling-point refrigerant, 184
Low-energy nuclear reactions (LENR),
        355–357
Low head hydro power plant, 142
Low-speed windmills, 73
Lumot Lake, 143
Lysekil Project, 290

M

Makai Ocean Engineering, 317, 318
Makban Geothermal Power Plant, 178, 184
Manifold absolute pressure (MAP) profile, 343
Manure, 96–97
MAP, see Manifold absolute pressure (MAP)
        profile
Maria Cristina Falls, 147
Martinique, 318
Masuda, Yoshio, 281
Masuda buoy, 281, 281
MCFC, see Molten carbonate fuel cell (MCFC)
McKubre, Michael, 357

Measuring instruments
   solar energy, 32–33
   units of measure, 405–406
   wind energy, 75–77
Mechanical three-cup anemometer, 75
Medium-capacity hydro power plant, 142
Methane-producing microbes, 104
Methanol fuel cell, 216–217
MFC, *see* Microbial fuel cell (MFC)
Micro-algae, 97
Micro-capacity hydro power plant, 142
Micro-hydro power units, 11, *11*
Microbial fuel cell (MFC), 223–224
Mills, Randell L., 355, 356
Mindanao, 145
Mini-hydro power plant, 142
Mitchell turbine, 131
Molten carbonate fuel cell (MCFC), 220–221, **225**
Mono-crystalline silicon solar PV system, 50
Montgolfier, Joseph Michael, 134
Moray, John E., 365
Moray, T. Henry, 359, 365
Motor efficiency, 369
Mount Makiling, 178
Mount Mayon, 178
MRI scanner, 344
MSW, *see* Municipal solid waste (MSW)
Multi-bladed windmill, 71, *71*, 72, 75
Multi-crystalline silicon solar PV system, 50
Multi-use type pumped storage hydro power
   plant, 145–147
Municipal sewage sludge, 96
Municipal solid waste (MSW), 95–96

N

N-machine, 21, 359–360, 366–367
NAAQS, *see* National Ambient Air Quality
   Standards (NAAQS)
Nagel, David J., 357
NAHB, *see* National Association of
   Homebuilders (NAHB)
Nannochloropsis gaditana, 379
Nano-hydro power plant, 142
NASA Helios unmanned aviation vehicle
   (UAV), 229, *229*
National Ambient Air Quality Standards
   (NAAQS), 390, **391**
National Association of Homebuilders
   (NAHB), 88
National Energy Technology Laboratory
   (NETL), 233

National Laboratory for Sustainable Energy, 69
National Renewable Energy Laboratory
   (NREL), 59, 69
Natural circulation solar water heater, 49, *50*
Natural Energy Laboratory of Hawaii
   Authority (NELHA), 317
NAUTILUS, 319
Naval Research Lab Spider Lion, 229–230
Neap tides, 241
NEB, *see* Net energy balance (NEB)
NELHA, *see* Natural Energy Laboratory of
   Hawaii Authority (NELHA)
NER, *see* Net energy ratio (NER);Nuclear
   electrical resonance (NER)
Nernst equation, 199
Net energy balance (NEB), 115, 394, 395–396, 398
Net energy ratio (NER), 115, 394, 395, 398
New Source Performance Standards (NSPS), 390
Newton, Isaac, 361
Newton's law of force, 65, 123
Newton's law of motion, 7
Newton's law relationship, 10
Newton's second law, 130
NEXA TM 1.2kW, 231
NEXT TM 1200, 232
Niagara Falls, 147
Nieper, Hans, 21, 350, 361, 371
Nissum Bredning, 290
Nitrogen dioxide (NO_2), 390, **391**
NMR, *see* Nuclear magnetic resonance (NMR)
NO_2, *see* Nitrogen dioxide (NO_2)
Nocera, Daniel G., 24
Non-condensing geothermal power cycle,
   174–175, *175*
Non-directional windmill, 74
Normal conductive gradient, 154, *157*
NREL, *see* National Renewable Energy
   Laboratory (NREL)
NREL maps, 69, *70*
NREL-USDOE formula for profitability of wind
   turbine farm, 89
NSPS, *see* New Source Performance Standards
   (NSPS)
Nuclear electrical resonance (NER), 366
Nuclear magnetic resonance (NMR), 366
Nuclear Power Corporation of India (NPCIL), 367

O

Oak Ridge National Laboratory (ORNL), 233
Ocean Energy buoy, 290–291
Ocean Power Technologies, 290

Ocean thermal energy conversion (OTEC)
　　systems, 19–20, 297–325
　actual conversion efficiency, 20
　applications of OTEC systems, 302, *302*
　base load electricity production, 305
　basic data for ideal OTEC system, *308*
　basic OTEC system, 299, *299*
　closed-cycle power generation, 309–315
　component parts, *300*
　cost, 320–321
　estimating heat absorbed by ocean water
　　bodies, 300–301
　future directions, 305
　heat of evaporation calculations, 316–317
　hybrid systems, 316
　ideal Carnot cycle efficiency, 301
　number of households service by OTEC
　　system, 317
　open-cycle power generation, 315–316
　projects currently under way for OTEC
　　systems, 317–319
　T-s diagram, *308*
　technical limitations, 319–320
　temperature profiles, 299–300
　theoretical pumping power, 304
　thermodynamic cycle, 306–309
　volume of ocean water needed to generate
　　MW of power, 303–304
　world's OTEC resources, *298*
Oil crops, 97–98
Okinawa, Japan, 318
One-bladed windmill, 75
Online LCOE calculator, 386
Open access, 397
Open-cycle OTEC system, 315–316
Orkney, Scotland, 291
Oscillating water column (OWC), *269*, 276–277
Oscillating wave surge converter, *269*, 276, 277
Osmotic energy, 189; *see also* Salinity gradient
　　energy
Ossberger turbine, 131
OTEC systems, *see* Ocean thermal energy
　　conversion (OTEC) systems
Overshot water wheel, 130
Overtopping device, *269*, 277–278
OWC, *see* Oscillating water column (OWC)
Oxen, 333

P

P-V diagram, *162, 170*
Pacific Northwest Generating
　　Cooperative, 291

Pacific Northwest National Laboratory
　　(PNNL), 233
Pacific Ring of Fire, 154, 156
PAFC, *see* Phosphoric acid fuel cell (PAFC)
Page, John, 87
Parabolic dish-electric transport system, *54*
Parabolic dish-steam transport and conversion
　　system, *53*
Partial evaporation, 177
Parts per million (ppm), 390
Pascal's law, 272
PDF, *see* Probability density function (PDF)
Peak load hydro power plant, 141–142
Pelamis wave energy converter, 291
Pelton runner, 130–131, 132
PEM fuel cell, *see* Proton exchange membrane
　　(PEM) fuel cell
PEMFC commercial fuel cell module by
　　Ballard, 231, *231*
Pendulator, 276, 277
Penstock, *148*
Pentland Firth, 242
Persistent seaweed, 98
PF, *see* Power factor (PF)
Philippine Geothermal Incorporated, 178
Philippines, 178–180
Phosphoric acid fuel cell (PAFC), 218–219, **225**
Photoperiod crops, 40
Photosynthesis, 4, 5
Piezoelectric motors, 344
Piezoelectric power, 20, 338–344
　alternative to strain gauges, 343
　applications of piezoelectricity, 340
　automotive industry, 343
　calculating energy produced from
　　piezoelectric element, 341, 342
　charging cell phones, 342–343
　dance floors, 341, 344
　detection of pressure variations due
　　to sound, 343
　high-voltage power sources, 340–343
　lead zirconate titanate crystal, 20
　most common applications (cigarette lighter/
　　charcoal igniter), 339
　naturally occurring crystals with
　　piezoelectric behavior, 339
　other naturally occurring materials, 339–340
　piezoelectric devices as sensors, 343
　piezoelectric devices as tiny actuators,
　　343–344
　piezoelectric motors, 344
　piezoelectricity, defined, 20
　potential future applications, 344

runways, racetracks, highways, 341
solar PV cells, 345
sophisticated medical devices (CT/MRI scanners), 344
synthetic ceramics, 340
synthetic crystals and langasite, 340
visual representation of, *339*
walking, 20, 340
Pilot anaerobic digester with floating gas holder, *106*
Piston power for surface attenuator, 274–275
Piston-type wave systems, 272–274
Plant biomass, 7–8
Plug-flow reactors, 105
Point absorber buoy, *269*, 269–274
Pollutant emissions, 390–393; *see also* Air pollution; Greenhouse gas (GHG) emissions
Pons, Stanley, 21, 350, *351*, 352, 353, 355, 356
Port Kembia, 289
Port MacDonnell, 289
Potassium oxide conversion, 3
Potter, M., 223
Power coefficient curve, *70*, 71
Power factor (PF), 368–370
Power law, 86
Power units of measure, 405
PowerBuoy, 290
PowerPal, *11*
ppm, *see* Parts per million (ppm)
Praceique, Bochaux, 269
Precognition, 361
Premier Rams, 150
Pressure-related osmosis (PRO), 195–198, *199*
Pressure units of measure, 405
Principle of transmission of fluid pressure, 272
Principles of Ultra Relativity, The (Seike), 365
PRO, *see* Pressure-related osmosis (PRO)
Probability density function (PDF), 77, 81
Propane-based fuel cell, 225–226
Proton exchange membrane (PEM) fuel cell, 215–216, **225**
Protonex, 231
Psychokinesis, 361
Pump/turbine specific speed, 131–132
Pumped storage hydro power plant, 142–147
Puthoff, Harold, 367
Pyranometer, 33, *33*, 42, *43*
Pyrheliometer, 33, 42
Pyrolysis, 107–108

Q

Quad/yr., 121
Quantum Vacuum, 368

R

Ralston eutropha, 24
Rankine cycle, 13, 161, 162, *162*, **167**, 306
Rankine cycle theoretical conversion efficiency, 14
Rapeseed, 100
Rapeseed methyl ester (RME), 100
Rated wind speed, 84, *85*
Rayleigh distribution, 77–80, 285
RBD oil, *see* Refined, bleached, or deodorized (RBD) oil
Reaction turbine, 130
Recirculating type pumped storage hydro power plant, 145, *146*
RED, *see* Reverse electro-dialysis (RED)
REDStack, 200
Reedsport, Oregon, 290, 291
Refined, bleached, or deodorized (RBD) oil, 99, 100
Reforestation, 93, 116
Refrigerant-grade anhydrous ammonia, 316
Regenerative fuel cell, 221–222
Reignwood Group, 319
Reliability, 88
Renewable energy
  advantages/disadvantages, 2–3
  alternative energy systems with better pollution rates, 398
  biomass energy, *see* Biomass energy
  cold fusion and gravitational energy, *see* Cold fusion and gravitational energy
  conversion efficiencies, 22–23
  cost, *see* Economics
  defined, 1
  fuel cells, *see* Fuel cells
  geothermal energy, *see* Geothermal energy
  humans and piezoelectrics, *see* Human, animal, and piezoelectric power
  hydro power, *see* Hydro power
  importance, 23–24
  intermittent nature of resources' availability, 384
  OTEC systems, *see* Ocean thermal energy conversion (OTEC) systems
  pollutant emissions, 390–393
  primary goal, 398
  "renewable mode" of usage, 23

salinity gradient, *see* Salinity gradient
  energy
social costs, 396–397
solar energy, *see* Solar energy
sustainability issues, 394–396, 398
tidal energy, *see* Tidal energy
wave energy, *see* Wave energy
wind energy, *see* Wind energy
Renewable Energy Act of 2008, 397
Renewable energy identification number
  (RINS), 9
Renewable Fuel Standard Program, 114
Residence time, 107
Reverse dialysis, 198
Reverse electro-dialysis (RED), 198–201
Reverse fuel cell (RFC), 221
Revolution in Technology, Medicine and Society:
  Conversion of Gravity Field Energy
  (Nieper), 21
Reykjavik, Iceland, 13, 181
Reykjavik Municipal Heating Project, 181
RFC, *see* Reverse fuel cell (RFC)
Rife ram, 138–139, 150
Ring of Fire, 154, 156
RINS, *see* Renewable energy identification
  number (RINS)
RME, *see* Rapeseed methyl ester (RME)
Rockfer, 150
Rossi, Andrea, 357–358
Ruminants, 8, 104
Run-of-river hydro power plant, 147, *148*
Runways, racetracks, highways, 341

S

Saccharification, 8
Saccharomyces cerevisiae, 102
SAFC, *see* Solid acid fuel cell (SAFC)
Saga University, 318
Salinity gradient energy, 15, 189–209
  advantages/disadvantages, 193–194, 202–203
  barriers to large-scale development, 204–205
  brine, 198
  desalination treatment plant, 198
  economics/cost comparisons, 203–204
  El Paso solar pond, 190, *190*, 193
  energy of sea water for desalination, 194–195
  energy required to boil salt water, 195
  environmental and ecological barriers,
    205–206
  factors affecting performance and feasibility,
    202–203
  halocline, 190

Nernst equation, 199
  potential energy, 204
  pressure-related osmosis (PRO), 195–198, *199*
  reverse electro-dialysis (RED), 198–201
  sensible heat equation, 191
  solar pond, 190–193
  specific applications or locations, 201–202
  theoretical and technical potential of salinity
    gradient, 204, **205**
  theoretical Carnot cycle efficiency, 192–193
  total daily energy absorbed by solar pond,
    191
  typical salinity gradient conversion
    system, *192*
  wastewater treatment facility, 198
Salter, S. H., 280
Salter's duck, 280, *280*
Salton Sea field, 158
SAM software, 59
Sano, 150
Satellite imagery, 156
Savonius windmill, *71, 72,* 74
Schlumf, 150
Sea Power Company, Ltd., 291
Seabased Industry AB, 291, 292
Search for Free Energy, 368
Seaweed, 98
SECA, *see* Solid State Energy Conversion
  Alliance (SECA)
Seike, Shinichi, 365–366
Semi-diurnal tides, 239
Sensible heat equation
  OTEC systems, 300
  salinity gradient, 191
Sharqawy, Mostafa, 198
Significant wave height, 285
Sihwa Lake power plan, 258, **259**
Siltation, 149, 150
Single-basin double-cycle generation system,
  252–253
Single-basin ebb cycle generation system,
  248–250
Single-basin tide cycle generation system,
  250–252
Single flash power generation cycle, *176,* 177
Slagging and fouling, 109–112
Slagging factor, 110
SME, *see* Soybean methyl ester (SME)
SO_2, *see* Sulfur dioxide (SO_2)
Social costs of renewables, 396–397
SOFC, *see* Solid oxide fuel cell (SOFC)
Soil warming, 181
Solar altitude angle, 35, *36*

Solar azimuth angle, 35
Solar chemical conversion, *4*, 5
Solar collector, 35, *36*
Solar collector system sizing, 55–57
Solar constant, 30
Solar cooling, 3
Solar declination angle, 35–37, 39
Solar dryer, 47–49
Solar drying, 3
Solar electric conversion, *4*, 5
Solar electrochemical conversion, *4*, 6
Solar energy, 3–6, 29–62
    available solar radiation on particular
        location, 42–45
    battery storage, 24, 55
    clearness index, 43–44
    conversion efficiency of solar PV systems,
        22–23
    daylight hours, 39–40
    diffuse solar radiation, 31, *32*, 42
    direct/beam radiation, 31, *32*, 42
    economics of solar conversion devices, 57–59
    extraterrestrial solar radiation, 30, 40–42
    geometric nomenclatures, 35–40
    high-temperature solar heat engine, 54–55
    measuring instruments, 32–33
    reduction in solar collector area with
        increased efficiency, 378–379
    SAM software, 59
    solar constant, 30
    solar dryer, 47–49
    solar PV system, *see* Solar photovoltaic (PV)
        system
    solar refrigerator, 45–47, *48*
    solar thermal collector system, 53, *54*
    solar thermal electric power system, 52, *52*
    solar time, 33–35
    solar water heater, 49, *50*
    sunrise/sunset times, 39
    thermodynamic pathways, 3–6
Solar incidence angle, 37–38
Solar noon, 35
Solar photovoltaic (PV) system, 50–52
    amount of energy required, **388**, 396
    battery storage, 24, 55
    BOS components, 382
    BOS cost, 383
    breakdown of installed cost of residential
        system, **383**
    conversion efficiency, 22, 378, *379*
    cost/pricing, 380, 381, 383, **383**
    environmental damage, 398
    estimating time to recover energy usage, 396

greenhouse gas (GHG) emissions, **388**, **389**
life cycle analysis (LCA), 387–389
major components, 55
major solar PV designs, 50, 387
piezoelectrics, 345
schematic of simple solar PV home
    system, *51*
sizing a PV system, 55–57
solar PV cost vs. balance of systems, 383
Solar pond, 190–193
Solar PV system, *see* Solar photovoltaic (PV)
    system
Solar pyranometer, 33, *33*, 42, *43*
Solar radiation, 349
Solar refrigerator, 45–47, *48*
Solar thermal collector system, 53, *54*
Solar thermal conversion, 3, *4*
Solar thermal electric conversion (STEC), *4*, 5
Solar thermal electric power system, 52, *52*
Solar thermal power system with distributed
    collectors, 53, *53*
Solar thermal power system with distributed
    collectors and generators, 53, *54*
Solar thermochemical conversion, 3–4, *4*
Solar thermomechanical conversion, *4*, 5
Solar time, 33–35
Solar village power system, 50, *51*, 52
Solar water heater, 49, *50*
Solid acid fuel cell (SAFC), 220
Solid oxide fuel cell (SOFC), 219–220, **225**
Solid polymer fuel cell (SPFC), 222
Solid-state alkaline fuel cell, 218
Solid State Energy Conversion Alliance
    (SECA), 233
Solids retention time (SRT), 105
Son of Energy in Which the Earth Floats, The
    (Moray), 365
Sonic 3D anemometer, 76, *76*
South Korea, OTEC systems, 319
Southern California Edison Company, 54
Southern Power Company, 112
Soybean methyl ester (SME), 100
Space heating and cooling, 181
Space power generator (SPG), 367
Space vortex theory (SVT), 367
Specific heats, units of measure, 406
SPFC, *see* Solid polymer fuel cell (SPFC)
SPG, *see* Space power generator (SPG)
Spider Lion fuel cell, 229–230
Spring tides, 239
Sri International, 290
SRI International, 357
SRT, *see* Solids retention time (SRT)

Standard time, 34
Statkraft PRO power plant, 196, 197
STEC, *see* Solar thermal electric conversion (STEC)
STEC plus electrolysis, *4*, 5
Steijn, Rob, 257
Stoichiometric combustion equation, 108, 109
Straight condensing geothermal power cycle, *175*, 175–177
Strain gauges, 343
Submerged pressure differential system, *269*, 278–279
Sugar and starchy crops, 98
Sugarcane bagasse, 109
Sulfur dioxide (SO_2), 390, **391**, 392–393
Sun path diagram, 36
Sunderbans, India, 258
Sunrise/sunset hour angle, 39
Sunrise/sunset times, 39
Super hydro power plant, 142
Surface attenuator, *269*, 274–276
Surface azimuth angle, 35, *36*
Sustainability issues, 394–396, 398
SVT, *see* Space vortex theory (SVT)
Swansea Bay, 258
Sweden, wave energy projects, 290
Sweet sorghum, 98
Swiss ML converter, 368
Synthesis gas (syngas), 107, 108
Synthetic ceramics, 340
Synthetic crystals and langasite, 340
Synthetic Genomics, Inc., 379
Systems Advisor Model (SAM) software, 59

T

T-RLF, *see* Tewari reactionless generator (T-RLG)
T-s diagram
   geothermal energy, *162*, *163*, *170*
   OTEC system, *308*
Table of conversion units, 405–406
Tachyon field energy, 361–362
Tachyon particles, 361
Tanner, H. A., 218
Technova Corporation, 356
Temperature profile in Earth's core, 154–158
Tesla (unit of measure), 361–362
Tesla, Nikola, 21, 22, 350, 360
Tesla switch, 368
Tewari, Paramahamsa, 21, 367
Tewari reactionless generator (T-RLG), 367
Texas, 159–160

Theoretical Carnot cycle efficiency, 192–193
Thermodynamic cycle
   geothermal energy, 162–163, *166*
   OTEC system, 306–309
Three-bladed windmill, 74, 75
3D ultrasonic anemometer, 76, *76*
Three Gorges Dam, 9
Tidal barrages, 257, **259**
Tidal capacity factor, 260
Tidal current technologies, 244
Tidal energy, 16–18, 239–263
   actual power and energy from tides, 17–18
   attractive forces between earth and moon, *17*
   capacity factor, 260
   controlled release of elevated reservoir water, 247
   Coriolis effect, 241
   cost of tidal energy systems, 258, **259**
   double-basin system, 253–256
   dynamic tidal power (DTP), 257–258
   energy produced in elevated stored water, 247
   environmental concerns, 259–260
   environmental impact assessment, 260
   how tidal energy works, 245
   hybrid tidal system, 258
   land drainage, 260
   neap tides, 241
   power generated from elevated stored water, 246
   release time for stored water, 246
   schematic of tidal power generation system, *245*
   semi-diurnal tides, 239
   single-basin ebb cycle power generation, 248–250
   single-basin tide cycle power generation, 250–252
   single-basin two-way power generation, 252–253
   spring tides, 239
   tidal barrage, 257, **259**
   tidal current technologies, 244
   tidal height variation, *240*, 241–242
   tidal "humps," 239
   tidal lagoon, 258
   tidal power generation schemes, 247–256
   tidal range technologies, 244
   tidal stream generator (TSG), 256–257
   water storage volume calculation, 246
   worldwide potential for use of tidal energy, *240*, 242–243
Tidal fence, 257

Tidal "humps," 239
Tidal lagoon, 258
Tidal range technologies, 244
Tidal stream generator (TSG), 256–257
Tidal type pumped storage hydro power
    plant, 147
Tide filling, 253
Tip-speed ratio (TSR), 70, *70*, *71*
Tipping fee, 95
Tiwi Geothermal Power Plant, 178, *179*, 184
Tokamak, 358
Ton refrigeration effect, 170
Torrefaction, 107
Trade winds, 63
Transesterification, 100, *100*
Trichoderma reesei, 102
TSG, *see* Tidal stream generator (TSG)
Tub wheel, 130
Turgo wheel, 131, *131*, 132
Two-bladed windmill, 71, *71*, *72*, 75

U

UAV, *see* Unmanned aviation vehicle (UAV)
Ultra relativity, 365
Ultrasonic devices, 343
Undershot water wheel, 130
Unipolar generator, 359
United Kingdom, wave energy projects, 289
United States, wave energy projects, 290
Units of measure, 405–406
Unmanned aviation vehicle (UAV), 229–230
Upwind windmills, 73
Urey, Harold Clayton, 349
U.S. DOE national labs, 233
U.S. Energy Information Administration, 24
U.S. Renewable Fuel Standards, 9
Utility-scale land-based 80-meter wind
    map, *70*
Utility scale wind turbine, 87

V

Valley dam hydro power plant, 147, *148*
Valone, Tom, 368
Vapor-dominated geothermal power plant, 159
Vapor refrigeration system, 168, *169*
Vapor-to-liquid mass ratio, 163
Vegetable oil, 97
Vertical axis windmills, 74
Vestas, 85
Vibrations, 345; *see also* Piezoelectric power
Volcanoes, 156

W

Walking, 20, 340
Water energy and food nexus group (WEF), 397
Water hyacinth, 149
Water power-generating devices, 129–133
Water transfer type pumped storage hydro
    power plant, 147
Water wheels, 130
Watt-peak circulation, 56, 58
Watts, James, 125, 328
Wave contouring raft, 276, *276*
Wave Dragon, 277, 290
Wave energy, 18–19, 265–295
    calculating power from wave per meter
        of crest head, 266
    economics, 287–288
    environmental impacts, 286
    estimated value of constant in wave
        equation, 19
    estimating power for given length of wave,
        267–268
    future/prospective wave energy projects,
        288–292
    hose pump, 279–280
    Masuda buoy, 281, *281*
    oscillating water column (OWC), *269*,
        276–277
    oscillating wave surge converter, *269*,
        276, 277
    overtopping device, *269*, 277–278
    piston power for surface attenuator, 274–275
    piston-type wave systems, 272–274
    point absorber buoy, *269*, 269–274
    Salter's duck, 280, *280*
    significant wave height, 285
    submerged pressure differential system, *269*,
        278–279
    surface attenuator, *269*, 274–276
    theoretical hydraulic power from waves,
        271–272
    typical hydraulic circuit for wave generators,
        281–285
    wave contouring raft, 276, *276*
    wave energy converter (WEC), 265
    wave power equation (wave energy flux
        equation), 266
    world's wave power potential, 268–269
    year-round distribution of wave power,
        286–287
Wave energy converter (WEC), 18, 265
Wave energy flux equation, 266
Wave hydraulic power system, 281–285

Wave piston system, 289
Wave point absorber, 269, 269–274
Wave power equation, 266
WaveBob project, 291
Weber units, 360
WEC, *see* Wave energy converter (WEC)
WEF, *see* Water energy and food nexus
         group (WEF)
Weibull distribution, 64, 80–84, 286
Weir, 147, 148
Westinghouse, George, 360
Whitehurst, 134
Wilk, Howard J., 356
Wind aero-generators, 63–64
Wind capacity factor, 87–88
Wind design parameters, 84–88
Wind energy, 6–7, 63–92
    actual energy and power from wind, 69–72
    actual power from wind, 72
    air density, 64
    availability and reliability, 88
    basic energy and power calculation from
         wind, 65–68
    capacity factor, 87–88
    cut-in wind speed, 84, 85
    cut-out wind speed, 84, 85
    downwind windmills, 74
    economics of wind machines, 88–89,
         380–381, 382–383
    energy vs. power, 66
    high-speed windmills, 73
    horizontal axis windmills, 74–75, 84–85
    kinetic energy available from moving
         wind, 7
    life cycle analysis (LCA), 389
    low-speed windmills, 73
    measuring instruments, 75–77
    NREL maps, 69, 70
    NREL-USDOE formula for profitability
         of wind turbine farm, 89

    power coefficient curve, 70, 71
    rated wind speed, 84, 85
    Rayleigh distribution, 77–80
    reducing LCOE, 382–383
    technological advancements, 379
    theoretical power extracted from
         wind, 67
    typical generation of wind movement, 64
    underestimation of wind power, 68
    upwind windmills, 73
    vertical axis windmills, 74
    Weibull distribution, 80–84
    wind aero-generators, 63–64
    wind design parameters, 84–88
    wind speed and height, 86–87
    worldwide wind energy potential, 69
Wind maps, 69, 70
Windustry.org, 88
Wire size, 56
WMO, *see* World Meteorological Organization
         (WMO)
Wood power, 112–113
Work animals, *see* Animal power
Work horse, 328, 332–333
World Meteorological Organization
         (WMO), 76
Worldwide wind energy potential, 69

Y

YSZ, *see* Yttria-stabilized zirconia (YSZ)
Yttria-stabilized zirconia (YSZ), 219

Z

ZAFC, *see* Zinc-air fuel cell (ZAFC)
Zenith, 35, 36
Zenith angle, 35, 36
Zero point energy, 367, 368
Zinc-air fuel cell (ZAFC), 222–223